Turfgrass Chemicals and Pesticides
A Practitioner's Guide

Turfgrass Chemicals and Pesticides
A Practitioner's Guide

Matt Fagerness

Rodney Johns

McGraw-Hill

New York Chicago San Francisco Lisbon London Madrid
Mexico City Milan New Delhi San Juan Seoul
Singapore Sydney Toronto

The McGraw·Hill Companies

Copyright © 2004 by The McGraw-Hill Companies, Inc. All rights reserved. Printed in the United States of America. Except as permitted under the United States Copyright Act of 1976, no part of this publication may be reproduced or distributed in any form or by any means, or stored in a data base or retrieval system, without the prior written permission of the publisher.

1 2 3 4 5 6 7 8 9 0   DOC/DOC   0 9 8 7 6 5 4 3

ISBN 0-07-141079-1

The sponsoring editor for this book was Larry S. Hager and the production supervisor was Pamela A. Pelton. It was set in Stone Sans by Lone Wolf Enterprises, Ltd.

Printed and bound by RR Donnelley.

 This book is printed on recycled, acid-free paper containing a minimum of 50% recycled, de-inked fiber.

Information contained in this work has been obtained by The McGraw-Hill Companies, Inc. ("McGraw-Hill") from sources believed to be reliable. However, neither McGraw-Hill nor its authors guarantee the accuracy or completeness of any information published herein and neither McGraw-Hill nor its authors shall be responsible for any errors, omissions, or damages arising out of use of this information. This work is published with the understanding that McGraw-Hill and its authors are supplying information but are not attempting to render engineering or other professional services. If such services are required, the assistance of an appropriate professional should be sought.

DEDICATION

To my loving wife Angela, who has supported this project from the beginning. She has been instrumental in helping me turn a vision into reality.

MF

PREFACE

When first asked to participate in authoring this text, I had to ask myself "How can the issue of turfgrass chemicals and pesticides be addressed to best serve the widest range of audience?" The answer to this quickly became clear. There are currently no other texts on the market that address the diversity of topics found in this one. There are many that specifically address groups of pests in great detail but none that cover this assortment of topics in an introductory manner. With that in mind, the purpose of this text is to provide a solid overview of the many facets associated with chemical and pesticide use in both turfgrass and landscape systems. Topics include the basics of chemical and pesticide use, weeds, diseases, insects, and the procedures associated with safe and responsible use of these materials.

Many texts out there that touch on some of these topics become wrapped up with technical information. While technical detail is sometimes necessary, it has been an underlying goal of this text to present topics in an accurate but conversational manner. This is for the benefit of the reader who wants the core information they need without needing a thesaurus or extensive academic background in the subject. The information presented herein is detailed enough to be suitable for trained professionals in the field of turfgrass and landscape management and introductory enough to be suitable for persons new to the field or with relatively little experience. As you proceed through the text, I encourage you to approach topics with an open mind and constantly evaluate your own level of knowledge and experience in them. Each chapter will end with a list of questions to both review covered material and to allow you to assess how the presented material pertains to your situation. This setup will allow you to get the most meaning out of these topics and be able to apply new knowledge immediately. The pursuit of knowledge in this area includes many key topics. The process of moving through these topics is an educational journey and one that I personally welcome you to. Let's get started!

—Matt Fagerness

ACKNOWLEDGMENTS

I would like to acknowledge the hard work and devotion of all the contributors to this text. Their provision of written material, illustrations, and specialized knowledge has helped increase both the scope and the impact of this text, for the benefit of all readers. To these individuals, I offer my personal and professional gratitude for a job well done.

MF

ABOUT THE AUTHORS

Matt Fagerness is currently the Extension Turfgrass Specialist at Kansas State University. A resident of Manhattan, Kansas, he received his Ph.D. from North Carolina State University with specialization in weed science and turfgrass growth regulators. Dr. Fagerness has been involved in both the applied and academic aspects of the turfgrass industry since 1989, including experience in golf course maintenance and turfgrass research/outreach.

Rodney Johns served in the golf course business for 10 years before starting Arki-Tec Landscaping and Sales Company, LLC in 1995. A resident of Canton, Missouri, Mr. Johns holds various degrees in plant science/horticulture and turf and greenhouse management.

CONTRIBUTORS

Ross Brown
Golf Course Superintendent, Shadow Lake and
Shadow Pines Golf Clubs, New York

Greg Hodges, Ph.D and Amanda Hodges, Ph.D.
Department of Entomology, University of Florida

Robert Wolf, Ph.D.
Department of Biological and Agricultural Engineering
Kansas State University

ART CONTRIBUTORS

I would like to acknowledge the contributions of art, pictures, and illustrations from each of the contributors to this text. Additionally, I would like to thank the University of Illinois and Kansas State University for allowing use of additional artwork in Chapter 10 of this text.

MF

Contents

CHAPTER 1
Introduction to Turfgrass Chemicals and Pesticides ... 1
History of Chemical and Pesticide Use in Turf: Early Times to 1950 .. 5
History of Chemical and Pesticide Use in Turf: 1950 to 1975 12
History of Chemical and Pesticide Use in Turf: 1975 to Present 16
Public Perception of Turfgrass Chemicals and Pesticides 19
Future Trends 25
Summary Questions 28

CHAPTER 2
Basic Chemical and Pesticide Principles 29
Pesticide Formulations 32
Chemical and Pesticide Safety Information—Using an MSDS Form .. 58
Understanding the Chemical/Pesticide Label 70
Environmental Impact of Chemicals and Pesticides 75
    Alternative Fates of a Pesticide 77
    Pesticide Mobility 80
    Economic and Aesthetic Pest Thresholds 82
Summary Questions 89

CHAPTER 3
Weeds Found in Turf and Landscapes 91
Weed Life Cycles: Annuals, Biennials, and Perennials 94
Weed Biology: Broadleaf Weeds, Grasses, and Sedges 114
Broadleaf Weeds Commonly Found in Turf and Landscapes 117
Grassy Weeds and Sedges Commonly Found
in Turf and Landscapes 139
Weed Ecology and Cultural Ways of Controlling Weeds 163
Summary Questions 170

CHAPTER 4
Herbicides Used in Turf and Landscape 171
Herbicide Families 173
   Preemergence and Postemergence Herbicides 181
   Selective and Nonselective Herbicides 185
   Contact and Systemic Herbicides 189
Interpreting and Following Herbicide Recommendations 192
Herbicide Use and Turf Establishment 196
Herbicide Resistance 201
Summary Questions 208

CHAPTER 5   by Ross Brown
Disease Problems and Fungicides Used in Turfgrass .. 209
Disease Problems in Turf 210
Fungicide Families and Modes of Action 234
Control Strategies for Turfgrass Diseases 241
Summary Questions 250

CHAPTER 6 by Rodney Johns
Disease Problems and Fungicide Use in the Landscape 251
Disease Control 256
Plant Health 262
Monocultures 266
Some Common Disease Problems 271
Control Strategies 276

CHAPTER 7
Insect Problems and Insecticides Used in Turf 279
Insect Problems in Turf 280
    Ground Pearls 284
    Mealybugs 286
    Leafhoppers 288
    Spittlebugs 290
    Chinch Bugs 292
    Mole Crickets 294
    White Grubs 298
    Sod Webworms 301
    Armyworms 305
    Cutworms 309
    Frit Flies 312
    Crane Flies 314
    Wireworms 316
    Billbugs 316
    Fire Ants 320
    Mites 323

Insecticides 324
Control Strategies for Herbivorous Insects 328
Summary Questions 334

CHAPTER 8 by Rodney Johns
Insect Control in Trees and Landscapes 335
Troublesome Insects 343
Evasive Species 353
Insect Control Issues 357

CHAPTER 9
Other Chemicals and Pesticides Used in Turf
and Landscapes 361
Adjuvants Used in Chemical and Pesticide Applications 362
Plant Growth Regulators 371
Materials and Strategies for Rodent and Burrowing Pest Control .. 380
Aquatic Plant Management 384
Control of Birds and Other Herbivores 393
Control of Nuisance Pests 398
Summary Questions 401

CHAPTER 10
Chemical and Pesticide Application
Equipment and Calibration 403
Equipment Used for Applying Chemicals and Pesticides 404
   Manual Sprayers 404
   Power Sprayers 407
   Power Sprayer Pumps 408
   Power Sprayer Tanks 413
Power Sprayer Agitation and Flow Control 415

Power Sprayer Strainers 418
Power Sprayer Hoses and Fittings 419
Power Sprayer Booms and Nozzles 421
Spray Guns 432
Calibrating Sprayers 434
  Calibrating Manual Sprayers and Spray Guns 435
  Calibrating Power Sprayers 436
  Selecting Nozzles for Hand-Held Booms 445
  Completing Sprayer Calibration 452
Sprayer Maintenance 455
Spreaders Used for Chemical and Pesticide Applications 457
  Making Applications Using Granular Spreaders 461
  Calibration of Granular Spreaders 461
Summary Questions 464

CHAPTER 11 by Rodney Johns
Pesticide Safety 465
Testing and Licensing 468
Safety 470
Types of Pesticides 482
Application Equipment 487
Disposal of Chemicals 491
Summary Questions 495

CHAPTER 12
**Best Management Practices and
Integrated Pest Management 497**
Reduced Pesticide and Chemical Use Through Strategic Plantings .. 499
  Turfgrass and Landscape Design Influences on Pesticide Use ... 502
Integrated Pest Management Principles 505

Interpreting Pesticide Risk 507
Developing an Integrated Pest Management Plan
for Turfgrass and Landscapes 512
Concluding Remarks 522
Summary Questions 523

Index **525**

Chapter 1

Introduction to Turfgrass Chemicals and Pesticides

The development of turfgrass as a component of both landscaping and our overall culture has seen tremendous strides over the past 100 years. Community parks and gardens used to be the means by which the general public could appreciate nature when they could not travel to natural places of beauty. Today, like no other time in history, the concept of greenspace has taken on very personal undertones. The turf industry encompasses and supports many different parts in what has become a multi-billion dollar industry. There are people involved in the turf industry from areas like commercial lawn care, grounds and athletic field management, golf course operations, and the needs of the general public itself (Figures 1.1 through 1.3). The magnitude of today's turf industry is evidenced visually by the extent and quality of managed turf. The turf industry is also economically massive, accounting for billions of dollars in gross receipts every year.

Critics will belabor the fact that so much is spent for what is seen as an unnecessary luxury. However, let us not ignore the fact that personal greenspace represents on a small scale the daily freedoms we often take for granted. Having a nice lawn and nice landscaping represents pride of ownership. This stems from an appreciation of natural beauty and, above all, tranquility in an otherwise busy world.

2 Turfgrass Chemicals and Pesticides: A Practitioner's Guide

FIGURE 1.1 Professional lawn care in action.

FIGURE 1.2 View of a maintained football field.

FIGURE 1.3 Colbert Hills Golf Course, Manhattan, Kansas.

4 Turfgrass Chemicals and Pesticides: A Practitioner's Guide

FIGURE 1.4 DDT being used for insect control during World War II.

The green industry therefore supports peace of mind, yet it does not put food on our tables like conventional agriculture. This pursuit of peace of mind has allowably become more complex over the years, due to advancements in the technology which supports this vast industry. We now have automated mowers in place of grazing animals, synthetic fertilizers in place of manure, sprinklers to help supplement

PESTPOINTER

As members of this industry, we have a responsibility to use the chemicals available to us with discretion, accountability, and precision. We also have a fundamental duty to serve the needs of those who seek protection for their greenspace and to address these needs.

natural rainfall, plant material which can better withstand the harshness of nature, and pesticides to help protect the greenspace which we create.

Somewhere between the cavalier pursuit of pesticide use without regard to safety or label instructions and the feverish criticism of any chemical or pesticide application lies the focus of the green industry. As members of this industry, we have a responsibility to use the chemicals available to us with discretion, accountability, and precision. We also have a fundamental duty to serve the needs of those who seek protection for their greenspace and to address these needs with technical knowledge, accurate but practical explanation, and timeliness. Proper management of the turfgrasses we aim to protect is the first step towards optimizing the use of chemicals and pesticides. With that in mind, turfgrass management, although not the specific focus of this book, will necessarily be discussed as it pertains to management of chemical inputs.

HISTORY OF CHEMICAL AND PESTICIDE USE IN TURF: EARLY TIMES TO 1950

The earliest documented cases of pesticide use date back well over two thousand years when armies would apply salt to the agricultural fields of those they conquered, rendering the fields useless. While crude and malicious, this represents the first example of directed herbicide use. Other pesticide use through the progression to the modern day was scattered and largely based upon casual observation. The early use of tobacco smoke or extracts for insect control was likely based upon visual observation, rather than on the understanding that nicotine has insecticidal properties. Most early pesticides bore potential human hazards, typically containing materials such as lead, mercury, cyanide, or arsenic. While not up to today's safety standards, these early pesticides were indeed functional and improved the quality of life for our ancestors.

The early decades were an important time for the development of pesticides used in production agriculture. While observations and early research showed the benefit of many materials, the advent of motorized farm equipment and airplanes made necessary the development of pesticides that could be efficiently applied on a larger scale. The first aerial pesticide applications were made in the early 1920s, usually with dusts containing nicotine or arsenic-based compounds for insect control. As newer and less toxic insecticides were developed, the use of these early materials gradually waned. The first organophosphate insecticide was discovered in the late 1930s, helping to project insecticide development into the modern era. The onset of World War II and United States involvement in the Pacific necessitated large quantities of insecticides that could be used to help prevent malaria. DDT (Figures 1.4 through 1.6) was introduced to

PESTPOINTER

While not up to today's safety standards, these early pesticides were indeed functional and improved the quality of life for our ancestors.

Introduction to Turfgrass Chemicals and Pesticides 7

FIGURE 1.5 Aerial application of DDT.

FIGURE 1.6 DDT being used for mosquito control on a public beach.

American scientists in 1942 and rapidly became a standard for combating this serious problem.

Some of the major milestones in the history of pesticides from 1900 to 1980 are shown below:

Year	Discovery/Milestone
1906	Federal Food, Drug, and Cosmetic Act takes effect.
1909	First experimental use of nicotine sulfate as an insecticide
1910	Federal Insecticide Act takes effect.
1912	First recommendation of zinc arsenite as an insecticide
1917	First dry carrier (dust) application of nicotine sulfate
1921	First use of an airplane for insecticide dust application
1922	First commercial use of calcium cyanide as an insecticide
1925	First experimental testing of selenium materials for insect control
1931	Discovery of thiram, the first sulfur-based fungicide
1932	Introduction of methyl bromide as a soil fumigant
1932	First reported use of the plant growth regulators, ethylene and acetylene, in fruit crops
1936	First reported use of the wood preservative pentachlorophenol
1938	Discovery of TEPP, the first organophosphate insecticide
1938	First use of *Bacillus thuringiensis* (BT) as a bacterial insecticide
1939	Discovery of the insecticidal properties of DDT
1941	First use of aerosol insecticides
1942	First experimental shipment of DDT to the U.S.
1942	Discovery of 2,4-D, the first phenoxy broadleaf herbicide

1944 First use of 2,4,5-T for vegetation control

Introduction of chlordane, the first persistent chlorine-based insecticide

Introduction of the first carbamate herbicide, propham

First American commercial use of organophosphate insecticides

Introduction of toxaphene and its widespread use in American agriculture

1947 Legalization of the Federal Insecticide, Fungicide, and Rodenticide Act (FIFRA)

1948 Introduction of dieldrin and aldrin, two highly persistent insecticides

Introduction of malathion, one of the safest organophosphate insecticides

Discovery of captan as a fungicide

1953 Discovery of the insecticide, diazinon

Amendment to the Food, Drug, and Cosmetic Act of 1906 establishes pesticide tolerance levels.

Introduction of Deet, one of the first successful insect repellents

Introduction of carbaryl, the first successful carbamate insecticide

Introduction of gibberellic acid for stimulation of growth in horticultural crops

Introduction of the triazine herbicide, atrazine and the nonselective herbicide, paraquat

1960 Introduction of trifluralin, the first dinitroaniline preemergence herbicide

1961 Introduction of the fungicide, mancozeb

1962 Rachel Carson publishes *Silent Spring*, attacking the use of DDT and other pesticides.

1965 Introduction of aldicarb, the first soil-applied insecticide and nematicide

1966 Introduction of carboxin, the first systemic fungicide

1969 The state of Arizona enacts policy to eliminate the use of DDT.

The USDA publishes the Mrak report, which outlines policy for restricting the use of persistent pesticides.

1970 The US Environmental Protection Agency (EPA) is officially formed.

1971 Introduction of glyphosate, the first systemic and nonselective herbicide

1972 Enactment of the Federal Environmental Pesticide Control Act (FEPCA)

1973 The EPA eliminates the domestic use of DDT.

1974 The EPA establishes the first set of worker safety standards for reentry into pesticide-treated areas.

1975 Elimination of the insecticides aldrin and dieldrin for all uses except for termite control

1976 Elimination of the use of mercury-based pesticides by the EPA

1978 Training program completed for commercial pesticide applicators.

The EPA first issues a list of restricted use pesticides.

1981 Enactment of the Comprehensive Environmental Response Compensation and Liability Act (Superfund) for major toxic cleanup projects

The World War II era was not only significant for the development of insecticides. A byproduct of the tire production process, critical during a time of war for production of military vehicles, was discovered in 1942 to selectively kill certain weeds in turfgrass. This new material, discovered to be 2,4-D, proved revolutionary for the agricultural and the green industry. An analog of 2,4-D, 2,4,5-T was discovered two years later and was rapidly developed for use as an aerial brush control herbicide. Once the war ended in 1945, both DDT and 2,4-D were rapidly incorporated into production agriculture and the green industry and fostered a new enthusiasm for new and even better pesticides (Figure 1.7).

HISTORY OF CHEMICAL AND PESTICIDE USE IN TURF: 1950 TO 1975

The 25 to 30 year period following the end of World War II saw both tremendous strides in the proliferation of pesticides and also tremendous awakenings in the perception of them. Materials like DDT spawned a revolution in insecticide development. Dozens were introduced in the 10 years following World II, including some still used today like Deet, malathion, and diazinon. The effectiveness and diversity of these new insecticides afforded luxuries to agriculture that had never been seen before. New herbicides like 2,4-D also revolutionized weed control in grass crops like corn or wheat, and offered new weed control options for the young green industry. The rapid increase in demand for these new pesticides was also significant to scientists who researched pest control and to pesticide manufacturers. Heavy emphasis was placed on development and investigation of new pesticides that would capture some of the insatiable demand from agricultural consumers. By 1960, herbicides like atrazine and trifluralin gave agriculturists and turf managers alike new options for grassy weed control. New fungicides that were significant for both agricultural and human health uses include:

- *Captan*
- *Mancozeb*
- *Benomyl*
- *Thiabendazole*

FIGURE 1.7 Early advertisement for DDT.

While the rapid development of pesticides during this period translated into immediate crop yield increases, this benefit also overshadowed the potential consequences of their use.

Government regulation of pesticides was not always via a common agency as it is today. The United States Department of Agriculture (USDA) and the Food and Drug Administration (FDA) each held responsibility for ensuring responsible but optimal production and distribution of food crops. However, swayed by the tremendous yield increases and subsequent export capabilities from United States agriculture, the potential negatives of pesticide use were not given due priority. The Miller Amendment of the Food, Drug, and Cosmetic Act of 1906 established acceptable pesticide residue thresholds for food crops. However, little attention was given to the environmental effects of pesticides, allowing manufacturers and agricultural consumers considerable freedom. The eyes of the nation and of the world were soon to be opened.

In 1962, Rachel Carson (Figure 1.8), a wildlife biologist, published *Silent Spring* and forever changed the world's view of pesticide use. People detached from the world of agriculture were introduced to the idea that pesticides could indeed contaminate the environment and have harmful effects on living organisms, including humans. While some members of the scientific community challenged perceived

PESTPOINTER

The effectiveness and diversity of these new insecticides afforded luxuries to agriculture that had never been seen before. New herbicides like 2,4-D also revolutionized weed control in grass crops like corn or wheat, and offered new weed control options for the young green industry.

inaccuracies in Carson's opus, *Silent Spring* had an effect that the author likely would never have anticipated. Her untimely death two years after publication of the book preceded much of the social and political upheaval that could later be credited back to the publication of *Silent Spring*. While social opinion and political groups rallied around the content of the book to challenge the pesticide industry and demand change, the pesticide industry remained somewhat unfazed. Release of new pesticides and promotion and use of existing ones continued in the agricultural industry, albeit with a new social conscience challenging its pursuits.

By the late 1960s government finally took action against the use of pesticides. Arizona placed a moratorium on the use of DDT, the pesticide most heavily criticized in *Silent Spring*, in 1969. The USDA developed regulations concerning the use of persistent pesticides the same year and published the Mrak Report, a document which laid

FIGURE 1.8 Rachel Carson, author of *Silent Spring*.

PESTPOINTER

The USDA developed regulations concerning the use of persistent pesticides and published the Mrak Report, a document which laid groundwork for environmental protection policy and eventually led to the formation of the Environmental Protection Agency (EPA) in 1970.

groundwork for environmental protection policy and eventually led to the formation of the Environmental Protection Agency (EPA) in 1970. While the EPA's directive was not extreme, it did set forth to create greater accountability within agriculture and the pesticide industry with which it had become so closely tied.

HISTORY OF CHEMICAL AND PESTICIDE USE IN TURF: 1975 TO PRESENT

Environmental awareness has been the underlying influence on turfgrass chemical and pesticide use in the past 25 to 30 years. With the advent of the United States EPA in 1970, pesticides began to be regulated by the "new sheriff in town." Not only were existing pesticides subjected to greater scrutiny, mandates of extensive preliminary environmental and toxicity studies were applied to new pesticides slated for release to the consumer market. This increased testing has become the standard for pesticide registration and release, but its immediate effect was seen via fewer released products on the market and the loss of those that had been available. Chemical companies understandably bear the cost burden of testing materials they are developing, so only those determined to have a sufficient market share upon release are seen as fit for full development. Along with increased required testing during pesticide development has come the foresight that more benign materials are those which will best pass the scrutiny of these tests. Thus, new chemistries are now pursued which do not have the potentially toxic or contaminating effects of their predecessors. This evolution has been a necessary one, but has also resulted in pesticides that may not pack the punch that older materials once had.

Another trend, which has evolved in the more recent past, is the use of reduced rates of new or existing pesticides. While a component of this trend is proactive environmental responsibility, another has been mandated for some pesticides. Certain pesticides have greater risk associated with higher use rates so determining effective use strategies at low rates has allowed some materials to remain registered and therefore marketable. Most new pesticides and other turfgrass chemicals are designed for low use rates, such that their toxicity and contamination potential are necessarily low. Greater diversity among available pesticides in the modern day has also been spawned by the appearance of herbicide resistance. Continued use of similar pesticides can result in the proliferation of pests that have natural adaptations to them. While many pesticides that were prone to the development of resistant pests have been eliminated for this or for other reasons, increased pesticide rotation has also become a

PESTPOINTER

Certain pesticides have greater risk associated with higher use rates so determining effective use strategies at low rates has allowed some materials to remain registered and therefore marketable.

standard practice to avoid the onset of resistance. Both pesticide rotation and lower use rates of pesticides have helped reduce the potential for environmental contamination.

Not all current trends in the green industry relative to pesticide use concern the pesticides themselves. As will be discussed in a later chapter, the 1960s and 1970s spawned the concept of Integrated Pest Management (IPM), which later was incorporated into the larger theme of Best Management Practices (BMPs). These management philosophies were not immediately accepted on a broad scale and the first practitioners were often labeled by their peers as being "closet" environmentalists. However, IPM has become a recognized standard for modern turfgrass pest management and is now expected of turfgrass managers. The implementation of IPM has resulted in great strides toward minimizing pesticide input and maximizing the public reputation of turfgrass managers.

One of the most recent and also most controversial strides taken in the arena of pesticide management is the ongoing development of herbicide-resistant turfgrasses. The technology is certainly appealing to the practitioner. Now biotechnology is used in every aspect of new turf development and pesticide production. The future will allow the unprecedented use of nonselective herbicides like glyphosate on sensitive species like Kentucky bluegrass or creeping bentgrass. While the

---

PESTPOINTER

Now biotechnology is used in every aspect of new turf development and pesticide production. The future will allow the unprecedented use of nonselective herbicides like glyphosate on sensitive species like Kentucky bluegrass or creeping bentgrass.

18 Turfgrass Chemicals and Pesticides: A Practitioner's Guide

FIGURE 1.9 Public sentiment often does not support genetically modified organisms (GMOs).

technology is innovative, it does have its critics. General criticism exists worldwide over the use of genetically modified organisms (GMOs) for production agriculture. The primary basis for this criticism is that GMOs are in defiance of nature and that scientists are foolishly "playing God" (Figure 1.9). Other concerns, such as development of highly resistant weeds and sensitivity of neighboring plants, have made acceptance of herbicide resistant crops and turf a gradual process, but it remains a significant breakthrough that will be a part of our future.

PUBLIC PERCEPTION OF TURFGRASS CHEMICALS AND PESTICIDES

Public perception of chemical and pesticide use has, with the technology itself, evolved dramatically over the course of history. From ancient times until the early to mid 20th century, pesticide use was predominantly limited to the agricultural sector. The proportion of the world's population devoted to agriculture was much higher in these times so the influence of pesticides was seen to promote and maintain personal livelihood, even though available pesticides at the time were, by today's standards, highly toxic. The anti-pesticide sentiment, which is well entrenched in today's society, was not evident in those days and would have had little practical impact, considering the strong vocal influence of the agricultural community and the relative lack of knowledge regarding the effects of pesticides on either human toxicity or the environment. It is probable that human casualties resulted from early pesticide use, considering many contained such dangerous compounds as:

- *Lead*
- *Mercury*
- *Cyanide*
- *Arsenic*

However, because of the comparatively imprecise means of determining cause of death in those days, the public perception of pesticides as beneficial to crop production and grower livelihood, and the less efficient means of transmitting information, continued without significant opposition. Imagine how the public's view of agricultural pesticide use in the late 1800s or early 1900s might have been different if they had the media and cyberspace access we have today.

Public perception of pesticide use has changed most dramatically since World War II for a number of different reasons. First, following the war, a pesticide revolution occurred which had dramatic impact on pest control. Two materials had the greatest causal influence on this revolution: DDT and 2,4-D. While they targeted different pests, their immediate impact made possible the rapid expansion of acreage used for agricultural pursuits. With expanding acreage for existing farms came a corresponding decrease in the number of people who considered agriculture their livelihood. This was not a trend isolated to the World War II era. When the United

States achieved independence, better than 95 percent of the population considered agriculture their means of livelihood. The Industrial Revolution, which began in the mid 1800's, brought thousands off the farm and into the cities where new means of earning a living represented new promise for a better life. This trend has continued to the present day where fewer than 5 percent of the American population is directly involved with production agriculture. The detachment of the majority of the population with agriculture has correspondingly led to decreased familiarity with practices such as pesticide use. It is easy to criticize or jump onto the anti-pesticide bandwagon when the vast majority of critics are not themselves practitioners who consider pesticide use a part of their livelihood.

Another significant step in the evolution of public perception of pesticide use occurred in 1962 with the publishing of Rachel Carson's *Silent Spring*, one of the first books which addressed the potential dangers of pesticide use, both to human beings and the natural environment. Although many scientists now concede that there were numerous inaccuracies in *Silent Spring*, the effect on the public was irreversible. Like in no other time previously, the public voice began to stir and question the effects of agriculture and pesticide use. The timing of *Silent Spring* could also not have been more perfect for its intended impact. The 1960s were and always will be known as a decade where the public learned they could seize a cause and implement change through vocal activism and challenging the status

---

PESTPOINTER

The detachment of the majority of the population with agriculture has correspondingly led to decreased familiarity with practices such as pesticide use.

PESTPOINTER

Silent Spring and numerous other books that followed were the foundation on which a new generation of environmental activists built their cause upon to stimulate their perceived need for change.

quo. *Silent Spring* and numerous other books that followed were the foundation on which a new generation of environmental activists built their cause to stimulate their perceived need for change. Government responded to this new vocal opposition to pesticides:

- *States like Arizona banned the use of DDT in agriculture in 1969.*
- *The USDA was developing policy to reduce and eventually eliminate the use of persistent pesticides.*
- *One year later, the EPA was formed to specifically regulate the use of all pesticides.*
- *By 1973, DDT was effectively banned in all 50 states, although it is still widely used today in production agriculture around the world.*

Chemical and pesticide use in turfgrass has been affected by the same trends just described for agriculture. Turfgrass management from the 1800s through the mid-1900s was a proportionately tiny component of total agricultural interests. Few chemical products were specifically targeted for turf use and public perception of turfgrass was certainly not what it is today. Golf, the unequaled catalyst for new technological development in turf, was a very young sport at the time and considered elitist by the great majority of the public. Since most people did not have access to the game, golf course management was of little interest to the public, as were the pesticides used on golf courses. The scientific advancements that produced a wave of new pesticides for agriculture also resulted in their use on turf.

Following World War II, golf began to gain more widespread acceptance among the general public and, as it did, more people wanted well-maintained turf on their own property. The scientific community responded as turfgrass began to develop rapidly as a recognized agronomic discipline in the 1950s. More pesticide products were specifically tested on turf and, along with improved turfgrass varieties and maintenance practices, turfgrass science rapidly improved golf course turf and turf used for a variety of commercial and residential purposes. Public perception of turf pesticide use was also heavily influenced by the tumultuous 1960s. If reaction to *Silent Spring* and the environmentalist movement that followed has forever changed agriculture, so too has been the case for turfgrass.

Practitioners in the green industry have an additional challenge not faced by agriculture, however. Agriculture has seen significant declines in numbers of practitioners and criticism of pesticide use comes from miles away from the farm. Turfgrass has become ubiquitous, as urban population centers are today home to millions of acres of maintained turf, be it in the form of golf courses, athletic fields, or lawns. With turf's development, literally, in our own back yards, has come heavy criticism of pesticide use that we as the public are more

---

PESTPOINTER

The scientific advancements that produced a wave of new pesticides for agriculture also resulted in their use on turf.

> ## PESTPOINTER
>
> The close proximity of turfgrass to the majority of the population has made environmentalist opposition to pesticide use an easy concept to sell to the public.

likely to come into physical contact with. Pesticides are easily obtainable over the counter and determining the extent of their use remains challenging. The close proximity of turfgrass to the majority of the population has made environmentalist opposition to pesticide use an easy concept to sell to the public. Who wouldn't be more attentive when, instead of details regarding farm use of pesticides hundreds of miles away, the public's attention is drawn to the turf their children are playing on?

Aggressive environmental lobbying and manipulation of malleable public opinion are today's enemies of the green industry, as turf has become a scapegoat for a myriad of urban environmental problems. Despite concerted efforts from the green industry to implement IPM programs, use reduced rate pesticides, and to proactively manage pesticide use responsibly, public concerns over pesticide use practices by turfgrass managers persist. Scientists and representatives from the turfgrass industries in many states have begun lobbying for the cause of the green industry, but there is still an uphill battle to fight. Armed only with persuasive but fact-based arguments in support of turfgrass management practices, lobbyists for the green industry have benefited us all. However, until the different sectors of our industry can unite and present a collective voice of support for what we do and why, the public will continue to be swayed by the passionate cries of foul play that come from influential environmentalists.

FUTURE TRENDS

What lies in store for chemical and pesticide use in the green industry remains open to speculation. Recent trends indicate that the industry will likely lose many more currently available products than will be replaced by new chemistry and new products. Table 1.1 lists some of the potential problems that may exist with turfgrass pesticides, what tests indicate these problems, and the likely fate of the pesticide. In most cases, reasons for a particular material to lose registration are a function of research. Research results showing that a pesticide may have negative characteristics are legitimate problems that should be addressed by eliminating production and use of that material. Chemical companies can ill afford to continue production and marketing of potentially hazardous pesticides when these discovered hazards are just the fuel the anti-pesticide groups seek to sway public perception. Most pesticides that have problematic characteristics are now identified as such before they are ever released for sale, so the loss of available pesticides usually is pertinent to older materials that were not tested as extensively during their development. The current process for development and release of a new pesticide now requires sometimes 10 years or more and costs millions of dollars. Only those materials that can endure the rigorous testing over

---

PESTPOINTER

Most pesticides that have problematic characteristics are now identified as such before they are ever released for sale, so the loss of available pesticides usually is pertinent to older materials that were not tested as extensively during their development.

TABLE 1.1 Reasons why pesticides are banned or eliminated.

Reason	Research That Substantiates the Problem	Result
Persistence	Animal metabolism studies, soil longevity studies, ability of microbes to degrade pesticide	Product deregistration, discontinued product development
Toxicity	Acute toxicity studies on animals, pesticide accumulation in animal tissues, effects of pesticides on animal reproduction	Product deregistration, discontinued product development
Production costs/ Market share	Inability to formulate pesticide into a usable product, chemical company market analysis and sales projections	Discontinued product development
Mobility	Drift studies, soil leaching studies, surface runoff studies, pesticide accumulation in wells, lakes, and ponds	Product deregistration, discontinued product development
Unacceptable injury to desired plants	Drift studies, surface runoff studies, efficacy trials on target pests and crops	Product deregistration, discontinued product development, limited crops for which the pesticide is labeled

this period are deemed worthy of such elaborate costs. Therefore, the risk involved with newly released pesticides is expectedly very low.

Of greater concern to the green industry than research-based elimination of some pesticides is the collective and persistent voice of environmentalists who condemn pesticide use on principle and without substantiation. This vocalized opinion is not one that can be underestimated, as proponents of pesticide elimination are not bound by the same fact-based code of ethics that the scientific community is held to. As such, environmentalists can use isolated examples and passion to fuel public fear, skepticism, and other damaging sentiments that conflict with the objectives of the green industry. Pesticides are not the only components of the green industry which have become media targets. Other products such as synthetic fertilizers and ice-melting materials have also been identified and targeted as possible environmental contaminants and therefore have come under serious criticism. We also experience opinion from the news media that offers general criticism of lawn care and landscaping, the proponents of which condemn the overall monetary and resource consumption of our industry. Turf and landscape professionals, therefore, are and will be faced with growing distrust by the consumers who rely on them.

While this trend may be disturbing or even alarming to the practitioner who relies on pesticides to maintain quality and pest-free greenspace, we have no choice but to embrace this trend as the reality of our profession. However, the positive sides to this are also numerous and should offer encouragement to green industry professionals. The homeowner and/or member of the public have every right to be skeptical of pesticide use by the green industry. As members of this great industry, we are however armed with a valuable weapon: knowledge. We are expected to have practical and technical expertise in both turfgrass management and in the selection and use of pesticides within turf. When this knowledge can be demonstrated and passed along to the consumer, the likelihood of distrust or skepticism diminishes. Sure, there are and will continue to be a subset of the public consumer base who simply will not favor the use of pesticides. Frankly, there is little that can be done to sway their opinion, but knowledgeable practice on our part does not exacerbate their concerns.

Our greatest contribution of knowledge is to the majority of the consumer base who, despite the negative light in which pesticides are often mentioned, still favor their use to maintain quality turf. When we can demonstrate to this majority that we do indeed have the expertise they believe us to have, this only reinforces the strength of the green industry and therefore our professional credibility. A colleague once presented this idea to me in a very enlightening way. Consider the members of the public consumer base who are willing to pay hard earned money for professional lawn and landscaping services or for private country club memberships. Most have significant levels of education, which have allowed for their financial success (e.g. doctors, attorneys, accountants, business executives, etc.). It only follows that they would expect knowledge and professionalism from those who are providing technical service relative to turfgrass management. If we as members of the green industry cannot demonstrate expertise and the ability to rationalize what we are doing, why should they extend their trust to us?

The future continues to be bright for the green industry, but there is growing expectation that we have a level of knowledge which

---

PESTPOINTER

Our greatest contribution of knowledge is to the majority of the consumer base who, despite the negative light in which pesticides are often mentioned, still favor their use to maintain quality turf.

exceeds that of our consumer base. It is therefore our responsibility to maintain that edge and to have the knowledge which is expected of us. The contents of this text are designed as a practical but thorough technical resource for present and future members of the green industry, such that its strength and credibility will continue forward in an ever-changing world.

SUMMARY QUESTIONS

- *What has been your philosophical approach to selecting and using pesticides?*

- *Have you experienced negative feedback from clients regarding pesticide use at your facility or on the job?*

- *What has your approach been when you have been challenged by questions or criticism of your pesticide use practices?*

- *What is your opinion of those who challenge or criticize what you do for a living?*

- *How would you rate your knowledge base when it comes to explaining the need for or the theory behind your pesticide applications?*

- *How does your base of knowledge stack up to the expectations of your clients?*

Remember: There is no right or wrong answer to any of these questions but they are worth considering when you evaluate your role as a turfgrass and pesticide practitioner and can help you understand the point of view your clients may be coming from.

Chapter 2
Basic Chemical and Pesticide Principles

Understanding the behavior of pesticides, why some are potential environmental contaminants, why some are toxic, and why so many people fear them, goes well beyond the scope of the practitioner. The old saying goes, "Familiarity breeds contempt." On the flip side, things we don't fully understand tend to cause fear and skepticism. Many practitioners and even more of the general public do not understand the science that is behind pesticides and other turfgrass chemicals. Because of this lack of understanding, pesticides are popular targets for public concern and environmentalist scorn. As a member of the scientific community, I can empathize with the bad rap pesticides get because they are simply not understood. As a turfgrass specialist who must present scientific information in an understandable format, I also recognize that the information out there is often complicated, boring to read, or otherwise cumbersome. Members of the green industry need to know why this knowledge gap exists and how best to overcome it.

Pesticide manufacturers employ Ph.D.-educated biologists, pesticide scientists, toxicologists, and chemists. Their roles within these companies extend to depths of detail that far exceed what the practitioner needs to know in the workplace or what the end consumer

PESTPOINTER

Practicing members of the green industry represent a link between the end consumer and the companies that develop, manufacture, and market pesticides.

would ever want to know. Practicing members of the green industry represent a link between the end consumer and the companies that develop, manufacture, and market pesticides. Therefore, the exhaustive amount of detailed information that is available for a particular pesticide needs to be condensed down into a user-friendly format that has the essential core elements to it. That is what herbicide labels and Material Safety Data Sheets (MSDS) represent. The information contained in a four-page pesticide label and an MSDS of similar length represents years of research and millions of dollars spent to bring a pesticide out of the laboratory and onto the shelf. So, the next time you find yourself scratching your head over the contents of one of these forms, remember that what you see is the highly condensed version. Scary perhaps, but this information is vital to applying chemicals safely and properly.

The perception the public has of pesticides often does not take into account the science behind these materials. Commercial practitioners and consumers alike take into account the basic instructions for use and perhaps some basic safety information. What is lost in the shuffle is that these instructions and safety guidelines establish a reputable basis for a pesticide to be marketed. Were there not this information, a pesticide could never be manufactured and sold. Directions for use represent the results from dozens of scientific studies investigating how a particular pesticide should be best used. Safety guidelines may imply danger but, if a pesticide is marketed, it has endured rigorous testing mandated by the EPA and thus has

PESTPOINTER

While any pesticide can be dangerous if used improperly, many household products can be much more dangerous than pesticides.

EPA's "stamp of approval." Many environmentalists refer to pesticides as "poisons" and would have the public believe that pesticide manufacturers have released these materials on society like the plague. While any pesticide can be dangerous if used improperly, many household products can be much more dangerous than pesticides (Figure 2.1). The reason why pesticides are subjected to a greater

FIGURE 2.1 Many common household products are more dangerous than most pesticides.

amount of criticism is not because they are scrutinized less by the EPA. It is simply because the public knows less about them.

This chapter is designed to help bridge the gap between the complexities of pesticide chemistry and the applications of these materials in the consumer market. A detailed understanding of chemistry, soil science, toxicology, or plant physiology is not required. We can leave that to chemists and researchers. However, by being able to grasp some basic principles and to directly see how they apply to the process of selecting and applying pesticides, the practitioner will be better able to comprehend their own actions and to explain them to their clients. The goal herein is to provide ample detail but to present it in such a way that it has meaningful relevance to the workplace. Let's get started!

PESTICIDE FORMULATIONS

The physical form that we see for pesticides is often a far cry from how they would naturally occur. All pesticides are fundamentally composed of an active ingredient, which is the chemical compound directly responsible for acting against a particular pest. The means by which a pesticide or chemical active ingredient specifically affects its target is called a mode of action. Mode of action for specific pesticides and other chemicals will be discussed further in later chapters. The

PESTPOINTER

All pesticides are fundamentally composed of an active ingredient, which is the chemical compound directly responsible for acting against a particular pest.

> **PESTPOINTER**
>
> Inert ingredients to help pesticides get into plant tissues may include water, organic solvents, additives, or even fertilizer.

way a pesticide active ingredient is packaged so that it has shelf life and so that it can be efficiently applied is via a pesticide formulation. The formulation includes the active ingredient and all other inert ingredients in a pesticide product. Inert ingredients to help pesticides get into plant tissues may include water, organic solvents, additives, or even fertilizer. All pesticides are formulated. None occur naturally in a form in which they could be efficiently applied to plants. This makes sense since most active ingredients in turfgrass chemicals and pesticides are synthetic, originating in the laboratories of the manufacturers. Even natural or organic products must be formulated so that they can be effectively packaged and distributed to consumers for use in the workplace. While many pesticide and chemical products originate in the laboratory, the role of the chemists is often preceded by a market analysis. The market analysis is highly critical to determine whether or not a developed product will be sufficiently marketable to justify the costs of discovery, development, and manufacturing.

Most pesticide manufacturers actually have two different types of chemists. The first type is a chemist who searches for good active ingredients. These chemists take a chemical compound and modify it until it has the properties they are looking for. Active ingredient chemists may spend years searching for a compound that is usable for pesticide purposes. Often, a compound they are interested in will have properties they are looking for but may not be stable. They must then change the chemical properties so that the compound doesn't break down. Otherwise, the compound wouldn't be worth

> **PESTPOINTER**
>
> Active ingredient chemists may spend years searching for a compound that is usable for pesticide purposes.

producing and marketing. Another challenge faced by active ingredient chemists is the reproducibility of the process involved in creating a compound. Pesticide manufacturers must be able to produce a compound efficiently in order for it to be cost effective. Much like for the automobile industry, a pesticide needs to be set up for assembly line production to be profitable. Every year, thousands of promising compounds developed by these chemists are dismissed because the chemical procedures required to produce the compound are too cumbersome or too expensive.

The discovery of an active ingredient is certainly an important part of developing a pesticide, but it is really only the beginning. Once an active ingredient compound is identified, it then must be formulated. Formulation chemists must take the active ingredient and package it in a form that is usable to the practitioner. This process of developing a suitable formulation is not easy and is another reason why some active ingredients never become products available

> **PESTPOINTER**
>
> Formulation chemists must take the active ingredient and package it in a form that is usable to the practitioner.

> **PESTPOINTER**
>
> Unstable compounds may not have adequate shelf life if they are packaged in a liquid formulation. These materials are often packaged as dry formulations such as powders or granules so that they retain stability until the practitioner uses them.

for purchase and use. The characteristics of an active ingredient compound heavily influence how it is formulated. For example, many active ingredient compounds are not soluble in water. For a pesticide that needs to be applied as a liquid, an insoluble active ingredient can pose a major problem to the formulation chemist. The chemist must determine how to package the active ingredient or the product will not be marketable. Another factor that influences pesticide formulations is the stability of the active ingredient. Unstable compounds may not have adequate shelf life if they are packaged in a liquid formulation. These materials are often packaged as dry formulations, such as powders or granules, so that they retain stability until the practitioner uses them. Marketing can also influence the choice of a pesticide formulation. If a manufacturer wants to be able to sell a combination product (e.g., a herbicide on a fertilizer carrier), the active ingredient needs to be properly formulated to accommodate this purpose. Just what different formulations are out there?

All matter in the universe can occur as a solid, liquid, or gas. Formulations can generally be broken down into these same three categories. Gaseous formulations are the least common for turf pesticides. They would include aerosols, foggers, and fumigants. Aerosols and foggers are fairly common pesticide formulations for in-home use (Figure 2.2) or use by exterminators but aren't common for turf

FIGURE 2.2 Insect repellent, a common household aerosol pesticide.

applications since they don't allow for adequate coverage of the affected areas. Fumigants are sometimes used in turf situations. An excellent example is methyl bromide, which is often used for golf course renovation. Methyl bromide effectively sterilizes the soil, killing all existing vegetation and even seeds. However, the fumigation process in turfgrass systems is cumbersome and is usually restricted to high profile areas like putting greens. Plastic covers must be secured over the target area to keep the methyl bromide gas trapped in once it is inserted. Fumigation can also be dangerous. Methyl bromide is highly toxic to applicators so its use must be handled with extreme caution. Without a doubt, the demands of applying pesticides to larger areas of turf make necessary more user-friendly formulations like solids or liquids.

Solids and liquids are the most common formulations for pesticides and chemicals used in agriculture, including turfgrass systems. Some common formulations for turfgrass pesticides and chemicals are listed in Table 2.1. Solid or dry formulations are manufactured, packaged, and sold in the dry form but may be applied as a dry material or in a liquid spray solution. Dry formulations that are applied dry, or sold in a ready-to-use form, include dusts, baits, and granules. All three of these formulations have low concentrations of active ingredient since they are ready to apply as packaged.

Dusts are not commonly used for turfgrass applications but have utility in smaller areas like gardens or ornamental landscapes (Figure 2.3). They are usually applied to cover the surfaces of plants that a particular pest is prone to target. Many fungicides and insecticides come in dust formulations but dusts are rare for herbicides or other chemicals. Dusts used to be more common for broad scale applications of materials like DDT (Figure 2.4) but this method is no longer considered safe or practical.

PESTPOINTER

Dusts are advantageous in that they are easy to use and treated plants are easy to identify. However, dusts are not useful for application to larger areas and the fine particle size can represent a hazard to the applicator if the dust is inhaled.

TABLE 2.1 Common dry and liquid formulations used for turfgrass chemicals and pesticides, or for facilities pest management.

Formulation	Abbreviation	Dry/Liquid	Method of Application	Description
Dust	D	Dry	Ready to use	Very fine particles.
Bait	B	Dry	Ready to use	Larger pieces, designed for pests to eat.
Granule	G	Dry	Ready to use	Pellet size particles.
Water dispersible granule	WG/WDG	Dry	Liquid spray	Similar in appearance to granules, break down and suspend in spray solution.
Water-soluble granule	WSG	Dry	Liquid spray	Similar in appearance to granules, dissolve readily in spray solution.
Wettable powder	WP	Dry	Liquid spray	Fine powdery material, suspends but doesn't dissolve in spray solution.
Soluble powder	SP	Dry	Liquid spray	Similar to wettable powder but dissolves in spray solution.
Dry flowable	DF	Dry	Liquid spray	Similar in concept to wettable powder but measured by volume, not by weight.
Flowable	F/FL	Liquid	Liquid spray	Thick, concentrated liquid that is diluted in spray solution.
Emulsifiable concentrate	EC	Liquid	Liquid spray	Organic liquid with ingredients to allow for "oil and water" to mix effectively.
Microencapsulated concentrate	M/MC/MEC	Liquid	Liquid spray	Active ingredient particles are encased in a polymer coating so they can suspend in spray solution.
Solution concentrate	SC	Liquid	Liquid spray	Active ingredient is packaged in a solvent that naturally mixes well with water.
Liquid	L/SL	Liquid	Liquid spray	Concentrated liquid solution that is easily diluted in water.
Ready to use spray	RTU	Liquid	Liquid spray	Diluted liquid solution for immediate use on target pests.

Basic Chemicals and Pesticide Principles 39

FIGURE 2.3   A dust insecticide coats the surface of a treated ornamental plant.

FIGURE 2.4   Dust application of DDT in mid-20th century agriculture.

Dusts have specific characteristics:

- *They are easy to use.*
- *Treated plants are easy to identify.*
- *They are not useful for application to larger areas.*
- *The fine particle size can represent a hazard to the applicator if the dust is inhaled.*

Baits are also not common for application to large areas but can be useful for household or maintenance building management of small animal pests such as rodents. Baits have specific pros and cons:

- *They are easy to use.*
- *Evenness of application is not a concern since ingredients in the bait attract the pest to it.*
- *It can be uncertain if the target pest has been eliminated.*
- *If the pest dies in an inaccessible location, odors can become a problem.*
- *Baits can also be hazardous to children or pets if they are placed in areas where they can be found and eaten.*

For turfgrass facilities like golf course maintenance shops, traps or use of domestic predators like cats have become preferred ways of managing the types of pests that baits would ordinarily target.

Among ready to use dry formulations, granules are the most common for turfgrass applications. Granules have an inherent advantage over other formulations (Figure 2.5):

- *They are similar in appearance to many fertilizers.*
- *They can be applied without the need for separate or specialized equipment like liquid sprayers.*

Basic Chemicals and Pesticide Principles   41

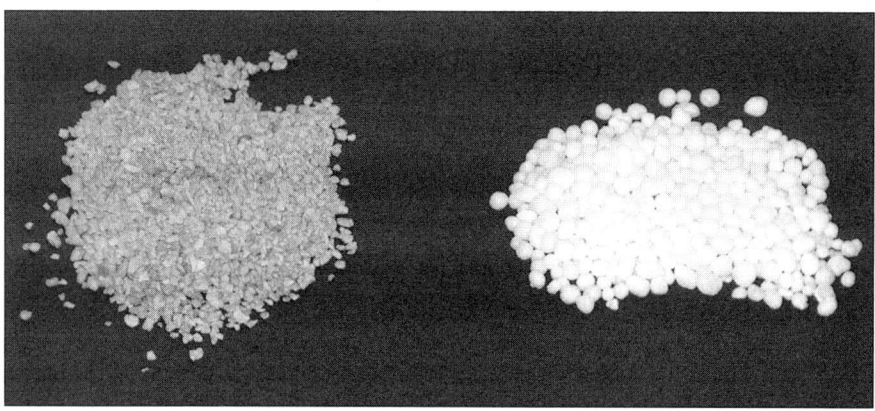

FIGURE 2.5   Particles of a common granular pesticide (left) are similar in size to those for synthetic fertilizers (right).

- *Most turfgrass operations already have spreaders for fertilizer applications so granular pesticides represent a user-friendly format for application (Figure 2.6).*
- *Granular pesticides contain lower concentrations of active ingredient since they are not diluted before they are applied, like many other formulations.*
- *The inactive or inert part of a granule is usually clay or an organic material that binds and stabilizes the active ingredient until the granule is applied. This type of formulation is ideal for pesticides that might be less stable in a liquid formulation.*

Upon application, exposure of the granule to water and other climatic elements gradually releases the active ingredient so that it can reach its target pests. While granules have many advantages, there are also some drawbacks. Care must be taken that granules be applied evenly since they naturally do not provide the type of coverage that can be expected from a liquid application. Because active ingredient concentrations in granules are low, larger amounts of the formulated product are needed for an application. From a manufacturer's standpoint, the costs of producing granules are higher

42 Turfgrass Chemicals and Pesticides: A Practitioner's Guide

FIGURE 2.6 Application of granular pesticides requires similar equipment as for fertilizer applications.

than for other formulations, so the user-friendly format granules offer often comes with higher product costs to the consumer.

Dry formulations that are applied as liquids in a spray solution include water-dispersible granules, wettable powders, soluble powders, and dry flowables. Unlike ready to use dry formulations like granules, these dry materials are packaged and sold in a more concentrated form with higher proportions of active ingredient. For example, granular pesticides used in turf usually have 5 percent or less active ingredient and rarely more than 10 percent. By contrast, dry formulations like wettable powders typically have 50-90 percent active ingredient with the expectation that the material will be diluted in the spray tank.

There are cost benefits to using the more concentrated materials as they usually cost less to manufacture, but these materials require more effort to mix and apply properly. Most solid formulations designed for liquid applications require a weighing scale to accurately define how much needs to be placed in the spray tank.

Water-dispersible granules (WDG) are packaged in a form that is similar in appearance to ready-to-use granules (Figure 2.7). They are sometimes also labeled as wettable granules (WG) or water-soluble granules (WSG) but the idea is similar. These products are all designed for application in a liquid spray solution. Whether this type of product is a WDG, WG, or WSG depends upon how soluble the material is in water. WDG and WG formulations do not readily dissolve in water, but the particles break down and are suspended in the spray solution. Because they don't actually dissolve, the spray solution needs to constantly mixed or agitated to keep the particles suspended evenly. Otherwise, they will tend to settle to the bottom of the spray tank. WSG formulations look much like WDG or WG materials but dissolve readily in water and don't require as much agitation once the spray solution has been initially mixed. These formulations are advantageous because the particle sizes are larger and are thus fairly

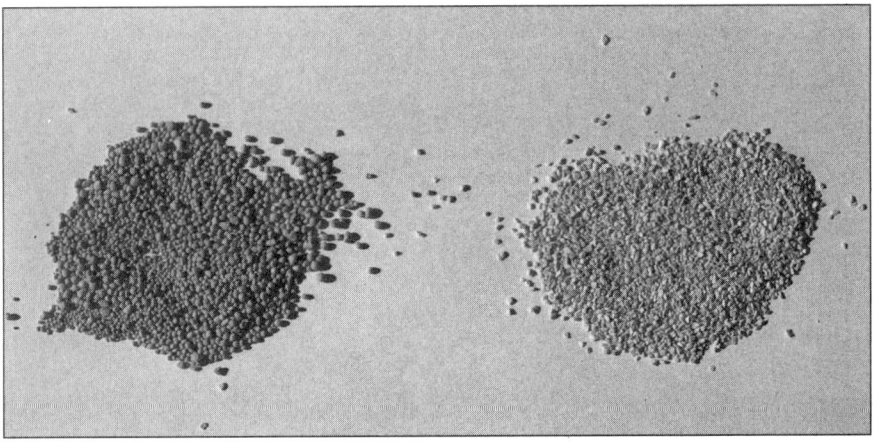

FIGURE 2.7 Particles from two different wettable granule (WG) products.

PESTPOINTER

Water dispersible granules (WDG) are packaged in a form that is similar to appearance to ready to use granules. They are sometimes also labeled as wettable granules (WG) or water-soluble granules (WSG) but the idea is similar.

easy to handle and measure without spillage. The large particle size also offers safety because particles are not prone to drift during handling and mixing and don't pose risk of inhalation to the person preparing for an application. Because these formulations are designed for liquid applications, they must be stored in a dry location and kept dry until they are placed into a spray solution.

Powders are another common class of dry formulations intended for use with liquid spray applications. As the name implies, powders feature much smaller particle sizes than do wettable granules (Figure 2.8). There are two primary types of powders, wettable powders (WP) and soluble powders (SP). The difference between them, as was the

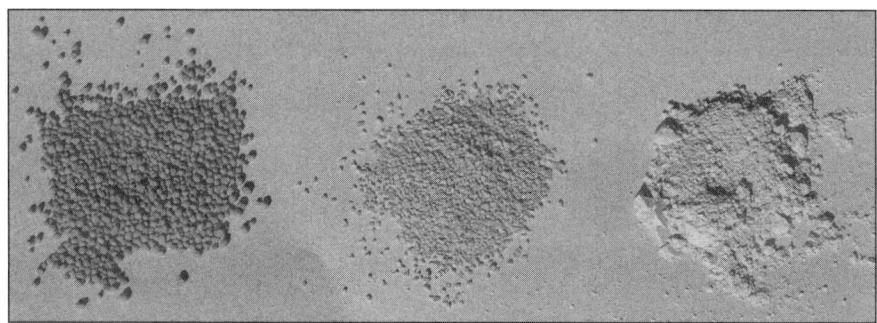

FIGURE 2.8 Powder particles (right) are much finer than those from wettable granules (middle/left).

case with the liquid-applied granules, has to do with how the product behaves in water. WP formulations do not dissolve readily in spray solution so they require agitation to maintain an even concentration. Think of wettable powders as kicking up some silt while wading in a stream. Initially, the water would become foggy with silt particles dispersed in it. However, in time, the water becomes clear again as the silt particles settle back to the bottom. Wettable powders behave much like this. If the spray solution is not agitated, the particles will settle out and the spray solution will contain less pesticide. The leftover powder residue can also contaminate later applications with the same spray equipment if it is not adequately cleaned. Soluble powders are similar in appearance to wettable powders but dissolve readily in water and do not require as much agitation.

Because powders are much finer particles, they do pose risk of inhalation during the mixing process. Many manufacturers have helped address this risk by packaging powders or other dry formulations in dissolvable packets that can be inserted into the spray tank without having to directly handle the product (Figure 2.9). These packets typically contain defined amounts of the pesticide, based upon recommended application amounts, so they can help reduce the time required in preparing for an application. A dry flowable (DF) formulation is similar to the powder formulations in appearance but is unique because the amount required for mixing is measured by volume rather than by weight. This simplifies the mixing process by making it more like that for liquid formulations without the problem with messiness that liquids can sometimes present.

FIGURE 2.9 The herbicide Plateau® is one that is sometimes packaged in dissolvable pouches to minimize product handling.

Liquid formulations are also very common for turfgrass pesticides and chemicals:

- *Liquids are always applied in a liquid spray but their composition can vary considerably.*
- *How liquid formulations vary depends greatly on how readily the active ingredient dissolves in water.*
- *Highly soluble materials are easy to formulate, as they can simply be packaged and sold as a concentrated water-based solution or as a diluted, ready to use product.*

Many common residential or "over the counter" pesticides are formulated in the ready to use or RTU fashion because these products require no mixing and are easy for the untrained applicator to use (Figure 2.10). Water-based concentrates, sometimes abbreviated L or SL, may or may not be colored but characteristically do not change their appearance in spray solution, other than perhaps appearing more diluted and thus paler in color (Figure 2.11). Water-based formulations don't require any agitation to avoid product settling but a thorough initial mixing ensures an even concentration of the spray solution.

The other liquid formulations used for turfgrass pesticides and chemicals are not water-based, usually because the active ingredient does not dissolve in water. These formulations must somehow package the insoluble active ingredient so that it can be applied in a water-based spray solution. Sound tricky? Formulation chemists have cleverly come up with means to get this difficult job done. Examples include:

- *Flowables*
- *Emulsifiable concentrates*
- *Microencapsulated concentrates*
- *Solution concentrates*

Flowables, abbreviated F or FL, are best described as liquid versions of their cousins, the dry flowables or wettable powders.

Basic Chemicals and Pesticide Principles   47

FIGURE 2.10   Ready to use (RTU) formulations are very user-friendly.

Essentially, the active ingredient is suspended in spray solution with the help of the other ingredients in the formulation. The packaged formulation appears as a thick liquid (Figure 2.12) and, like dry flowables or wettable powders, will settle out if the spray solution is not consistently agitated. Flowables often contain higher amounts of active ingredient than other liquid formulations because of their thick, concentrated form. While this increases the use efficiency for the consumer, flowables are difficult to manufacture and therefore not as common as other formulations.

FIGURE 2.11 A true water-based formulation, shown diluted in water (left) and undiluted (right).

Basic Chemicals and Pesticide Principles   49

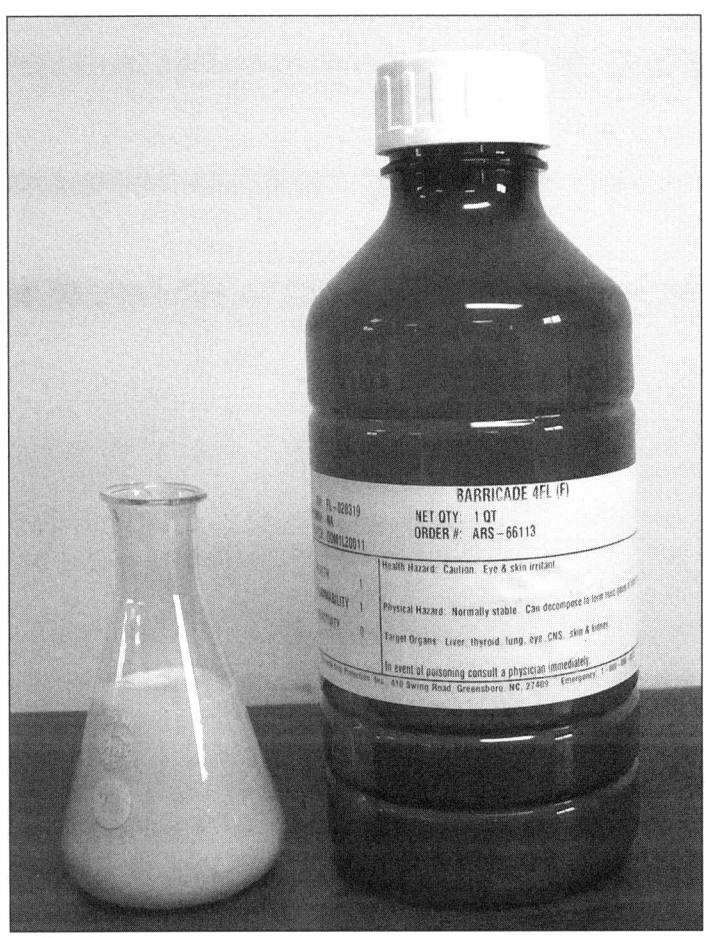

FIGURE 2.12   Barricade 4FL®, a liquid flowable herbicide product.

Emulsifiable concentrates (ECs) are very common liquid formulations for pesticide active ingredients with low water solubility. In many cases, active ingredients may not dissolve in water but will readily dissolve in an organic liquid. Petroleum compounds like gasoline are organic liquids, as are laboratory solvents like toluene or xylene. An EC contains active ingredient that is dissolved in one of these solvents, rather than water. However, another hurdle must be crossed because organic liquids themselves don't mix well with water. Think of an oil slick floating on water as a visual reference. To solve this problem, chemists add what is called an emulsifying agent to the formulation. The emulsifying agent allows the active ingredient and organic liquid component to be evenly dispersed in water, forming a mixture of two different types of liquid known as an emulsion. A common household example of an emulsion is chicken soup broth, which is fundamentally water but has visible oil droplets suspended within it. ECs are usually colored liquids in the packaged form but their reaction when put into water is striking. Once an EC is combined with water, white clouds immediately form in the water (Figure 2.13) and, as the new emulsion is mixed, the spray solution becomes evenly white (Figure 2.14). Once an even emulsion between water and the EC is formed, it remains stable and is not prone to separate or settle out.

FIGURE 2.13 An emulsifiable concentrate (EC) forms thick clouds when first added to water.

Basic Chemicals and Pesticide Principles 51

FIGURE 2.14 Once mixed thoroughly, an EC formulation results in a white spray solution.

A microencapsulated concentrate (MC, M, or MEC) is similar to an EC in that there is an emulsifying agent in the formulation to help the tiny pesticide-containing particles disperse evenly in the spray solution. Because there is a similar ingredient as with the EC, MC formulations tend to form white liquids when mixed with water. The difference is that the active ingredient in an MC is usually even less dissolvable in water than that for an EC. This situation can be remedied but would require greater quantities of more toxic organic liquids to get the job done. To combat this problem, MC formulation chemists encase microscopic portions of the stubborn active ingredient in a tiny plastic shell that releases the pesticide once the spray application is made. Concerns regarding the environmental effects of organic liquids used in EC formulations have made the MC formulation a common alternative or replacement for EC chemistry. However, as you might suspect, the microencapsulation process can be expensive for the manufacturer and may lead to higher consumer costs.

A final category of liquid formulations is the solution concentrate (SC). SC formulations are less complicated than ECs or MCs but also are not feasible for as many active ingredients, limiting their availability. The logic is similar to an old law of mathematics that states that if A=B and B=C, then A must equal C. Here's how a SC formulation works. The active ingredient does not dissolve in water, just like for flowables, ECs, or MCs. However, the active ingredient does dissolve in a liquid that, in turn, can be dispersed evenly or dissolved in water. SC formulations are often already milky liquids in the packaged form (Figure 2.15) so they don't undergo the radical visual change from jug to tank that we see for EC or MC formulations. Finding the right liquid that can serve this "mediator" role is challenging and is why SC formulations are not more common. However, in some cases, this ideal scenario is feasible and can result in both a more cost efficient manufacturing process and the use of less hazardous ingredients.

Now that we've discussed the types of formulations that one might encounter in the workplace, it is necessary to discuss how the formulation is identified to the consumer and what it means to the applicator or to someone preparing a spray solution for application. The formulation is often identified or indicated as an attachment to the name of the product (Figure 2.16). This attachment will include a number and then the abbreviation for the formulation. What exactly does it mean to the practitioner? The meaning of the number will vary, depending upon whether a material is dry or a liquid. For dry materials like granules or wettable powders, the number indicates

Basic Chemicals and Pesticide Principles 53

FIGURE 2.15 Packaged EC formulations are often colored liquids (left) while solution concentrates (SC) are often milky liquids (right).

the percent of active ingredient that is contained in the formulated product. For example, a 2 G formulation is a granule that contains 2 percent active ingredient while a 50 WP is a wettable powder that contains 50 percent active ingredient. This same reasoning also holds true for SP, WG, WSG, and WDG formulations.

Liquid formulations have a different numbering scheme. Some liquids don't have an identified number or formulation abbreviation in their title (Figure 2.17). Usually, these are materials that are simply diluted in water and for which the manufacturer recommends a particular volume of the product that is to be applied to a given size area or for a known quantity of spray solution. Other liquids specify in their title a number and a formulation abbreviation (Figure 2.18). In these cases, the number indicates how many pounds of active ingredient are contained in one gallon of the formulated product. For example,

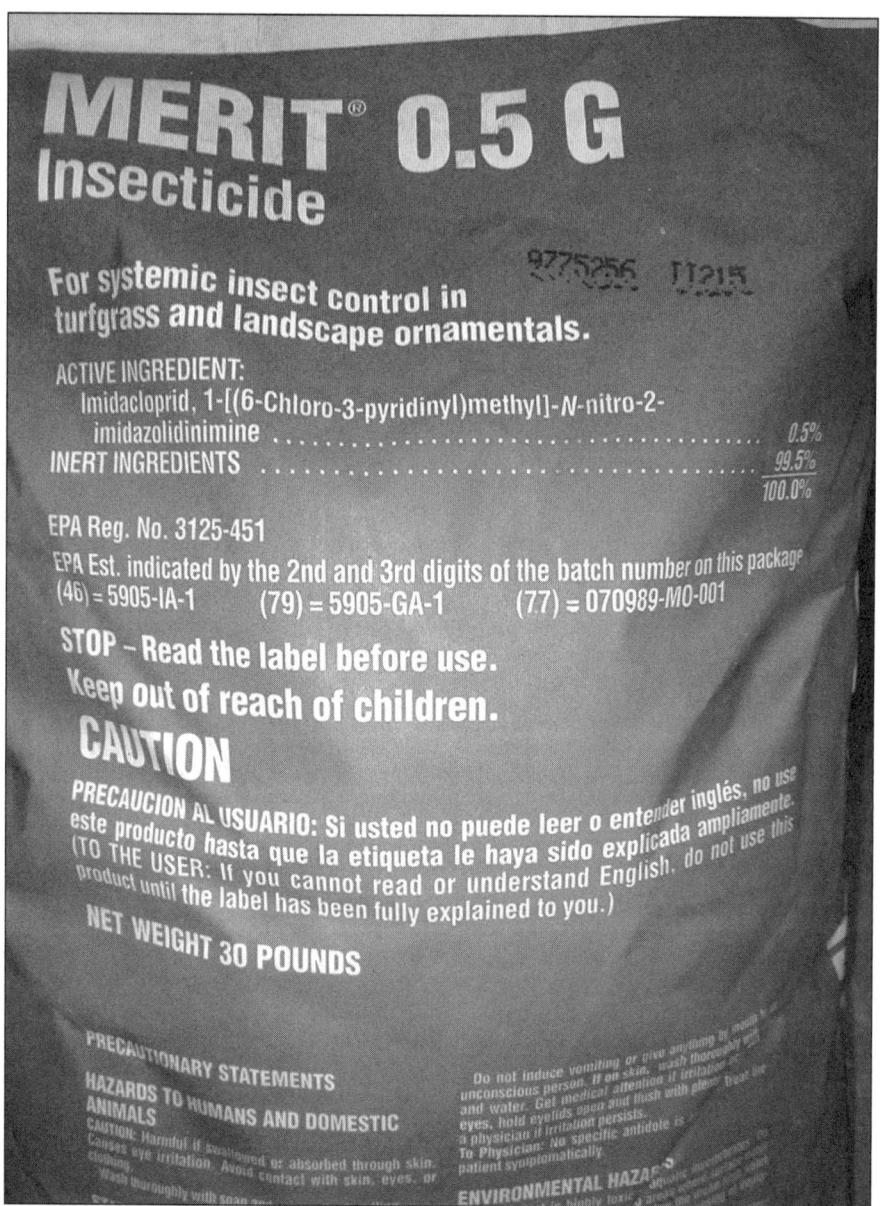

FIGURE 2.16   The granular insecticide Merit® 0.5G includes reference to the formulation in the title.

Basic Chemicals and Pesticide Principles 55

FIGURE 2.17 Many water-based pesticides don't include a reference to the formulation.

FIGURE 2.18 Liquid products created with organic solvents usually include the formulation in their title.

PESTPOINTER

Knowing what the numbering scheme for formulations means can help you properly calculate the correct amount of a formulated material to use for a given application.

a 3.3 EC is an emulsifiable concentrate containing 3.3 pounds of active ingredient per gallon while a 2 SC, as pictured in Figure 2.18, is a solution concentrate with 2 pounds of active ingredient per gallon of product. Most product labels will indicate recommended amounts of product for a given size area, but they usually also indicate recommended amounts of active ingredient for that same area. Knowing the numbering scheme for formulations means can help you properly calculate the correct amount of a formulated material to use for a given application. More on proper calculations for pesticide and chemical applications will be covered in a later chapter.

Many pesticides or other chemicals are formulated in more ways than one to offer flexibility to the consumer. Is there an advantage to using one formulation instead of another? The answer to that question is usually based upon the specific needs at your facility:

- *Granular formulations can usually be applied to larger areas in less time, thus influencing labor costs associated with pesticide applications.*

- *Granular pesticides can also be advantageous in residential lawn care where close proximity of sensitive landscape plantings to turf may increase the risk of drift injury with liquid applications.*

- *Granular formulations are also the most common for combination products containing pesticides and fertilizer.*

Despite these advantages, granular materials are not always the best choice for a pesticide application.

Many pesticides are not available in a granular form so they must be applied as a spray solution. Liquid applications such as this also have their advantages. Sensitive turf areas such as putting greens are not well suited for granular pesticide applications as granules can affect playability of the surface, are highly visible, and can damage turf by introducing a concentrated amount of pesticide to a small area. Liquid applications are essentially discreet in that there is little evidence of the application once it has been made. Liquid applications also result in more even coverage to the target area, decreasing the chances of unintended injury. The mode of action, or the means by which a pesticide's active ingredient targets a pest, can also be better suited to a liquid application. Herbicides that are absorbed by leaves of target weed species are usually more effective when applied as liquids since there is more even pesticide coverage on the leaves. Many fungicides function best when they are able to coat turfgrass leaves and thus be evenly present when the disease-causing organism attempts to feed on the turf plant. These types of materials are also best applied as liquids.

Lastly, the toxicity of many pesticides may point to using a liquid application. Innocent bystanders like birds may consume granular insecticides and be unintentionally harmed or killed. Liquid applications may be justified in such a case to minimize the impact

PESTPOINTER

Many fungicides function best when they are able to coat turfgrass leaves and thus be evenly present when the disease-causing organism attempts to feed on the turf plant.

of the pesticide on organisms it does not target. While all of these issues pertaining to either solid or liquid applications need to be considered, cost issues and equipment limitations also need to be a part of the decision process for selecting formulations.

CHEMICAL AND PESTICIDE SAFETY INFORMATION—USING AN MSDS FORM

Material Safety and Data Sheets, better known as MSDS, are required to be included with pesticides at the time of sale. Most commercial turf and landscape facilities or organizations keep an active database of MSDS, both to satisfy regulations from groups like OSHA or EPA and to serve as a handy reference in case the information is needed in the workplace. Just about any chemical or material used will feature an MSDS. All chemical reagents used in laboratories have MSDS and there is even one for water. Sound crazy? Well, the truth of today's workplace is that we must have precautionary strategies for most any type of situation to minimize the occurrence of problems. We can no longer presume that common sense is adequate to prevent mishaps and other problems. The MSDS is a highly detailed document with information most of us will rarely have to use or reference. Nonetheless, we must have a basic understanding of what information the MSDS does contain so, when the time does come, you can use it appropriately.

PESTPOINTER

The EPA classifies pesticides with one of three signal words: caution, warning, and danger.

Just what does an MSDS contain? Unlike pesticide labels, which can feature information often requiring six to eight pages of text, the MSDS is a detailed but condensed (usually two pages) document that allows for similar presentation of data, be the topic material water, a chemistry kit reagent, or a pesticide. MSDS have 16 general sections, each of which will be discussed briefly in this section. The intent is to decipher what may seem to be complicated terminology, so that you can better understand what the information really means to you as the practitioner.

All MSDS have a brief heading which identifies the trade name, the formulation, and the type of material (e.g. a herbicide). Also included in the header is the name of the manufacturer, along with an address and phone number so they can be contacted in case of a problem or emergency. Immediately following the heading is Section 1, which provides basic information about the material. Let's use a turf fungicide as an example (Figure 2.19). Section 1 of a herbicide MSDS would provide the trade name, the given name and concentration of the active ingredient, the chemical name of the active ingredient, and the name of the fungicide family to which the active ingredient belongs. The EPA also classifies pesticides with one of three signal words: caution, warning, and danger. The level relevant to the fungicide in Figure 2.19 is caution. Most herbicides also fit into the caution category, since they tend to be less toxic than other pesticides. Herbicides with more toxic properties and many other fungicides are given the signal word warning while the most toxic pesticides, including many insecticides, are given the signal word danger to maximize caution used in their handling and use. Section 2 lists all of the components that go into the pesticide product. This, of course, includes the active ingredient but also the other constituents that make up the pesticide formulation. Remember that all pesticides are formulated so a product may only contain a small proportion of the active ingredient. Identification of the other ingredients is important since they may also have potentially toxic properties that should be listed in an MSDS.

Sections 3 through 8 on an MSDS cover the hazards associated with exposure to and use of a pesticide and how to deal with

> **HERITAGE FUNGICIDE**
> Syngenta Crop Protection, Inc.
> Post Office Box 18300
> Greensboro, NC 27419
> In Case of Emergency, Call 1-800-888-8372
>
> **1. PRODUCT IDENTIFICATION**
> Product Name: HERITAGE FUNGICIDE
> Product No.: A12704A
> EPA Signal Word: Caution
> Active Ingredient(%): Azoxystrobin Technical (50.0%)
> CAS No.: 131860-33-8
> Chemical Name: Methyl (E)-2-{2-[6-(2-cyanophenoxy)pyrimidin-4-yloxy]phenyl}-3-methoxyacrylate
> Chemical Class: A beta-methyoxyacrylate fungicide
> EPA Registration Number(s): 100-1093 (formerly 10182-408)
> Section(s) Revised: 3, 5, 7, 16

FIGURE 2.19 Introductory portion of the MSDS for Heritage fungicide.

potential problems. Included in these sections are proper first aid strategies, fire-fighting recommendations, methods for cleaning up spills, proper pesticide storage, and proper protective equipment for handling and use. This information can be critical when a problem arises. For example, Sections 3 and 4, which cover the symptoms associated with exposure to the pesticide and first aid measures, often contain useful information that can be passed along to a doctor in the event of a medical problem. As a practitioner, this information can expedite the medical care a person may require and, in an extreme case, could help save a life.

Fire fighting recommendations, listed in Section 5 of the MSDS, not only indicate whether or not the pesticide is flammable but also indicate what sort of fire extinguisher is proper to put out a fire and what personnel evacuation procedures should be followed. Spill control recommendations provide optimum ways to handle and minimize the spread of product spills. Access to either type of information can be essential in the event of an emergency to minimize both property damage and personal injury. Handling and storage recommendations are useful to manufacturers, transporters, and end users alike. Making sure that anyone who may encounter a pesticide knows how to handle or store it properly and what hygiene practices are necessary to avoid exposure maximizes worker safety. The recommendations for personal protective equipment (PPE) on an MSDS are often for the benefit of the manufacturer or for personnel involved in transport of the pesticide. Additional PPE recommendations for application of a pesticide are usually included in the product label, which is discussed in the next section of this text.

Sections 9 and 10 on an MSDS cover physical and chemical properties that are pertinent to the practitioner:

- *A description of the appearance*
- *Melting point*
- *Boiling point*
- *Physical density*
- *Solubility in water*
- *Volatility or vapor pressure*

  The scale used ranges from 0-4 where:

- *0 represents the least amount of hazard potential*
- *1=slight hazard potential*
- *2=moderate hazard potential*
- *3=high hazard potential*
- *4=severe hazard potential*

Aside from the appearance description, this information might not seem practical but it does offer some insight as to why the product is recommended to be stored in a certain way. Whether they are chemically stable or not, it is recommended that most products be stored in a dark and dry location where the effects of water, sunlight, and heat or cold are minimized. These storage conditions promote maximum shelf life of a product, many of which are not used in their entirety during a single application period. Section 10 of an MSDS briefly comments on the chemical stability of a product and indicates whether or not it will react with other chemicals. For instance, oxidizing agents like nitrogen fertilizers or flammable materials may pose hazards if brought into contact with certain pesticides. Section 10 points out that these types of materials should not be stored with or near the pesticide described on the MSDS.

Sections 11 and 12 of the MSDS cover the toxic potential for the subject pesticide. Section 11 addresses the potential for toxicity to humans who may be handling or using the pesticide. As you might expect, humans are not the test organisms used to determine this information. Usually, scientists will use small mammals like rats in tests to determine toxicity. Examples of information in this section are shown in Figures 2.20 through 2.22. Acute or immediate toxicity is the concentration at which the pesticide can cause injury or death when eaten or swallowed, absorbed through the skin, or inhaled. The units used to express acute toxicity are either an LD50 or an LC50, which indicate the lethal dose or lethal concentration of a pesticide, respectively, at which 50 percent of a tested population of organisms are killed. Note that the oral LD50 for the example herbicide and fungicide (Figures 2.20, 2.21) are greater than 5,000 mg/kg while the oral LD50 for the insecticide (Figure 2.22) is only 173 mg/kg. This tells us that the insecticide is much more toxic if consumed because a much lower quantity is required to kill 50 percent of a test population. Acute toxicity also addresses whether or not the pesticide is prone to cause eye or skin irritation. You can again see from the three example figures that the extent to which pesticides can irritate the eyes or skin can vary. Mutagenic potential is the ability of the pesticide to cause mutations. Reproductive hazard potential assesses the concentration at which a pesticide can harm either a developing fetus or the mother. Chronic toxicity assesses how a pesticide may build up in tissues and what, if any, organs are affected by the buildup. Carcinogenic potential identifies the potential for tumor formation and in what parts of the body. While most of the toxicity informa-

> **11. TOXICOLOGICAL INFORMATION**
> Acute Toxicity/Irritation Studies
> Ingestion: practically non-toxic
> Oral LD50 (Rat): >5,000 mg/kg body weight
> Dermal: slightly toxic
> Dermal LD50 (Rat): >2,000 mg/kg body weight
> Inhalation: Not Available
> Inhalation LC50 (Rat) : Not Available
> Eye Contact: mildly irritating (Rabbit)
> Skin Contact: mildly irritating (Rabbit)
> Skin Sensitization: sensitizing (Guinea Pig)
> Mutagenic Potential
> Prodiamine: None Observed
> Reproductive Hazard Potential
> Prodiamine: Fetal toxicity at high dose levels (rats); developmental and maternal toxicity observed at 1g/kg/day.
> Chronic/Subchronic Toxicity Studies
> Prodiamine: Liver (alteration and enlargement) and thyroid effects (hormone imbalances) at high dose levels (rats); decreased body weight gains.

FIGURE 2.20 Toxicological MSDS information for the herbicide product Barricade®.

tion in Section 11 refers to the pesticide active ingredient, there is also information there that identifies potential toxicity of the other ingredients that make up a pesticide product.

Section 12 follows up the human toxicity information with both ecological toxicity and environmental impact data. Ecological toxicity covers the potential impact for the pesticide to harm wildlife. For example, concentrations are often listed which are lethal to test fish species. Again, the term used is the LC50, which identifies the concentration (in this case, in a body of water) at which 50 percent of the test population of a fish species are killed. The lower the LC50, the more toxic the pesticide is to the test organism. LC50 values represent acute toxicity potential while chronic toxicity looks more at how a pesticide might accumulate over an extended period of time and cause long-term health problems in the test organism.

> **11. TOXICOLOGICAL INFORMATION**
>
> **Acute Toxicity/Irritation Studies (Finished Product)**
> Ingestion: Practically Non-Toxic
> Oral (LD50 Rat): > 5,000 mg/kg body weight
> Dermal: Slightly Toxic
> Dermal (LD50 Rat): > 2,000 mg/kg body weight
> Inhalation: Moderately Toxic
> Inhalation (LC50 Rat): > 4.67 mg/l air - 4 hours
> Eye Contact: Moderately Irritating (Rabbit)
> Skin Contact: Slightly Irritating (Rabbit)
> Skin Sensitization: Not a Sensitizer (Guinea Pig)
> Reproductive/Developmental Effects
> Azoxystrobin Technical: Shows weak chromosomal damage in mammalian cells at cytotoxic levels. Negative in whole animal assays for chromosomal and DNA damage at high dosages (> or = 2,000 mg/kg).
> In rabbits, no effect was observed up to the highest dose level (500 mg/kg/day). In rats, developmental effects were seen only at maternally toxic doses (100 mg/kg/day).
> Chronic/Subchronic Toxicity Studies
> Azoxystrobin Technical: In a rat 90-day feeding study, liver toxicity was observed at 2,000 ppm. This was manifest as gross distension of the bile duct, increased numbers of lining cells and inflammation of the duct.

FIGURE 2.21 Toxicological MSDS information for the fungicide product Heritage®.

Environmental fate of a pesticide is also covered in Section 12. This topic will be covered in greater detail in a later portion of this chapter. The environmental fate of a pesticide relevant to an MSDS would include information such as pesticide responsiveness to sunlight and how long it can remain in a stable form in soil. Sunlight can break down many pesticides, especially if they are mixed into a liquid solution. The duration a pesticide can last in solution before it breaks down gives us an idea of how stable it is. The process by which sunlight can break down a pesticide in solution is called photolysis. The unit used for this process is the half-life, or the time required for 50 percent of the pesticide to break down. Half-life is also important

> **11. TOXICOLOGICAL INFORMATION**
>
> Acute Toxicity/Irritation Studies (Finished Product)
> Ingestion: Moderately Toxic
> Oral (LD50 Rat): = 173 mg/kg body weight
> Dermal: Slightly Toxic
> Dermal (LD50 Rat): > 2,000 mg/kg body weight
> Inhalation: Slightly Toxic
> Inhalation (LC50 Rat): > 0.764 mg/l air - 4 hours
> Eye Contact: Severely Irritating (Rabbit)
> Skin Contact: Moderately Irritating (Rabbit)
> Skin Sensitization: A weak skin sensitizer.
> Reproductive/Developmental Effects
> Cypermethrin Technical: There were no cypermethrin-induced effects in fertility in two separate two-litter three (filial) generation studies in the rat.
> Chronic/Subchronic Toxicity Studies
> Cypermethrin Technical: NOEL (2-yr) for dogs 5 mg/kg, rats 7.5 mg/kg.
> Nervous system effects typical of pyrethroids (motor incoordination, gait abnormalities) in a range of repeated dose studies (dog and rat). Possible nerve fiber degeneration in 14-day study in rats.

FIGURE 2.22 Toxicological MSDS information for the insecticide product Demon.

when evaluating how long pesticides remain active in soil. Soil half-life gives us an idea of how long the pesticide can actively control pests or potentially impact the environment. For example, a herbicide applied on April 1 with a soil half-life of two months would be at maximum concentration at the time of application, a 50 percent concentration by June 1, a 25 percent concentration by August 1, and down to a 12.5 percent concentration by October 1. Temperature and rainfall extremes can affect the soil half-life but you get the idea.

Sections 13 through 16 of the MSDS address the regulatory issues, which govern the handling of a particular pesticide. Section 13 outlines any specific instructions for proper disposal of a pesticide product, especially those that supplement local and regional laws for pesticide disposal. Section 14 outlines any specific rules associated

> **PESTPOINTER**
>
> Much of this information might be deemed common sense, but remember that common sense can not be presumed and we must recognize that documentation of this information is a necessity for these products to be available to us.

with proper transportation of the pesticide. Much of this information might be deemed common sense but remember that common sense can not be presumed and we must recognize that documentation of this information is a necessity for these products to be available to us. Section 15 lists any additional regulatory information that may pertain to a pesticide while Section 16 is the safety net known as "Other Information," which includes anything not covered earlier in the MSDS. One useful piece of information often found in this last section is the National Fire Protection Association (NFPA) hazard rating for the pesticide (Figure 2.23). These ratings use a numeric scale for three hazard categories:

- *Health hazard*
- *Flammability hazard*
- *Chemical reactivity hazard*

    The scale used ranges from 0-4 where:

- *0 represents the least amount of hazard potential*
- *1=slight hazard potential*
- *2=moderate hazard potential*
- *3=high hazard potential*
- *4=severe hazard potential*

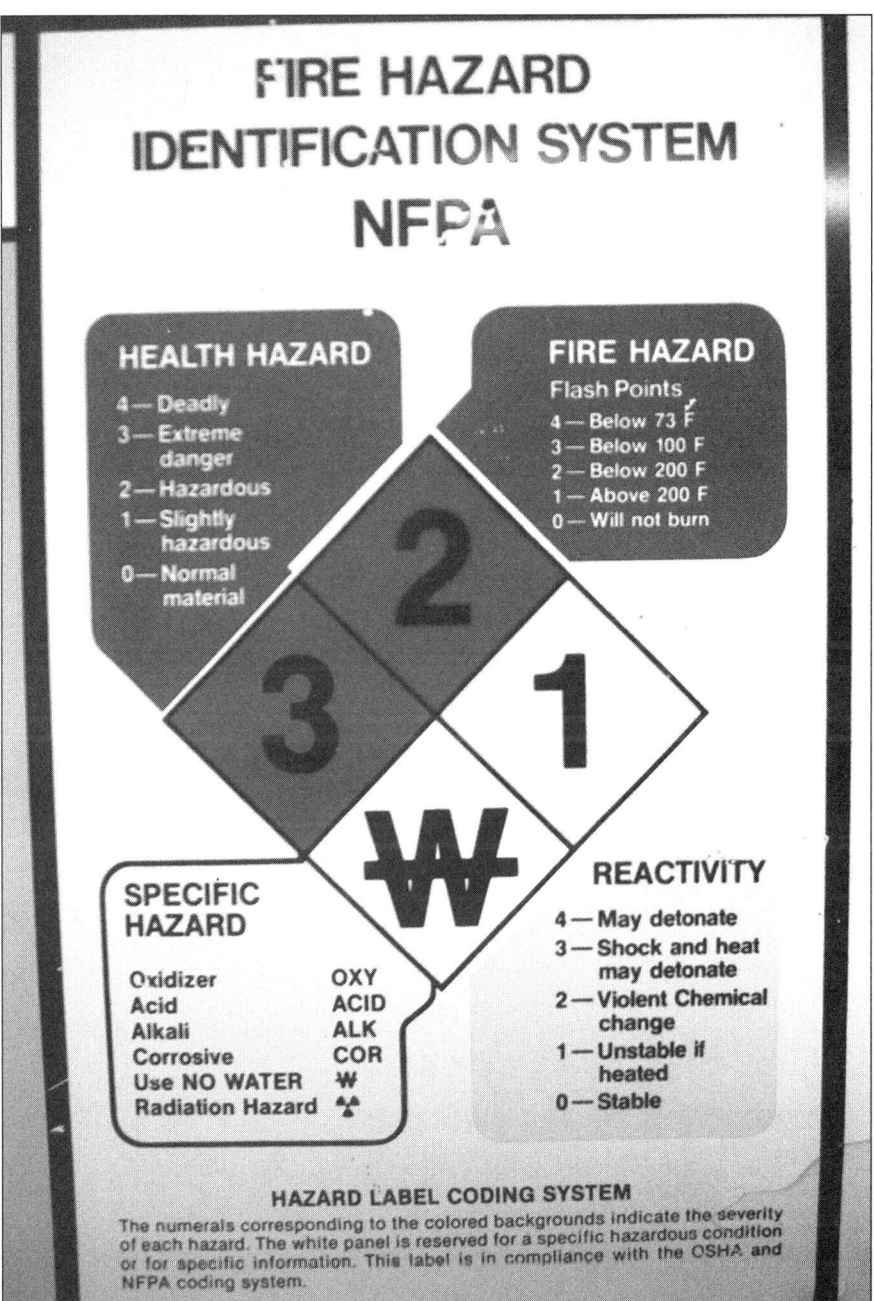

FIGURE 2.23 National Fire Protection Association (NFPA) hazard ratings posted show information for potentially hazardous chemicals.

I mentioned earlier that many herbicides are given a caution rating by the EPA, the lowest hazard category possible. The EPA looks primarily at toxicity or environmental contamination potential when rating pesticides. The hazard scale used by NFPA is similar but is more pertinent for proper storage and handling of a pesticide. A pesticide labeled caution by EPA might commonly be assigned a 1 (slight hazard potential) in all three hazard categories by NFPA while stored fuels might be assigned a 3 or 4 (high to severe hazard potential) in the same categories. Precautions such as these are commonly featured on the doors of pesticide storage buildings or rooms to indicate the hazard potential of the contents (Figure 2.24).

Remember that, while the MSDS might not seem as important or meaningful as the pesticide label to the average practitioner, it can be critical in the event of an emergency. Knowing where the MSDS for a particular pesticide is located and how to interpret its contents can help you make proper decisions in an emergency situation and, in some cases, might help save someone's life. This is important for both managerial personnel and for employees. Many organizations that employ people who use and therefore are exposed to chemicals include safety training as an early part of the employment process. Many of the readers of this text may require such training or have received it themselves. However cumbersome or boring the training may seem, a typical and critical part of it is knowing where to find safety documents like MSDS.

PESTPOINTER

Many organizations that employ people who use and therefore are exposed to chemicals include safety training as an early part of the employment process.

Basic Chemicals and Pesticide Principles 69

FIGURE 2.24 Precautionary statements posted outside a pesticide storage facility.

UNDERSTANDING THE CHEMICAL/PESTICIDE LABEL

The label for a chemical or pesticide is perhaps the most important reference document that exists for the benefit of the practitioner. MSDS are critical for safety issues but don't offer much in terms of how to actually use a particular product. Think of a pesticide label as both the warranty and the owner's manual for the product it references. Most questions that you may have regarding use of a pesticide will be answered somewhere within the contents of the label. It is the ultimate responsibility of the practitioner to apply a material properly but most potential problems can be avoided by following label specifications. As with the MSDS, the label contains a lot of detailed information. However, as a practitioner, you will find the label more meaningful and informative because it addresses the actual use of the pesticide. The label is typically a bigger document than an MSDS, sometimes requiring 10 or more pages. However, by systematically describing the contents of the label, it is my aim to help you make more sense out of it and allow you to effectively use the information contained within it for your specific needs in the workplace.

Just what is contained in a product label? Perhaps a better question to ask would be, "What is not contained in a product label?" Labels do vary somewhat but their diversity is attributable to the wide variety of turf chemicals and pesticides that are out there. There is some common ground among product labels and this discussion will focus on what is similar among the labels you might encounter in the workplace. First and foremost, the trademark name of the product is featured at the beginning of the label. Some product names will include reference to the formulation. Other products, particularly liquid formulations, may not indicate the formulation in the title but will usually feature a "fine-print" description of the formulation in the early portion of the label. Brief summaries of the uses for the product and, for pesticides, the sorts of pests they control are also listed at the beginning. The name of the active ingredient in the product and its concentration in the formulated product are indicated on all product labels.

Following the introductory segment, labels will feature condensed versions of the information that is also included in the MSDS. Examples of this are:

- *First aid recommendations*
- *Recommended personal protective equipment (PPE)*
- *Potential hazards to humans or animals*
- *Storage/disposal recommendations*
- *Potential environmental hazards*

Because this safety information is not as detailed as it is in the MSDS, the label should not be viewed as a complete substitute for the MSDS in making safety-related decisions. However, the presence of this information in the product label does serve as a convenient reference for the practitioner and can minimize, in many cases, the number of documents a practitioner needs to use when preparing for and making an application.

Product labels all include some sort of legal disclaimer offered by the manufacturer as a condition for sale and use of the product (Figure 2.25). While most of us, author of this text included, are not fond of fine print that is written in "legalese", this legal disclaimer does establish what rights you do and don't have as a practitioner. The disclaimer, regardless of what product label it is a part of, will

PESTPOINTER

Because safety information is not as detailed as it is in the MSDS, the label should not be viewed as a complete substitute to the MSDS for making safety-related decisions.

CONDITIONS OF SALE AND LIMITATION OF WARRANTY AND LIABILITY

NOTICE: Read the entire Directions for Use and Conditions of Sale and Limitation of Warranty and Liability before buying or using this product. If the terms are not acceptable, return the product at once, unopened, and the purchase price will be refunded.

The Directions for Use of this product should be followed carefully. It is impossible to eliminate all risks inherently associated with the use of this product. Crop injury, ineffectiveness or other unintended consequences may result because of such factors as manner of use or application, weather or crop conditions, presence of other materials or other influencing factors in the use of the product, which are beyond the control of SYNGENTA CROP PROTECTION, Inc. or Seller. All such risks shall be assumed by Buyer and User, and Buyer and User agree to hold SYNGENTA and Seller harmless for any claims relating to such factors.

SYNGENTA warrants that this product conforms to the chemical description on the label and is reasonably fit for the purposes stated in the Directions for Use, subject to the inherent risks referred to above, when used in accordance with directions under normal use conditions. This warranty does not extend to the use of the product contrary to label instructions, or under abnormal conditions or under conditions not reasonably foreseeable to or beyond the control of Seller or SYNGENTA, and Buyer and User assume the risk of any such use. SYNGENTA MAKES NO WARRANTIES OF MERCHANTABILITY OR OF FITNESS FOR A PARTICULAR PURPOSE NOR ANY OTHER EXPRESS OR IMPLIED WARRANTY EXCEPT AS STATED ABOVE.

In no event shall SYNGENTA or Seller be liable for any incidental, consequential or special damages resulting from the use or handling of this product. THE EXCLUSIVE REMEDY OF THE USER OR BUYER, AND THE EXCLUSIVE LIABILITY OF SYNGENTA AND SELLER FOR ANY AND ALL CLAIMS, LOSSES, INJURIES OR DAMAGES (INCLUDING CLAIMS BASED ON BREACH OF WARRANTY, CONTRACT, NEGLIGENCE, TORT, STRICT LIABILITY OR OTHERWISE) RESULTING FROM THE USE OR HANDLING OF THIS PRODUCT, SHALL BE THE RETURN OF THE PURCHASE PRICE OF THE PRODUCT OR, AT THE ELECTION OF SYNGENTA OR SELLER, THE REPLACEMENT OF THE PRODUCT.

SYNGENTA and Seller offer this product, and Buyer and User accept it, subject to the foregoing conditions of sale and limitations of warranty and of liability, which may not be modified except by written agreement signed by a duly authorized representative of SYNGENTA.

FIGURE 2.25 Sample legal disclaimer found on a pesticide product label.

place most of the responsibility for proper product use on the practitioner. This may seem unfair but most of the responsibility should be placed on the practitioner. Only the practitioner can choose when to apply a product, how to prepare for and make an application, and determine what results should be expected. Chemical manufacturers are responsible for providing the formulated product as indicated on the product label. Beyond the point of sale, the manufacturer's legal obligations are usually limited. These terms are stipulated in the product disclaimer and should be understood before a product is purchased.

What rights do practitioners really have? If a practitioner has reason to believe a product is tainted, contaminated, or damaged at some point prior to when the product is purchased, the disclaimer clearly states that the remedy for such a situation shall be either product replacement or a refund. Liability by the manufacturer usually ends there, although most manufacturers and/or distributors will generously adhere to these guidelines and proactively inform consumers that a batch of product may be defective. Otherwise, customers may lose faith in a product or any products offered by a particular manufacturer and choose to take their business elsewhere. Further legal obligations may exist, at the manufacturer level, if a practitioner actually uses defective or contaminated products. When these applications result in a negative outcome (e.g. turfgrass injury or death), the manufacturer may be liable. However, the practitioner must be able to prove that the product contamination was solely responsible for the undesirable injury. Ruling out other reasons for such injury, such as applicator error or environmental conditions, is essential for a practitioner to have a legal claim against the manufacturer. Proof of manufacturer negligence, usually knowledge of a product defect before the product is sold, is also a helpful and sometimes necessary component of a legal claim against a manufacturer. Proper documentation of each and every application made is thus critical for a practitioner to maintain their consumer rights.

The above-described components of a product label are given more discussion because they are common components of all labels. The remainder of most labels, and usually the biggest proportion, details how a product is to be used. This part of the label represents the most important information to the majority of practitioners. Because products vary so much, in terms of what pests they target and for what situations they are most useful, it is impossible to generalize this part of a product label in a few paragraphs. However, common examples of information contained in the instructional portion of a product label include:

- *Proper mixing instructions for a chemical or pesticide, either alone or tank-mixed with other materials*
- *How to determine compatibility of a chemical or pesticide with other products*
- *Whether or not a chemical or pesticide requires supplemental ingredients, such as adjuvants, for proper application*
- *Whether or not a liquid spray solution containing a particular pesticide or chemical requires special treatment like agitation*
- *What volume of water per unit area a pesticide or chemical should be applied in (liquid applications only)*
- *What amount (rate) of a pesticide or chemical should be applied per unit area and how this amount varies for different turfgrasses or for different levels of management*
- *What turfgrass species a pesticide or chemical can be safely applied to*
- *What alternative plant species (e.g. ornamentals) a pesticide or chemical can be applied to or in proximity to*
- *Areas where a pesticide or chemical should not be applied (e.g. near bodies of water, sensitive areas like putting greens)*
- *Recommended timing(s) for application of a pesticide or chemical*
- *Specific pests that a pesticide will target and to what extent*
- *Specific recommendations for how to treat or maintain an area following an application of a pesticide or chemical*
- *Environmental conditions in which a pesticide or chemical should, or should not, be applied*

- Duration of pesticide or chemical activity that should be expected and how that varies with specific amounts used or with different pests
- How much total pesticide or chemical can be safely applied during a growing season or calendar year
- How many total pesticide or chemical applications can be safely applied during a growing season or calendar year
- Geographical variations in recommendations for pesticide or chemical use (e.g. climatic differences, regulatory issues)

The above list covers much of what might be expected in the instructional portion of a product label but may not be all-inclusive. Therefore, it cannot be emphasized enough that a thorough understanding of the instructional portion of a product label is essential to your successful use of that product. As much as following label instructions can help an application, ignorance of or ignoring label instructions can also mean poor results. Being sure you understand the label yourself and also familiarizing your staff with label contents will help ensure that pesticide and chemical use are positive components of a total approach to turfgrass management.

ENVIRONMENTAL IMPACT OF CHEMICALS AND PESTICIDES

The environmental impact of pesticide applications is always a sensitive issue and is the basis for many of the criticisms that are levied against the use of these materials. As practitioners, understanding how pesticides may influence other organisms or the natural environment is a key component of responsible pesticide use. To understand the possible consequences of pesticide use, we must first understand what fates may lie in store for a pesticide after it is applied. Stated another way, where can a pesticide go once it leaves the sprayer or spreader? Table 2.2 lists some possible fates of a pesticide one it has been applied.

TABLE 2.2 Possible fates of turfgrass chemicals and pesticides.

Fate	Description
Plant metabolism	Pesticide is absorbed into plant tissue and broken down by plant enzymes.
Pest consumption	Target pests consume pesticide and metabolize it.
Soil absorption	Pesticide becomes bound to soil particles.
Organic matter absorption	Pesticide becomes bound to organic matter in thatch and soil.
Microbial degradation	Soil microorganisms like bacteria consume pesticide and metabolize it.
Photolysis	Pesticide is broken down by the energy contained in sunlight.
Chemical degradation	Pesticide is broken down via chemical reactions. with water and other compounds contained in the soil.
Volatility	Pesticide is lost to the atmosphere via evaporation from plant and soil surfaces or via drift.
Surface runoff	Pesticide moves with flowing water across turf, soil, or impermeable surfaces before it enters target plants or soil.
Leaching	Pesticide moves vertically (percolates) through soil, reaching ground water or subsurface drainage sources.

There are a number of such possible fates but it is also important to remember that rarely does one possible fate account for 100 percent of an applied pesticide. Combinations of these fates are more common so practitioners must try to maximize the desirable fates and minimize those that result in loss to the environment.

The most desirable fate of a turfgrass chemical or pesticide is usually uptake by the target plant or pest. How this happens can vary significantly, depending upon what type of chemical is used and what the target pest or plant is. Some pesticides are applied to target turfgrass plants, some are applied to target weeds, and some are applied with intent that they reach the soil. Most herbicides are applied as a broadcast application, meaning they will contact both turfgrass and weeds. Absorption or uptake into both kinds of plants must therefore be expected. Turfgrasses tolerant to herbicides that target broadleaf weeds are able to harmlessly metabolize the herbicide while the sensitive weeds are injured or killed when they metabolize the same material. In either case, uptake by plants diminishes the chance for the herbicide to experience alternative fates. However, some persistent herbicides can still experience other fates if they are removed from the target site with turfgrass clippings during a mowing event. Proper disposal of clippings must include consideration of what pesticides may be contained in them.

Some herbicides and many other pesticides are applied with pest consumption as the intended fate. For example, preemergence herbicides are applied to form a thin layer just beneath the soil surface so that emerging weeds grow into the layer and then absorb the herbicide. Many fungicides are applied to coat the surfaces of turfgrasses susceptible to a particular disease. When the pathogen attacks the turf, it also consumes fungicide and is thus controlled. Other pathogens and many insects feed on turfgrass roots so pesticides that target these pests must be situated in the thatch or soil so the pests will consume them. Whether or not these pesticides are actually absorbed into turfgrass tissues depends upon the specific characteristics of the pesticide. Those that are not taken up by turfgrass plants may be more subject to alternative fates, in addition to pest consumption.

Alternative Fates of a Pesticide

Something must happen to the proportion of the pesticide that does not get taken up or used by plants or pests in the target area. What are some of these possible alternative fates of a pesticide?

Pesticides can be subject to volatility or loss into the air. Volatility can happen in one of two ways: evaporation from the target surface or drift prior to reaching the target surface. Drift is the more common means by which pesticides are lost to the air and is an avoidable problem. High spray pressures and windy spray conditions are two common reasons why drift occurs. Evaporation from the plant or soil surface is less common because most pesticides applied as liquids are formulated to rapidly adhere to absorb into the plant or the soil. Some active ingredients are unstable and are thus chemically prone to dissociate into the air. However, this shortcoming is usually addressed by formulating these active ingredients in more stable forms like dry granules.

Both the characteristics of the target area and the chemical properties of the pesticide can influence the appearance of a pesticide in a non-target area. Site characteristics most relevant to this discussion include slope, proximity to sensitive areas, soil texture, and the type of vegetation that is present. Sloped areas are more prone to pesticide loss, usually surface runoff, than flatter areas because water from heavy rain or irrigation will often flow downhill faster than it can percolate into the soil. Turfgrass areas near sensitive areas like ponds or wetlands present more risk for pesticide contamination simply because of the close proximity. (Figure 2.26)

Conservative or avoided use of pesticides and alternative plantings like buffer strips are common solutions to this potential problem. Water may infiltrate into coarse soils like sands faster than others soils but coarse soil particles also do not bind pesticides as well. Therefore, turfgrass growing on sandy soils may be less prone to surface runoff but more prone to pesticide loss via leaching or percolation through the soil. The opposite is true for heavier soils like clays. Slow water infiltration rates may increase the likelihood of surface runoff with heavy rainfall or irrigation events. Vegetation can also affect the potential for pesticide loss. Turfgrass is a consistent vegetative surface that can intercept and take up a high percentage of applied pesticides. However, landscaped areas with less uniform ground cover may allow for more alternative fates for an applied pesticide.

Basic Chemicals and Pesticide Principles 79

FIGURE 2.26 A turfgrass area near a pond often includes a protective buffer strip with taller vegetation.

Pesticide chemical properties that most greatly influence potential for environmental contamination are persistence and mobility. Pesticides that have long half-lives and thus degrade more slowly have more opportunities to move away from where they were applied. Most pesticides naturally dissipate or degrade by several mechanisms. Plant uptake and use is one means by which a pesticide can be degraded. The influence of climate is also important. Solar radiation and water can both help break down a pesticide over time. Chemical reactions in the soil can also either break down a pesticide into smaller chemical components. Lastly, the influence of soil microbes can significantly affect a pesticide's longevity. All soils have populations of microscopic organisms like bacteria and fungi, which feed on organic matter in the soil. Food for these organisms can include plant residue like thatch or pesticide compounds that are in the soil. Persistent pesticides are those that can best resist or withstand these different possible forms of break down.

Pesticide Mobility

Pesticide mobility presents greater risk for environmental contamination than persistence, with the products that are available in today's market. Mobile pesticides are only considered safe (and therefore marketable) by the EPA if it can be shown they have low persistence. Persistent pesticides, by similar reasoning, can only be registered if they have low mobility. This decreases the likelihood that persistent pesticides will move away from the target area. However, let's consider the two primary ways pesticides can move away from a target site. The first is surface runoff, where a pesticide or chemical moves with flowing water across the surface of the ground, eventually arriving in streams, ponds, or other areas where drainage collects (Figure 2.27). Surface runoff can result in contamination with any kind of pesticide if the runoff occurs at the wrong time. The classic example of a situation resulting in surface runoff is a thunderstorm, which deposits heavy amounts or rainfall in a short amount of time. Pesticides that have recently been applied and/or have not moved into the soil are subject to surface runoff, regardless of their persistence. Sloped areas or areas adjacent to bodies of water are the most prone to surface runoff contamination.

The other way pesticides may move away from the target area is through leaching. Unlike surface runoff, leaching contamination requires percolation through the soil with water. Eventually, the contaminant reaches the water table or drainage pipes that can lead into collection areas. Leaching is less common than surface runoff but the likelihood of it happening increases with the use of mobile pesticides. By their nature, mobile pesticides more readily dissolve in water and do not bind well to soil particles, allowing them to percolate freely through the soil. Shallow water tables, large volumes of water moving through soil such as with heavy rainfall or irrigation, and coarse soil textures (e.g. sands) are all conditions that can favor leaching contamination. Golf course putting greens are common turfgrass environments that are prone to leaching because

FIGURE 2.27 Surface runoff can be accelerated on slopes and impermeable surfaces like pavement.

many are sand-based; they are regularly irrigated, and subsurface drainage is usually shallow. Use of mobile pesticides in this sort of environment would certainly risk pesticide contamination.

Although environmental contamination is a risk for any pesticide application, proper application techniques and sensitivity to the environment can minimize potential problems. Turfgrass systems are inherently less prone to environmental contamination because there is consistent ground cover, unlike for most field crops. Turf is an excellent interceptor and filter for materials applied to it. As practitioners, we must use our heads and keep our applications away from sensitive areas like bodies of water or impermeable areas like concrete. Beyond these common sense issues, adherence to pesticide application guidelines will take you a long way towards proper environmental stewardship.

Economic and Aesthetic Pest Thresholds

The need to control pests, be it through natural or chemical control measures, is often a matter of consumer opinion. One homeowner, for example, might have minimal expectations for their lawn and therefore tolerate a certain amount of weeds or damage from insects or diseases. Other homeowners might have extremely high expectations for their lawns and therefore have a so-called "no tolerance" approach to any pests their turf may be exposed to. Consumer sentiment can thus have a significant role in how the pesticide practitioner approaches a given turf situation and how the pests are best managed. Money or budgetary issues can also play a key role in the control of pests. Whether it is a home lawn, an athletic field, or a golf course, the amount of funds available for the purchase of pesticides or to pay to have an application made often dictates the level of pest control that is achievable.

Many would perceive that golf courses have the greatest ability to budget for pest control and pesticide use. However, it is more likely that homeowners have the highest amount of flexibility when it comes to pest control. There are no patrons who might complain about pest problems in a home lawn. There also aren't supervisors who may dictate the amount of funds available for pest control pursuits. Whether the limitations to pest control are due to money, aesthetics, or both, the concept of an acceptable pest threshold becomes very important and warrants further discussion.

What is a pest threshold? It might best be defined as "the maximum tolerable level of infestation or injury, by a particular pest, before the infestation or injury becomes objectionable". Pest thresholds differ among the different kinds of pests and the plant systems within which they occur. The agricultural threshold for pests is most often associated with the amount of injury to plants that they cause. However, pest thresholds can also be based upon their potential impact on humans. For example, homeowners or grounds managers might be concerned about bee or wasp damage to tree fruit but are also concerned about the likelihood of bee stings if the insects are sufficient in number. Some weeds, like thistles or sandburs, can be potentially injurious to humans, livestock, or pets, which necessitates a certain threshold for these types of weeds. Despite the potential human impact of some pests, the majority of pest thresholds relevant to agriculture and to turf are still related to either economics or aesthetics.

The concept of acceptable pest thresholds is not a new one but is something that has evolved over the course of history. Early agriculturists did not have access to the pest control methods that we experience today. Farms were small and harvested crops were more intended to ensure survival rather than to achieve economic gain. The destructive impacts of pests could, at times, be severe, and put people at risk of survival. As far back as biblical times, we read of reported locust swarms that could decimate crops and the food source for many people. Under these circumstances, pest thresholds were very low because the stakes were high.

The modern era of agriculture has changed our views of acceptable pest thresholds. Reliable pesticides are available to help control highly destructive pests. Larger farming operations have created surpluses of food crops, such that significant pest problems aren't individually prone to cause starvation or other societal tragedies. However, while this is the case in more developed parts of the world, there are still many places where drought or major pest invasions can have drastic effects. In the third world, pest thresholds are still very low because of the risk of pests to human survival. Back home, modern pest thresholds are more based upon economics.

High yielding cropping systems diminish the per acre value of harvested crops and create a narrower margin between a profitable crop and one which breaks even or results in an economic loss for the grower. Pest thresholds in these cases are established with more precision, and are based upon specific estimates of yield loss due to a particular pest. Let's take a closer look at how thresholds are established for different pests and how these thresholds are modified when we look at turf, rather than conventional agriculture.

Insect damage can influence crop yields in a very visible manner, based upon crop injury. Farmers and turf managers alike must determine how much damage can be tolerated before yields or aesthetics are compromised. Insect populations can be monitored to an extent but many damaging insect pests are migratory, such as locusts or mole crickets, and damage can occur rapidly if control measures are not employed. Growers or turfgrass managers must therefore establish, in advance, tolerance levels for insect pests. These levels are based upon the type of crop, reproductive and behavioral characteristics of the insect, and costs of controlling the pest. Many insect pests have higher threshold levels for large-scale grain cropping systems, due to the high costs of treating the pest. Fruit crops often have very low thresholds for insect pests because any damage caused by the pest is sufficient to reduce the value of the crop. Wormy apples aren't going to be attractive at the marketplace to the end consumer so costs of treating orchards for insects are easily justified.

Turf insect thresholds are similar to that for fruit crops because of the high aesthetic premium placed on turf. However, there is variability in insect thresholds among different types of turf. Homeowners or commercial landscape managers may be able to tolerate some grub damage without needing to apply insecticides. However, thresholds in commercial sod production for insects like sod webworm may be very low, due to the impact of these pests on quality of and ability to sell the sod. Golf course putting greens have very low thresholds for insects, due to their high visibility on the golf course and the immediate visual impact of pest damage (Figure 2.28). Problems with putting greens are the top reason why superintendents lose their jobs so great care must be taken to avoid insect damage on greens.

Diseases are also pest threats that can significantly damage target plants. Be they caused by fungi, bacteria, or viruses, diseases can influence yields of agricultural crops or the aesthetics of ornamental plants such as turfgrass. Whether diseases require treatment or not is at the discretion of the grower, be they a farmer, turf manager, or homeowner. Once again, the concept of a pest threshold

Basic Chemicals and Pesticide Principles   85

FIGURE 2.28   Golf course putting greens are highly visible and have low thresholds for pest problems.

becomes very important. Some diseases can be tolerated in grain crops due to the high costs of treatment and acceptable damage levels caused by the disease. Fruit crops have very low disease thresholds for the same reasons low thresholds exist for insects. Moldy or scabbed fruit is not appealing to the consumer and therefore cannot be at all tolerated.

Turf disease thresholds will again vary by circumstance. Minor brown patch outbreaks in a home lawn may not warrant corrective treatment. When diseases threaten a highly visible golf course, the thresholds drop considerably. Low disease thresholds are why fungicides are one of the biggest costs for highly maintained golf courses. Disease thresholds highly influence the strategy selected for disease control. The primary two options are either preventive or curative. Preventive disease control means treating for a disease at or prior to when conditions are favorable for disease incidence, such that symptoms do not appear. Golf courses or high profile athletic fields with low disease thresholds would usually opt for a preventive control program so symptoms don't detract from the aesthetics and playability of the turf. Curative disease control means treating once disease symptoms have appeared, so as to avoid further damage. Many homeowners view disease from this curative standpoint. If disease appears, they then would decide whether or not to control it. Facilities with low disease thresholds unfortunately do not have this sort of flexibility.

Cost considerations also can influence disease thresholds. It is no secret that preventive disease control is much more costly than is curative control. Preventive control measures are also more reliable and avoid problems associated with visible disease symptoms. High fungicide costs can create dilemmas for turfgrass managers and force critical decisions to be made regarding approaches to disease control. Where budget constraints are a problem, high profile areas like putting greens usually maintain the lowest disease thresholds while less visible areas might be allotted a higher acceptable threshold for some diseases or be switched from a preventive to a curative disease control program. Disease thresholds can, in some cases, stimulate broad changes to a turf facility. Highly damaging diseases like gray leaf spot warrant significant preventive fungicide costs for golf courses that grow perennial ryegrass. Instead of maintaining low disease thresholds and incurring these annual expenses, many golf courses have switched to alternative species like zoysiagrass or Kentucky bluegrass that are not sensitive to this disease.

Weeds may be a more subtle threat among the possible pest groups. The threshold for weeds is usually based more upon the numbers of weeds present in a given area than the direct damage they cause to desired plants. The impact of weeds may not visually compromise the health of a desired plant species. However, over the course of a growing season, competition for space and soil resources can cause costly yield reductions. That is why weeds are the number one pest category in American agriculture, from the standpoint of pesticide use. Because weeds exact a more gradual toll on crop production, the intellectual and financial effort to determine the economic threshold for these pests has been considerable. Factors such as weed emergence, crop planting date, numbers of seeds a single weed can produce, and size of the weed can all influence threshold values and have been the subjects of numerous research studies. The true threshold established for weeds may differ considerably among different cropping systems.

A farmer sees weeds as plants that significantly decrease yield. This yield loss can pertain both to a current growing season and to those in the future. Some weeds are capable of producing hundreds of thousands of seeds per plant, ensuring years of difficulty. Velvetleaf is an example of such a weed and, not surprisingly, its acceptable threshold is very low because of its seed-producing capabilities. Some weeds are recognized as "noxious". State regulatory agencies identify such weeds and establish zero tolerance thresholds for them so that they can be kept in check. Weeds that are most capable of reducing yields, or otherwise interfering with crop harvest, typically have the lowest acceptable thresholds. These thresholds vary significantly among crop species, as management inputs and the herbicide control measures available differ from crop to crop.

A turf manager is more apt to see weeds as diminishing the beauty of a landscape. Because of practices like mowing, turf weeds are not as numerous as for other crops. However, those weeds which do thrive in turf often can tolerate being mowed so alternative strategies must be executed to control them. Much as for insects and diseases, thresholds for weeds in turf will depend on the specific turf and management intensity. Weeds would have a very low threshold on putting greens, both because they stand out visually and because they impact the playability of the surface. In general, the greater visibility a turf area has, the lower the acceptable threshold for weeds. A dandelion infestation on a professional baseball field would not be tolerated while a homeowner might be able to live with a few weeds here and there.

Acceptable weed thresholds in turf can be influenced by the cost of herbicides, the expected impact of herbicide treatment, and the consequences of allowing the weeds to remain. A good example of a threshold dilemma that weeds can create is volunteer bermudagrass. While bermudagrass is a desired turf species in many areas, it can be an aggressive invader of other species, particularly in the transition zone. Because of its aggression, many turfgrass managers would identify a very low threshold for bermudagrass. However, few reasonable selective control options exist for it, resulting in an increase in the acceptable threshold. The spreading nature of bermudagrass can permanently alter the species composition of a turfgrass area. At some point, a turf manager must choose to accept the infestation and perhaps choose bermudagrass as the desired turf, or aggressively pursue controlling it with no guarantee of success. The perennial nature of turfgrass systems is often subject to such difficult situations when it comes to weeds. Because weeds are such a difficult problem for so many turfgrass situations, they will be the focus of the next two chapters.

This chapter has been designed to introduce you to some of the many terms, concepts, and supporting documents that are associated with pesticide and chemical use in turfgrass management. Some of the key points that you should now be familiar with are:

- *The science that goes into discovery and development of a pesticide product*
- *Why pesticides are formulated*
- *The types of formulations that exist for the products you use on the job*
- *Advantages and disadvantages of different formulations*
- *Why a pesticide MSDS is important*
- *What information is contained in an MSDS and how it benefits the practitioner*

- *Why a pesticide label is critical to successful application of a product*
- *What information is contained in a product label and how to use it*
- *What possible fates exist for a pesticide, once it has been applied to turf*
- *What conditions promote undesirable fates for a particular pesticide application*
- *The importance of establishing pest thresholds*
- *What criteria are necessary for determining pest thresholds*
- *How pest thresholds differ for different pests and why*
- *How pest thresholds differ for different turf situations and why*

I encourage you to review parts of this chapter if some of the highlighted key points are still unclear. Subsequent chapters will expound upon the different topics discussed in Chapter 2 so a good understanding of the basics will make later chapters more meaningful to you as a practitioner.

SUMMARY QUESTIONS

- *What pesticide formulations are you familiar with? Do you work more with dry or liquid formulations in the workplace?*
- *When you mix for pesticide applications, do you measure by total product amount or active ingredient amount? If you've done both, which are you more comfortable with?*
- *How often have you referenced an MSDS for a pesticide? If you've used it, was it for information or out of necessity, such as for an emergency?*
- *How frequently do you use pesticide labels on the job? If not, do coworkers use them or do you rely on past experiences for determining how a pesticide product should be properly used?*

- Do you have a common storage area for chemical or pesticide labels and MSDS? Was knowing the location of these forms part of your training or is it a component of how you train employees?

- Does your facility have areas that are sensitive to pesticides (e.g. wetlands)? How do you treat these areas differently?

- Do you have established pest thresholds for your facility? How do they differ among different types of pests? What factor most influences your decision to use pesticides: type of turf, type of pest, or cost of the pesticide?

Chapter 3

Weeds Found in Turf and Landscapes

Most people, when they envision what the term "pest" means to them, will think of nuisance insects, rodents, or perhaps that annoying little neighbor kid. What turfgrass practitioners envision as a pest differs considerably. They may also think of insects, but of those that feed on turf. They usually think of diseases that can cause numerous management headaches, especially during summer. Lastly, turfgrass managers think of weeds when they think of what pests are. Many people will envision illegal plants when they hear the term "weed" (Figure 3.1). In this context, weeds are some of the most problematic pests found in managed turfgrass and landscapes. Surveys conducted at local, state, and national levels have shown that the majority of turfgrass professionals consider weeds to be the most troubling pest problem they deal with. How can this be? Weeds don't appear overnight, as some disease and insect problems do. Weeds don't have the potential to be lethal to turfgrass, as diseases and insects can be. So what is it about weeds that make them so notorious? The answer to that question will differ, depending on whom you talk to. Some common answers, however, might be:

- *Many weeds can't be prevented with pesticides, like diseases or insects.*

92 Turfgrass Chemicals and Pesticides: A Practitioner's Guide

FIGURE 3.1 Wild Cannabis sativa plants, growing in a natural setting.

- *Weeds are easy to spot and their presence draws criticism.*
- *More pesticide products are needed to control the spectrum of weeds I must deal with.*
- *Some weeds are highly invasive and the problem gets worse every year.*
- *Pesticides aren't available to control some weeds without harming my turf.*

Do any of these responses sound familiar or have particular pertinence to your situation? If so, you may be one of the many turfgrass professionals out there who view weeds as the worst of their pest problems. Each example response listed above represents a circumstance where weed control differs and is more challenging than is the case for other turfgrass pests.

Fundamentally, weeds are defined as plants out of place. Therefore, what plants are viewed as weeds will depend upon the context of the situation. Some turfgrass managers may view a particular plant as a desired turf while others may view the same plant as a weed. Determining what plants are weeds is at the discretion of the decision makers at a particular facility. Weed control is inherently more challenging than control of other pests because weeds are uninvited plants growing in a plant ecosystem. The biological similarity of weeds to the crop or landscape plants they invade makes managing and controlling them particularly difficult. This is because control strategies are more likely to cause harm to desired turf than control measures that target non-plant pests like diseases or insects.

The purpose of this chapter is to introduce the reader to some of the most common weeds they might experience in a turfgrass or landscape setting. The weeds discussed in this chapter may not be all-inclusive for your area but do represent the majority of the more common weed problems in turf and landscape settings. As with other pests, control of weeds may be a necessary but time consuming and

PESTPOINTER

Fundamentally, weeds are defined as plants out of place. Therefore, what plants are viewed as weeds will depend upon the context of the situation.

costly portion of your management budget. Because time and money are valuable commodities in any operation, we must focus on fundamentals before getting to the specifics of weed control. The core fundamentals of weed control are proper identification of weed species and their life cycles. Because these two issues are cornerstones of a solid weed management program, they will be discussed prior to any discussion of specific weed control strategies.

WEED LIFE CYCLES: ANNUALS, BIENNIALS, AND PERENNIALS

Understanding the life cycle of a weed is less complicated than for disease or insect pests but is absolutely critical to optimizing management and control of weeds. A life cycle for a pest is best described as the period of time between birth/germination and death from natural causes. The life cycles for plant pathogens that cause disease symptoms or for insects can vary considerably in duration for different species and can often occur multiple times in a given growing season. Weeds tend to be more basic, falling into one of four life cycle categories: summer annual, winter annual, biennial, and perennial. Proper recognition of a weed's life cycle not only helps in selection of a good control method but also assists with and sometimes dictates the timing of applied control measures. We'll now proceed through the basic life cycles and describe each in more detail.

PESTPOINTER

Weeds tend to fall into one of four life cycle categories: summer annual, winter annual, biennial, and perennial.

Annual weeds most closely resemble the life cycles for other pests. Insects and disease organisms, with few exceptions, have short life cycles and rely upon prolific reproduction for species persistence. Annual weeds, as the name implies, usually do not live more than twelve months, even though some annuals may live during portions of two calendar years. They characteristically rely upon seed production to fuel future generations of weeds. Seed production varies among species but can be highly prolific. Some weed species can produce a million or more seeds per plant! Thankfully, most common annual turf weeds are not this productive, in terms of seed production. Also, common cultural practices like mowing can reduce or prevent seed production. Because seed production is the primary means of reproduction for annual weeds, seeds produced from a previous generation tend to germinate readily when climatic conditions become suitable.

Annuals have two primary growth phases: a vegetative phase and a reproductive phase. The vegetative phase begins at the time of germination and represents the biggest portion of the plant's life cycle. During the vegetative phase, annual weeds can produce large amounts of plant biomass and tend to grow fairly rapidly. As such, they can be aggressive competitors with turf for nutrients, water, and other resources. The reproductive phase involves flowering and seed production and is triggered primarily by temperature and daylength. However, the reproductive phase can also be induced by environmental stresses such as drought. Drought-induced flowering and seed production are part of a survival mechanism that helps the plant fuel

PESTPOINTER

Annuals have two primary growth phases: a vegetative phase and a reproductive phase.

PESTPOINTER

Within the larger group of annual weeds that are found in turf and in other cropping systems, there are two primary subgroups: summer annuals and winter annuals.

later generations of weeds under adverse conditions. Seeds produced by annuals can germinate as soon as the following growing season but can persist for longer periods of time if conditions are adverse or if they are buried deeper in the soil by practices such as tillage. Within the larger group of annual weeds that are found in turf and in other cropping systems, there are two primary subgroups: summer annuals and winter annuals.

Summer annuals are so named because summer represents the peak or the central season that comprises their life cycle. They, unlike winter annuals, carry out their life cycle within a single calendar year. Most summer annuals germinate in late winter or spring but are capable of germinating well into the summer. A common example of this ability is late summer breakthrough of weeds like crabgrass (Figure 3.2). Preventive herbicides applied in spring can eliminate many of the weeds that are available for germination but some can persist until the herbicide wears down and sprout later in the summer. Herbicide application strategies must therefore account for these capabilities as much as possible.

When summer annual weeds actually germinate can vary considerably among species but is most always a direct function of soil temperature. Placing a month or date of germination to certain weeds is often unreliable as climatic conditions vary from year to year and certainly among different regions of the country. For example, summer annuals like crabgrass germinate in January or February in

Weeds Found in Turf and Landscapes 97

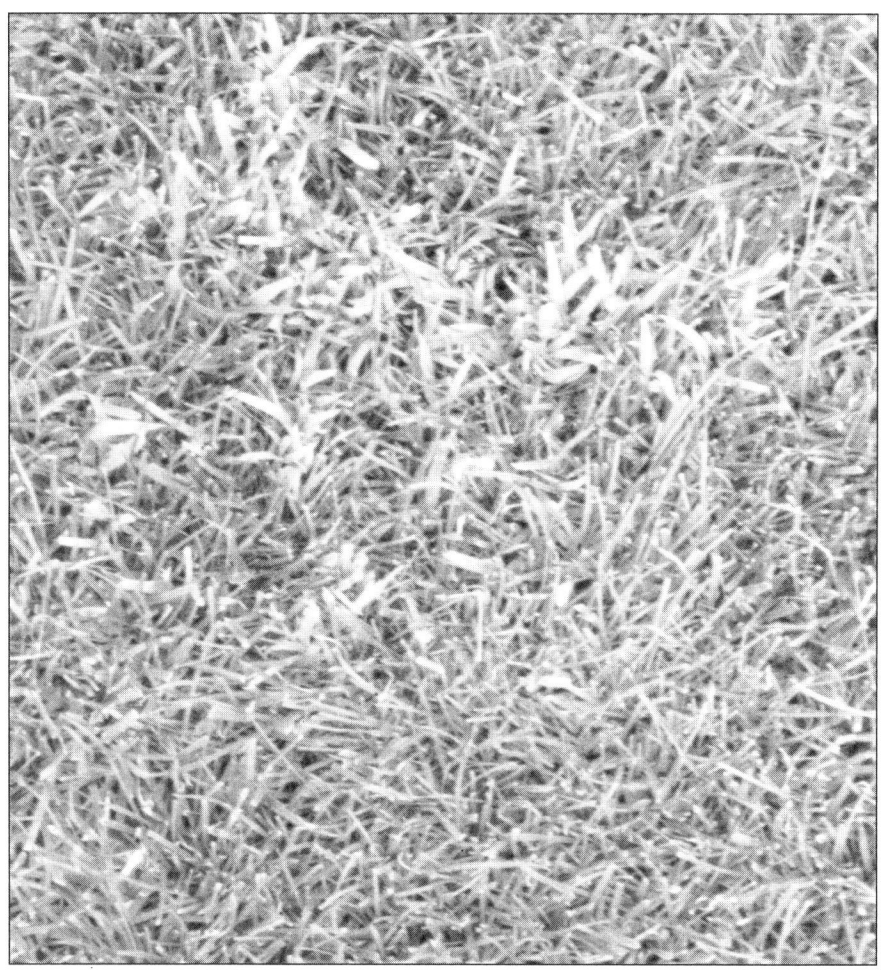

FIGURE 3.2 Crabgrass infestation in managed turfgrass.

many southern regions of the country but will not germinate until May or even June in some northern states. Many cropping systems rely upon model units called growing degree-days, which factor in temperature data from a particular year and help growers plan activities like planting. Such precision has not been pursued with most weed species so always be sure to consult peers or state-based extension literature if you are unsure when certain weeds emerge in your area. Summer annuals often remain in a seedling or immature stage for the first month or two after germination because air temperatures have not yet reached their seasonal peaks. Growth of summer annuals can become very rapid when it does peak and these weeds usually will grow at a rate much faster than for desired turf. As a group, summer annual plants can tolerate longer daylengths and higher temperatures than many other plants so they thrive during the peak of summer. This gives them a distinct competitive advantage if they are not managed or controlled.

A common feature among all summer annuals is induction of the reproductive phase in late summer or fall. Induction of this phase of growth is usually quite noticeable:

- *Broadleaf plants produce a flower.*
- *Grasses produce long stalks that will bear the seeds.*
- *General decline in the vigor of vegetative portions of the plant*
- *Leaves will tend to lose color*
- *Many broadleaf summer annuals will become rigid and stemmy (Figure 3.3). This trend is largely due to the plant's increased allocation of vital resources to its flowering and seed producing parts.*
- *Much of the plant's foliage will senesce or die back, much akin to the leaf dropping process seen with deciduous trees.*

This trend is sometimes preceded by purple coloring in leaves, which is caused by breakdown of the green chlorophyll pigment that gives actively growing leaves their color and greater expression of other naturally occurring plant pigments.

As for germination, specific timings for induction of the reproductive phase in summer annuals will vary among weed species but they will tend to vary less across regions of the country. This is because the summer annual reproductive phase is in part triggered by daylength. This mechanism helps promote flowering and seed production, even when conditions may still be fairly hot in many locations. Annual plants

Weeds Found in Turf and Landscapes   99

FIGURE 3.3   A mature prostrate spurge plant, with numerous branching stems.

have internal sensors that can detect the daylength they are exposed to. Once daylength drops to a certain level, the plant is triggered to initiate the reproductive phase. The daylength trigger helps signal the plant that colder weather is on the way and that the seed production process needs to begin. Different areas of the country experience greater daylength fluctuation. Practitioners in areas such as the

> **PESTPOINTER**
>
> Practitioners in areas such as the south where daylength fluctuations are less severe may experience a gradual reproductive phase for summer annuals while northern areas that see faster reductions in fall daylength may experience more rapid flowering and seed production with the same weed species.

south where daylength fluctuations are less severe may experience a gradual reproductive phase for summer annuals while northern areas that see faster reductions in fall daylength may experience more rapid flowering and seed production with the same weed species.

Temperature is also intimately related to induction of the reproductive phase in summer annuals. Flowering and seed production tend to begin once night temperatures begin to cool in late summer or early fall and are accelerated by brief freezing events like the first fall frost. This effect of temperature will again occur sooner and will be more rapid in northern climates while it may be more gradual as one moves south. Because of the more consistent daylengths and annual temperatures that exist in many subtropical or tropical climates, some traditionally annual weeds like goosegrass can assume more persistent characteristics and may behave more like perennials in these areas (Figure 3.4). Practitioners in Hawaii or parts of Florida may encounter this additional challenge in their weed management programs but such an occurrence is rare elsewhere.

Winter annuals are usually not active during winter months but, as is the case for their summer annual cousins, winter is the central season that the life cycle always includes. Winter annual plants include many weeds but also some crops like winter cereals. Many winter annuals germinate in the fall, pass through winter as a juve-

FIGURE 3.4 Goosegrass, an annual grass species that can be perennial in more tropical climates.

nile, and continue developing to maturity once conditions are again suitable in spring. However, fall germination is not always a requirement as members of a winter annual species can germinate in fall, early spring, or both. Some winter annuals, including winter cereal crops like wheat, must be planted in fall so the plant can experience winter vernalization, a necessary exposure to cold temperatures that is essential for flowering and seed production in the spring. Vernalization is not a requirement for all winter annual plants, especially wild plants like weeds that would suffer as species if they experienced mild winters and could subsequently not reproduce.

As a group, winter annual plants all prefer cool to moderate temperatures and so thrive outside the confines of summer conditions. The duration of a winter annual life cycle is comparable to that for summer annuals but will often occupy portions of two calendar years. Northern regions, which favor relatively short life cycles for summer annual plants, will recognize longer life cycles for winter annuals. Conversely, regions of the country like the south, where summer annual life cycles may be longer, will also tend to experience shorter life cycles for winter annuals. Subtropical or tropical areas have few winter annuals since there is an extremely short or nonexistent cool period in which these plants are able to thrive. In some cases, annual weeds can be either summer or winter annuals, depending upon the particular climate in an area.

Much like summer annuals, winter annuals will undergo both a vegetative growth phase and a reproductive growth phase. Even when winter annuals germinate in fall, they experience relatively little growth prior to spring, when growth conditions are much

---

PESTPOINTER

As a group, winter annual plants all prefer cool to moderate temperatures and so thrive outside the confines of summer conditions.

more suitable. The active vegetative growth phase for winter annuals can begin as soon as January in more temperate climates and can last well into the month of May in northern climates. Leaf and root development are the primary emphases of winter annuals during the vegetative phase of growth. Grasses can produce secondary shoots and leaves (called tillers) while broadleaf plants will tend to branch out and become more robust. Both broadleaf and grassy winter annuals in turf will remain fairly compact during the vegetative phase.

Winter annuals undergo a transformation to a reproductive phase that is biologically similar to that for summer annuals but which occurs at a different time of the year. Daylength and temperature are again the most important environmental stimuli that induce the reproductive phase. Depending upon conditions at a particular location, the reproductive phase for most winter annuals can occur as soon as March and as late as June. Typically, the longer days and warmer temperatures that indicate summer is around the corner will stimulate this transformation. However, early spring warm spells or dry spring conditions can induce the reproductive phase earlier than normal. Also, cool and moist conditions may delay the onset of the reproductive phase in winter annuals. The reproductive phase typically lasts four to eight weeks for most weeds but can be much longer for some species. Annual bluegrass is a winter annual grass that is known as a progressive seed producer, meaning it can consistently produce seed over an extended period when the conditions are right (Figure 3.5). In transition zone areas, annual bluegrass might produce seed for a conventional six week period and is limited in this regard by the hot temperatures that are customary in this climate. However, in a maritime climate like the Pacific Northwest, annual bluegrass is capable of producing seed over a three to four month period!

Many of the biological changes that accompany transition into the reproductive phase are similar between summer and winter annuals. Winter annual grasses can produce large quantities of seed but tend not to produce the large seed-bearing stalks that many summer annual grasses do. Many winter annual grasses are adapted to

FIGURE 3.5 Annual bluegrass in the flowering stage of growth.

coexist with managed turf. By virtue of this adaptation, they can often still produce seed while in a compact form that can tolerate closer mowing practices. Annual bluegrass is perhaps the best example of a winter annual grass species that invades and occupies turf. In many climates, annual bluegrass is so prolific that it is managed as the featured turf species. It can tolerate mowing heights as close as those associated with putting greens and can still produce flowers and seeds at these kinds of mowing heights. Because of this adaptation, management or control of annual bluegrass is often better served by trying to chemically suppress seed production. Winter annual broadleaf weeds in turf are often different from summer annual broadleaf weeds in that they are more compact in the vegetative phase but tend to grow much taller as they mature and flower (Figures 3.6 and 3.7). Common turf weeds like henbit and chickweed are both examples of this trend. Vertical growth during the reproductive phase subjects these weeds to practices like mowing but they are often very dense, allowing them to still reproduce effectively. By contrast, summer annual broadleaf weeds in turf (e.g. spotted spurge) tend to remain prostrate, even as they mature into the reproductive phase (Figure 3.8).

A less common life cycle for weeds is a biennial. The term translates roughly to "two years" and, true to the translation, these weeds usually have a two-year life cycle. These weeds will characteristically germinate in the first year and develop into a prostrate, vegetative

FIGURE 3.6 A juvenile or seedling henbit plant.

FIGURE 3.7 A mature henbit plant, which is taller and in the flowering stage.

form called a rosette (Figure 3.9). As for some winter annuals, biennials require a vernalization period (during winter) between the first and second year of growth in order to flower and reproduce. The second year of growth proceeds much like a winter annual, in that warmer spring temperatures and longer days will induce flowering and seed production in the plant. Comparatively few weeds feature this life cycle but they can often be confused with winter annuals since the flowering period is similar.

PESTPOINTER

The common ground that all perennials share is some ability to withstand all seasons.

Weeds Found in Turf and Landscapes 107

FIGURE 3.8 Prostrate growth is still present in spotted spurge, even at a more mature stage of growth.

The final primary life cycle for weeds is perennial. Many of the most common weeds in turf and landscapes fit into this category, which is unfortunate because many perennials are more difficult to manage and control than annual or biennial weeds. The term

FIGURE 3.9   A common rosette form found in the first year of a biennial life cycle.

perennial implies longevity and many vigorous perennials can indeed persist for many years. By definition, a perennial must be able to persist for more than two years. Some persist for little more than this required period of time while others might last indefinitely under adequate growth conditions. Managed turfgrasses are them-

Weeds Found in Turf and Landscapes 109

selves perennials, which of course is desirable from a maintenance standpoint. The common ground that all perennials share is some ability to withstand all seasons.

Most perennial weeds can and will produce seeds to further the development of the species (Figure 3.10). This serves the function of allowing perennial weeds to spread but doesn't result in the death of the original plant, as would be the case for annual weeds. What distinguishes perennials is the existence of some sort of storage structure that allows them to store food reserves and survive both summers

FIGURE 3.10 Perennial weeds like yellow nutsedge still produce seeds as a means of survival.

and winters. These structures will vary among biological groups of weeds and even among species in a similar group. Storage structures common in perennial weeds are listed with example species in Table 3.1.

Roots are common storage organs for all plants. They are therefore one of the most common means by which perennial weeds persist from year to year. A highly robust and vigorous root system is a common characteristic of most perennial grasses. Orchardgrass, as exampled above, relies exclusively upon its root system while other perennial grasses may have rhizomes, stolons, or both to supplement the storage capabilities of the roots.

Taproots are exclusive to broadleaf weeds and are thus present in both annual and perennial species. However, the taproot in an annual serves simply to optimize the plant's ability to grow during its life cycle while the same structure in perennials serves additionally as the primary food reserve storage area during winter months.

Rhizomes and stolons are generally more common in grasses than in broadleaf weeds but can exist in both types of weeds. Both structures are essentially lateral or horizontal stems. Rhizomes develop underground while stolons develop at the soil surface (Figure 3.11). Since either type of structure serves both to facilitate spread of the weed during the growing season and to store food reserves, they are almost exclusively found in perennial weeds. The short life cycle of annual weeds would make unnecessary the creation and devel-

TABLE 3.1 Common perennial weed storage structures with an example weed species.

Storage Structure	Example Species
Vigorous root system	Orchardgrass
Taproot	Dandelion
Rhizome	Bermudagrass
Stolon	Field bindweed
Bulb/Corm	Wild garlic
Nutlet	Yellow Nutsedge

Weeds Found in Turf and Landscapes   111

FIGURE 3.11   Patches of white clover develop from horizontal stolons.

opment of such structures. Rhizomes and stolons emerge from the original growing point of perennial weeds and spread outward. The extent to which a rhizome or stolon can spread is determined by whether the plant has determinate or indeterminate growth.

Determinate growth is more limited, with stolons or rhizomes able to progress for a distance and then produce a secondary shoot or "daughter plant" from a new growing point called a node. The node can generate both roots and new leaves. Determinate weeds produce only a single node from a particular rhizome or stolon and progress no further. An example of this kind of growth is found in the turfgrass species Kentucky bluegrass, which produces determinate rhizomes. Weeds with indeterminate rhizome or stolon growth can produce these same structures with many nodes, allowing for rhizomes and stolons that can extend for much larger distances from the original plant. Aggressive spreading species like field bindweed or bermudagrass have this indeterminate form of growth (Figure 3.12).

Bulbs, corms, and nutlets are all underground storage structures that are exclusive to monocot weeds other than grasses. They are

FIGURE 3.12 Perennial weeds like bermudagrass, which have indeterminate stolon and rhizomes, can aggressively encroach into established turf.

therefore commonly found in cultivated plants like irises or tulips but also in some weeds, like wild garlic or many of the sedge species. Because these structures are not found in grasses or broadleaf weeds, their relative importance pales in comparison to other perennial storage structures. Bulbs and corms are very similar in appearance and are more common in cultivated plants than in weeds. Weedy species like wild garlic tend to have much smaller bulbs than cultivated plants like tulips or edible garlic, most likely a function of years of breeding efforts with these domesticated plant species. Bulbs in weeds are round or elliptical in shape and store ample food reserves to produce new plant shoots when growing conditions are suitable (Figure 3.13). Nutlets are exclusive to perennial sedge species and are most common in yellow and purple nutsedge. They are produced as extensions from the roots of these species and, like bulbs, give rise to new shoots when conditions permit it.

Storage structures in perennial weeds make controlling these weeds more difficult than for annuals. Because new plant growth can arise from these structures, destroying a perennial weed means one must also destroy the storage organs. While some perennials have only a single means of storing food reserves, others have multiple means. Not coincidentally, those weeds that have more than one place to store food reserves are also those that are more difficult to manage and control. More on this topic will be discussed in the next chapter.

PESTPOINTER

Storage structures in perennial weeds make controlling these weeds more difficult than for annuals.

114 *Turfgrass Chemicals and Pesticides: A Practitioner's Guide*

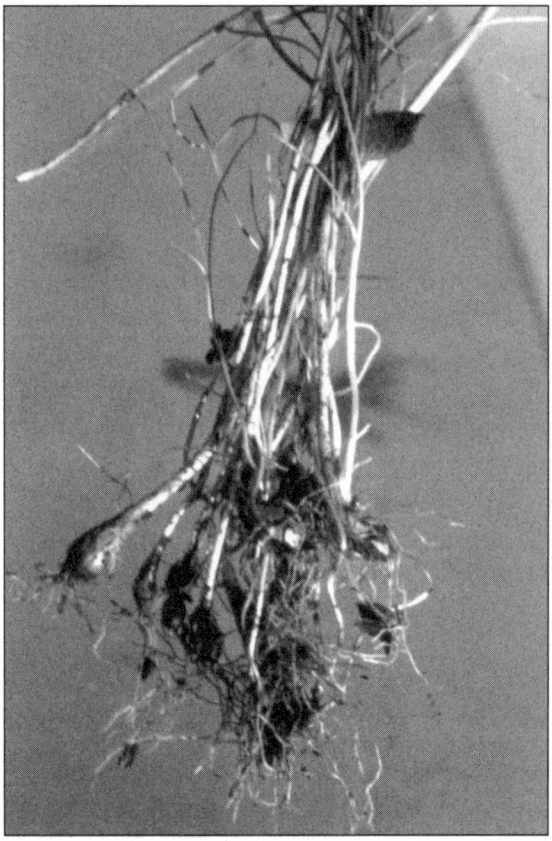

FIGURE 3.13   Bulbs from a wild garlic plant.

WEED BIOLOGY: BROADLEAF WEEDS, GRASSES, AND SEDGES

The biology of weeds is just as important to understand as the life cycle if weed control is to be optimized. As will be discussed in the next chapter, chemical weed control is founded upon using herbicides that are toxic to weeds but comparatively safe to turf. Much of the basis for this has to do with the biology of the target weeds. The more biological similarity a weed has to turf, the more difficult it will be to control. As we proceed through discussion of the dif-

> **PESTPOINTER**
>
> Weeds will tend to fall in one of three primary biological categories: broadleaf, grass, or sedge.

ferent biological groups that weeds can fall into, think of how these plants resemble or differ from turfgrass as these similarities/differences have tremendous bearing on what chemicals will effectively control weeds. Weeds will tend to fall in one of three primary biological categories: broadleaf, grass, or sedge.

Broadleaf weeds are a diverse group of plants that represent many taxonomic families. All of these families and species do fall within the dicot order of plant taxonomy. The term dicot is often used to describe broadleaf plants but actually refers specifically to the seeds of these plants. Dicot (actually dicotyledonae) is a Latin term that crudely translates to "two seed leaves". Essentially, seeds of broadleaf plants have two halves that develop into two seed leaves as the plant germinates. This is why young broadleaf seedlings have two leaves. Another practical way to visualize dicot seed structure is to think of a broadleaf seed crop like a peanut, which can be easily split into two halves. Taxonomy aside, broadleaf plants have numerous common identifying features that can easily separate them from other weeds like grasses or sedges. Examples of these features include:

- *Branching stems with leaves situated away from the primary stalk*
- *Netted (think of a maple leaf) vein arrangement in the leaves*
- *Taproot-based root system*
- *Larger and often showy flowers*
- *Leaves that are usually broad or more round in shape*

As might be expected from Mother Nature, there are always exceptions to the above examples. However, these characteristics can provide a fundamental basis for differentiating dicot or broadleaf plants from other plants that may occur as weeds in turf.

To the well-trained eye, grasses and sedges that are considered weeds in turf are fairly easy to differentiate from desired turf. However, because they have more visual similarities to many turfgrass species, they are sometimes more difficult to recognize than broadleaf weeds. As was discussed previously, broadleaf weeds fall into the dicot order of plant taxonomy. By contrast, grasses (including turfgrass species) and sedges fall into the monocot order of plants. By definition, the term monocot means the seed has one primary food storage compartment and therefore will sprout as a single leaf. However, this taxonomic distinction has less practicality for the practitioner so we will focus on some more recognizable characteristics of these plants.

Grasses and sedges, as a group, have thinner, more elongated leaves than broadleaf plants so they are often easily distinguishable in this regard. In addition to the general shape of the leaves, some other common identifying characteristics for grasses and sedges are:

- *Non-branching and compressed stems with leaves attached to the primary stalk*
- *Parallel vein arrangement in the leaves*
- *Fibrous or netted based root system*
- *Small and usually non-showy flowers*
- *A basal growing point (crown) near the surface of the ground*

Unlike broadleaf plants, grasses and sedges offer few exceptions to the guidelines highlighted above. However, there are also some fundamental differences between grasses and sedges, even though they often appear to very similar at first glance. Sedges are usually shinier and have more rigid leaves than grasses. When conditions are good for both types of plants to grow, sedges will tend to grow faster than grasses. Using a hypothetical example, one week after

PESTPOINTER

The biological differences between grasses and sedges often result in different herbicide recommendations for what will effectively control these plants as weedy invaders.

a lawn with sedge weeds in it has been mowed, the sedges will stand taller than the turfgrass, often by a significant margin. Sedge plants have triangular stems while grasses have round stems. Pulling up a plant from each group and rolling the stem in your fingers will help assist in identifying this biological difference. Another more visual difference is how leaves are arranged on the stem. Grasses will develop leaves that emerge from two sides of the stem while sedges develop leaves that emerge from three sides of the stem. Looking down from directly above a growing grass or sedge plant should help expose this difference to a practitioner (Figure 3.14). As will be discussed in the next chapter, these biological differences between grasses and sedges often result in different herbicide recommendations for what will effectively control these plants as weedy invaders.

BROADLEAF WEEDS COMMONLY FOUND IN TURF AND LANDSCAPES

Broadleaf weed control encompasses the greatest proportion of annual dollars spent on turfgrass and landscape pest control. This may or may not be the case in commercial turfgrass systems but is true when one considers expenditures from the residential consumer sector. Table 3.2 lists some of the most common broadleaf weeds found in turfgrasses and landscapes, sorted by their life cycles.

FIGURE 3.14 Note how the leaves on a yellow nutsedge plant emerge in three directions. Grasses, by contrast, only develop leaves in two directions.

While they may seem cumbersome, scientific names are important because common names can vary considerably across regions.

Because this text is not devoted exclusively to weed control, it is not feasible to discuss each of the weeds listed in Table 3.2 in detail. However, I will attempt to cover the spectrum of the weeds in the list as completely and as illustratively as possible. For weeds listed that aren't covered in the sort of detail that meets your needs, I encourage you to consult with your local turfgrass and/or weed control university expert for more guidance.

Annual broadleaf weeds are some of the most common weeds found in turfgrass and landscapes. Figures 3.15 through 3.22 depict some of the common summer and winter annual broadleaf weeds found in turfgrass and landscape systems. One of their common characteristics, although they and those not pictured from Table 3.2

TABLE 3.2 Common broadleaf weeds found in turf and landscape systems.

Weed (common name)	Weed (scientific name)	Life cycle
Carpetweed	Mollugo vercillata	Summer annual
Prostrate knotweed	Polygonum aviculare	Summer annual
Common purslane	Portulaca oleracea	Summer annual
Prostrate pigweed	Amaranthus blitoides	Summer annual
Common lespedeza	Lespedeza striata	Summer annual
Black medic	Medicago lupulina	Summer annual
Eclipta	Eclipta prostrata	Summer annual
Florida pusley	Richardia scabra	Summer annual
Spotted spurge	Euphorbia maculata	Summer annual
Purple cudweed	Gnaphalium purpureum	Summer/Winter annual
Catchweed bedstraw	Galium aparine	Summer/Winter annual
Prickly lettuce	Lactuca serriola	Summer/Winter annual
Common groundsel	Senecio vulgaris	Winter/Summer annual
Mayweed chamomile	Anthemis cotula	Winter/Summer annual
Horseweed	Conyza canadensis	Winter/Summer annual
Common mallow	Malva neglecta	Winter/Summer annual
Knawel	Scleranthus annuus	Winter/Summer annual
Hairy bittercress	Cardamine hirsuta	Winter/Summer annual
Common chickweed	Stellaria media	Winter annual
Henbit	Lamium amplexicaule	Winter annual
Purple deadnettle	Lamium purpureum	Winter annual
Small hopclover	Trifolium dubium	Winter annual
Lawn burweed	Soliva pterosperma	Winter annual
Yellow rocket	Barbarea vulgaris	Winter annual/Biennial
Wild carrot	Daucus carota	Biennial
Bull thistle	Cirsium vulgare	Biennial
Wild geranium	Geranium carolinianum	Biennial
Common yarrow	Achillea millefolium	Perennial
Mugwort	Artemisia vulgaris	Perennial
Florida betony	Stachys floridana	Perennial

Continued on next page

TABLE 3.2 *(continued)* Common broadleaf weeds found in turf and landscape systems.

Weed (common name)	Weed (scientific name)	Life cycle
English daisy	Bellis perennis	Perennial
Oxeye daisy	Chrysanthemum leucanthemum	Perennial
Chicory	Cichorium intybus	Perennial
Dichondra	Dichondra carolinensis	Perennial
Dogfennel	Eupatorium capillifolium	Perennial
Dandelion	Taraxacum officinale	Perennial
Mouseear chickweed	Cerastium vulgatum	Perennial
Birdseye pearlwort	Sagina procumbens	Perennial
Field bindweed	Convolvulus arvensis	Perennial
Birdsfoot trefoil	Lotus corniculatus	Perennial
White clover	Trifolium repens	Perennial
Ground ivy	Glechoma hederacea	Perennial
Healall	Prunella vulgaris	Perennial
Common blue violet	Viola papilionacea	Perennial
Yellow woodsorrel	Oxalis stricta	Perennial
Broadleaf plantain	Plantago major	Perennial
Buckhorn plantain	Plantago lanceolata	Perennial
Red sorrel	Rumex acetosella	Perennial
Curly dock	Rumex crispus	Perennial
Moneywort	Lysimachia nummularia	Perennial
Virginia buttonweed	Diodia virginiana	Perennial
Bulbous buttercup	Ranunculus bulbosus	Perennial
Slender speedwell	Veronica filiformis	Perennial
Poison ivy	Toxicodendron radicans	Perennial
Silvery thread moss	Bryum argenteum	Perennial

represent numerous plant families, is that they are fairly low growing in the vegetative phase. This feature allows these annuals to thrive, even under regular mowing practices.

Common summer annual broadleaf weeds represent many different plant families. However, there are groups of these weeds that

Weeds Found in Turf and Landscapes **121**

FIGURE 3.15 Carpetweed, a common summer annual broadleaf weed in turf and landscapes.

are taxonomically similar to one another and, where relevant, these weeds will be discussed together. Carpetweed is a summer annual with long, narrow leaves that emerge in a circular arrangement from the stem (Figure 3.15). Carpetweed, like many other summer annuals, has a prostrate growth form when in the juvenile phase of growth. Similar species are catchweed bedstraw, which also has the circular leaf arrangement and is best known for its tendency to stick to clothing, eclipta, which is a common weed in southern regions,

FIGURE 3.16 Prostrate knotweed is a low-growing species that is one of the earliest summer annual weeds to germinate.

FIGURE 3.17 Black medic is similar to clover but has an extended central leaflet that sets it apart.

Weeds Found in Turf and Landscapes   123

FIGURE 3.18   Spotted or prostrate spurge is usually identifiable by the dark watermark found on the leaves.

and horseweed (sometimes called marestail), which is a more upright growing species that thrives in lower maintenance or native areas (Figure 3.20).

Some of the best-known summer annual broadleaf weeds have earned their reputations by thriving in areas that are not as suitable for growth of turf and other plants. Among the most noteworthy of these is prostrate knotweed, a prostrate summer annual that can form dense patches, especially in highly compacted areas (Figure 3.16). This plant germinates very early in the growing

FIGURE 3.19 Prickly lettuce looks a lot like thistles but is less spiny to the touch.

season but is not easily identifiable until it is mature. The leaves of knotweed are small throughout its life cycle and it is perhaps best defined via the ochrea, a papery appendage that appears at the point where leaf-bearing stems develop. The ochrea is a feature only common to plants in the Polygonum genus, which also

Weeds Found in Turf and Landscapes 125

FIGURE 3.20   Horseweed is a common invader of low maintenance turfgrass areas or landscapes.

includes field crop weeds like smartweed and ladysthumb. This unique characteristic can thus be used to help differentiate knotweed from other low-growing weeds. A species often confused with prostrate knotweed is prostrate or spotted spurge. Spurge will also have a low and spreading growth habit and can similarly form dense patches in compacted areas (Figure 3.18). It is best told apart from knotweed by its leaves, which are both more round in shape and have a distinct dark watermark. Both of these species can be solid indicators of compacted soils that may need to be aerated before turf will again successfully grow. Other annual species that can develop in compromised areas include prostrate pigweed and puncturevine, the latter of which can be a hazard to applicators and equipment due to its sharp spines.

Some common broadleaf annual weeds are close relatives of the well-known white clover, which is itself a perennial. Such annual species include summer annuals like common lespedeza and black medic, and the winter annual hop clover. These species all look like clovers in that they have leaves with three smaller leaflets but can be distinguished from white clover upon closer inspection. Common lespedeza is a species that is similar in appearance to clover but that has noticeable veins that extend towards the perimeter of the leaflets. It also has purple flowers, instead of white as seen on white clover, and is more common in southern regions. Both black medic and hop clover have yellow flowers but black medic can be distinguished by the fact that its middle leaflet is born on a short stalk (Figure 3.17).

Black medic is biologically more similar to alfalfa than it is to clovers. Hop clovers are close relatives of white clover and have a similar prostrate growth habit, so use of flower color can be an excellent means of telling them apart from their perennial cousin.

Common winter annual broadleaf weeds in turf include chickweed and henbit. Common chickweed is a pale green winter annual with smooth, teardrop-shaped leaves (Figure 3.21). Its vertical growth during the later part of its life cycle can make it a visual distraction but mowing can help minimize the visual impact of this weed. A similar species is mouseear chickweed, a perennial that resembles common chickweed but is a much hairier plant. Henbit is a unique weed in that it has numerous identifiable characteristics. It has showy purple flowers in the mature form and, in sufficiently large populations, can thus have some ornamental attributes (Figure 3.22).

PESTPOINTER

Common winter annual broadleaf weeds in turf include chickweed and henbit.

Weeds Found in Turf and Landscapes 127

FIGURE 3.21 Common chickweed is a winter annual with small, teardrop-shaped leaves.

Henbit is a member of the mint family and thus has square stems, rendering it unique among weed species. Its leaves are also very closely attached to the stem, which differs from the leaf arrangement on most broadleaf weeds. The heavily lobed leaves of henbit are similar to those of ground ivy, a creeping perennial that is a common turf invader. Other less common winter annual broadleaf weeds

128 Turfgrass Chemicals and Pesticides: A Practitioner's Guide

FIGURE 3.22 Henbit is a winter annual with square stems, showy purple flowers, and a minty fragrance.

include common groundsel, mayweed chamomile, knawel, hairy bittercress, and common mallow, the last of which is readily identified by its branching, maple leaf-style leaf vein arrangement.

Figures 3.23 through 3.25 show some common biennial broadleaf weeds that are commonly found in turfgrass or landscapes. These weeds are again often confused for winter annuals because both will tend to flower in the spring. They also may be confused for perennial weeds, due to their persistence in the vegetative form during the first year of their life cycle. Since this life cycle includes so few weeds, proper recognition and identification of these species is critical to properly plan a weed control program.

Common biennial broadleaf weeds can be placed into two categories to assist with their identification. The first is the true weedy category, within which can be placed yellow rocket and bull thistle. Yellow rocket is a member of the mustard family but has no value as a source of edible mustard. The plant forms a dense, low-growing rosette in the first year of its life cycle and then undergoes a vertical growth spurt when it enters its reproductive phase (Figure 3.23). The

FIGURE 3.23 Yellow rocket, a broadleaf weed that is a winter annual in some climates but more commonly has a biennial life cycle.

yellow flowers and rapid growth spurt contribute to the naming of this weed species. The rosette form can tolerate mowing during the earlier portions of the life cycle. Bull thistle is one of several thistle species that have the biennial life cycle. It and other biennial thistles also form a dense rosette during the first year of their life cycle but are easily distinguishable from other weeds by the leaf spines common to most thistles (Figure 3.24). Bull thistle, like yellow rocket and other biennials, will bolt or grow rapid vertical in its second year and produce showy flowers before releasing seeds. The weedy nature of thistles is both due to their low, dense growth form at the juvenile stage and the spiny leaves that can be painful to the touch. A species resembling many thistles is prickly lettuce but this annual is less spiny than the thistles it is often confused with (Figure 3.19).

130 Turfgrass Chemicals and Pesticides: A Practitioner's Guide

FIGURE 3.24 Bull thistle, a biennial broadleaf weed, displays the classic rosette vegetative form during the first year of its life cycle.

The second category for biennials is an ornamental category, within which wild carrot and wild geranium can be placed. The reason for this categorization of these weeds is their use as or similarity to common ornamental plants. Wild carrot is often called Queen Anne's lace and is a common decorative addition to many floral arrangements. As a weed species, wild carrot can be easily identified by its umbrella-shaped flower head that gives the plant ornamental value. This shape of the flowering structure is also an identification aid for the perennial species,

Weeds Found in Turf and Landscapes 131

FIGURE 3.25 Wild geranium, a biennial broadleaf weed that has similar physical characteristics to its cousins in the flower and nursery industry.

common yarrow. The vegetative form is also very similar to the leaves of cultivated carrot or other broadleaf species like dogfennel. Wild geranium has no ornamental value but is very similar in appearance to the geranium varieties that are sold as flowering plants (Figure 3.25). Wild geranium leaves are smaller than those of their commercial cousins but the shapes of the leaves can help identify this plant.

Some of the most widely recognized and also most troublesome broadleaf weeds in turf and landscapes are perennials. Figures 3.26 through 3.32 depict some common perennial broadleaf weeds found in turf and landscapes. These are the most difficult broadleaf weeds to control effectively, due to their perennial storage structures and sometimes invasive characteristics. The next chapter will discuss in detail how control strategies must differ to specifically and effectively target these types of weeds.

Without a doubt, the most commonly recognized and the most widespread perennial broadleaf weed in turf is dandelion. Dandelion

FIGURE 3.26   Dandelion, one of the most commonly recognized perennial broadleaf weeds in the green industry.

FIGURE 3.27 Field bindweed is one of the most troublesome broadleaf weeds to manage, due to its invasiveness.

FIGURE 3.28 White clover, another of the most easily recognized perennial broadleaf weeds, can form large patches in managed turf areas.

Weeds Found in Turf and Landscapes 135

FIGURE 3.29 Yellow woodsorrel (often called oxalis) can often be mistaken for clover because of its similar leaf structure.

FIGURE 3.30 Broadleaf plantain is unique among broadleaf weeds in that it has parallel leaf veins, normally a characteristic of grasses.

FIGURE 3.31 Curly dock is one of the larger broadleaf weeds found in turf and landscapes and is often found in areas with lower maintenance.

Weeds Found in Turf and Landscapes **137**

FIGURE 3.32 Speedwells are a family of low-growing broadleaf weeds that can easily escape the routine practice of mowing.

has a low growth habit that gives it excellent tolerance to mowing. It exists primarily in the vegetative form but can flower prolifically in either fall or spring. Flowers are bright yellow and can regenerate rapidly if they are mowed off. Eventually, flowers develop into a seed producing puffball that relies on wind or young children for seed dispersal (Figure 3.26). Dandelion persists with the help of a deep taproot that allows it to survive stressful periods. Wind-dispersed seed can result in dense dandelion infestations if existing weeds are not controlled.

Another very common perennial broadleaf weed species is white clover. It has a very low growth habit and can thus also escape the effects of regular mowing. White clover has the classic trifoliate leaf structure common to many legume plants and often has a whitish watermark on the leaflets (Figure 3.28). It spreads with the help of stolons, and can form dense patches in turf when left uncontrolled. White clover produces small, white, ball-shaped flowers that additionally assist in its development. Other common perennial weeds in turf that are similar to clover include birdsfoot trefoil and yellow woodsorrel. The latter species, sometimes referred to simply by its genus name oxalis, also has the trifoliate leaf structure but leaflets are more heart-shaped and the plant produces yellow flowers (Figure 3.29). While the dense development of white clover can gradually choke out desired turfgrasses, its stolons do not render it highly invasive. A stoloniferous species that can be highly invasive is field bindweed. Its rapid spreading ability and ability to thrive in dry areas can make it extremely difficult to manage. Field bindweed has been a problem in field crops for years but has more recently become a growing issue in many turfgrass and landscape areas. Its arrow-shaped leaves make it relatively easy to identify (Figure 3.27).

Some perennial broadleaf weeds are more sensitive to mowing and can be more common in less maintained areas or in landscapes. Examples include plantains, curly dock, red sorrel, mugwort, Florida betony, and chicory. Plantains are unique among broadleaf weeds in that they have a parallel leaf vein arrangement that is more common in monocot weeds. Two common plantain species exist in turf and landscapes. Broadleaf (sometimes called blackseed) plantain has shorter, thicker leaves (Figure 3.30) while buckhorn plantain has longer, narrower leaves. Both species can tolerate some mowing but tend to thrive in areas receiving less maintenance. Curly dock is a species with large leaves that have curled edges (Figure 3.31). It and its close cousin red sorrel have little tolerance to mowing but can become very robust in landscapes or native areas adjacent to turf stands. Mugwort, Florida betony, and chicory are less common in most turf areas but are very common in both landscapes and the container nursery industry.

A low growth habit is a common feature of many perennial broadleaf weeds and the speedwell family is an excellent example of this trait. A common weedy species of speedwell in turfgrass is slender speedwell, which has small, lobed leaves and can form thick patches if left uncontrolled (Figure 3.32). Other small-leaved prostrate perennial broadleaf weeds include daisies, the southern species Virginia buttonweed, buttercup, dichondra, and blue violet. Dichondra and violet can both be identified by their leaves, which are roundish and cup towards the stems. Dichondra is often used as a ground cover in southern regions but its ability to form dense patches can make it a troublesome weed. Perennial violet is one of the most troublesome broadleaf weeds in landscapes, due to its persistence even when control measures are employed. An additional low-growing problematic weed is moss, which falls into its own unique taxonomic class. Because mosses are more primitive and don't have the characteristics of many higher plants, control options must also be unique. Mosses are common in moist, shaded areas and have become a growing problem on golf courses, especially on putting greens.

GRASSY WEEDS AND SEDGES COMMONLY FOUND IN TURF AND LANDSCAPES

Grasses and grass-like plants are often more challenging weed problems for turfgrass practitioners because they have more biological similarity to the turfgrasses we manage. These similarities can also make proper identification of these weeds difficult. Most broadleaf weeds have numerous identifiable characteristics and can often be identified based solely on leaf shape. The grassy weeds often do not offer such clues and it may require closer inspection or consultation with experts or peers to properly identify these types of weed species. There are generally fewer grassy weeds that turfgrass and landscape managers must contend with than broadleaf weeds but control options are also more limited. It is for this reason that this aspect of weed management is usually regarded as one of the most difficult for practitioners. Tables 3.3 and 3.4 list some of the most common weedy grass species and grass-like weeds, respectively, that turfgrass and landscape practitioners must contend with.

We will now cover some details concerning these weeds, placing them in common groups where it is applicable. Again, it will not be feasible to cover them all in detail but, as you encounter unknown grassy or sedge-like weeds, focus first on whether the plant is a grass or a different kind of monocot plant. From there, proceed to looking for structures that would point to the plant as being annual or

TABLE 3.3 Common grass weeds found in turf and landscape systems.

Weed (common name)	Weed (scientific name)	Life cycle
Field sandbur	Cenchrus longispinus	Summer annual
Large crabgrass	Digitaria sanguinalis	Summer annual
Smooth crabgrass	Digitaria ischaemum	Summer annual
Goosegrass	Eleusine indica	Summer annual
Yellow foxtail	Setaria glauca	Summer annual
Crowfootgrass	Dactyloctenium aegyptium	Summer annual
Broadleaf signalgrass	Brachiaria platyphylla	Summer annual
Annual ryegrass	Lolium multiflorum	Winter annual
Annual bluegrass	Poa annua	Winter annual
Tufted hardgrass	Sclerochloa dura	Winter annual
Little barley	Hordeum pusillum	Winter annual
Bermudagrass	Cynodon dactylon	Perennial
Orchardgrass	Dactylis glomerata	Perennial
Tall fescue	Festuca arundinacea	Perennial
Nimblewill	Muhlenbergia schreberi	Perennial
Dallisgrass	Paspalum dilatatum	Perennial
Bahiagrass	Paspalum notatum	Perennial
Roughstalk bluegrass	Poa trivialis	Perennial
Quackgrass	Elytrigia repens	Perennial
Torpedograss	Panicum repens	Perennial
Johnsongrass	Sorghum halepense	Perennial

TABLE 3.4 Common sedge or sedge-like weeds found in turf and landscape systems.

Weed (common name)	Weed (scientific name)	Life cycle
Annual sedge	Cyperus compressus	Summer annual
Texas sedge	Cyperus polystachyos	Summer annual
Spreading dayflower	Commelina diffusa	Summer annual
Yellow nutsedge	Cyperus esculentus	Perennial
Globe sedge	Cyperus globulosus	Perennial
Purple nutsedge	Cyperus rotundus	Perennial
Green kyllinga	Kyllinga brevifolia	Perennial
Slender rush	Juncus tenuis	Perennial
Wild garlic	Allium vineale	Perennial
Wild onion	Allium canadense	Perennial
Star-of-Bethlehem	Ornithogalum umbellatum	Perennial

perennial. These decisions will be critical to determining a proper course of action. If the identification of the plant remains a mystery, use your local extension resources to firm up the identity of the plant before pursuing control strategies.

Figures 3.33 through 3.39 show some common summer and winter annual grasses found in turf and landscapes. Because these grasses are annual, they can be much easier to control effectively than perennial grasses. However, as a group, they produce large quantities of seed and can be perennial problems for turfgrass managers.

Some of the annual grasses listed in Table 3.3 and pictured may be more familiar to you than others. Let's briefly cover some of the environments and/or areas where they are most common. Crabgrass is one of the most widespread and recognizable grassy weeds in turf (Figure 3.34). Its broader and shorter leaf shape makes it easily distinguishable from turf. The two most common species of this weed are large and smooth crabgrass. They may appear at first glance to be the same but large crabgrass has hairy leaves, upon closer inspection, while smooth crabgrass does not. The good thing is that

FIGURE 3.33 Field sandbur can pose particular difficulties because of the spiny burrs that can be a nuisance to those who encounter them.

FIGURE 3.34 Crabgrass is one of the most widespread summer annual grassy weeds and represents a large proportion of annual herbicide expenditures.

control strategies for the two species do not differ. Another similar species is southern crabgrass, which exists only in more tropical areas of the country. Crabgrass has excellent tolerance to mowing and can be an invader of a wide variety of turfgrass settings, from lawns to golf course putting greens.

Field sandbur is most notorious for its spiny burrs that can plague people and animals that encounter it (Figure 3.33). It does very well in hot, dry climates such as the Great Plains states. It is highly prevalent in waste areas or roadsides but is also common in managed turf. Goosegrass is a summer annual grass more common to golf course turf than in many other areas relevant for turfgrass and landscape managers (Figure 3.35). It tends to develop later in summer than crabgrass and prefers hotter climates, thus making it more prevalent in the transition zone and in southern areas. It's open and prostrate growth habit when mature makes it highly undesirable as a weed. Goosegrass thrives on compacted soils and can often indicate that compaction is a problem for a particular location. Foxtails are a

FIGURE 3.35 Goosegrass is easily distinguishable by its open-faced growth habit and "wheel spoke" center.

group of summer annual grasses that are common invaders of field crops and are becoming more prevalent weeds in turf and landscapes (Figure 3.36). The soft seedhead gives foxtails their common name and is a highly recognizable trait, although some species can handle reasonable mowing heights and still persist. Yellow foxtail is one of the more common foxtail species in turf but others include green or giant foxtail. Like for sandbur, they are most common in low maintenance areas but are a growing concern in managed turf, especially when new housing developments and golf courses expand into previously rural areas. Other summer annual grass species, like crowfootgrass and broadleaf signalgrass, are more prevalent in southern regions of the country.

Among the common winter annual grassy weeds in turf, annual bluegrass is the most recognized and most widespread species (Figure 3.37). It struggles in very hot climates so is usually limited to the transition zone or northern states. Its prolific growth in cool climates makes it often the turfgrass species of choice since removal of it from other species would be too great and too perennial a challenge. Two unique biotypes exist for annual bluegrass. One is a true annual that is a prolific seed producer while the other is a weak perennial that produces less seed and is most commonly found as a competitor of creeping bentgrass on golf course putting greens. The widespread management of annual bluegrass as a turf in the northern US has resulted in breeding efforts to select varieties that can be used more deliberately. Annual ryegrass is another winter annual grass that has a history of turf use, especially as a species for overseeding warm-season grasses in the southern US. Higher quality species like perennial ryegrass and rough bluegrass have replaced annual ryegrass for this purpose in the modern era. Often called Italian ryegrass, annual ryegrass is less of a problem in turf than in field crops but can often be found in turf areas such as sod farms, where crops used to be grown.

Other lesser-known winter annual grass weeds include little barley and tufted hardgrass. They are not as widespread as some of the other annual grasses mentioned earlier but they can still present management problems. Little barley is a wild cousin of the barley vari-

Weeds Found in Turf and Landscapes   145

FIGURE 3.36 Foxtails are traditionally thought of as field crop weeds, but are a growing problem in turf due to the progression of urban landscapes into rural areas.

eties grown commercially for grain (Figure 3.38). When uncontrolled, little barley can form dense, low-growing clumps and will eventually produce seedheads that resemble those from cultivated barley. This species is less pronounced in urban, developed areas and

146 Turfgrass Chemicals and Pesticides: A Practitioner's Guide

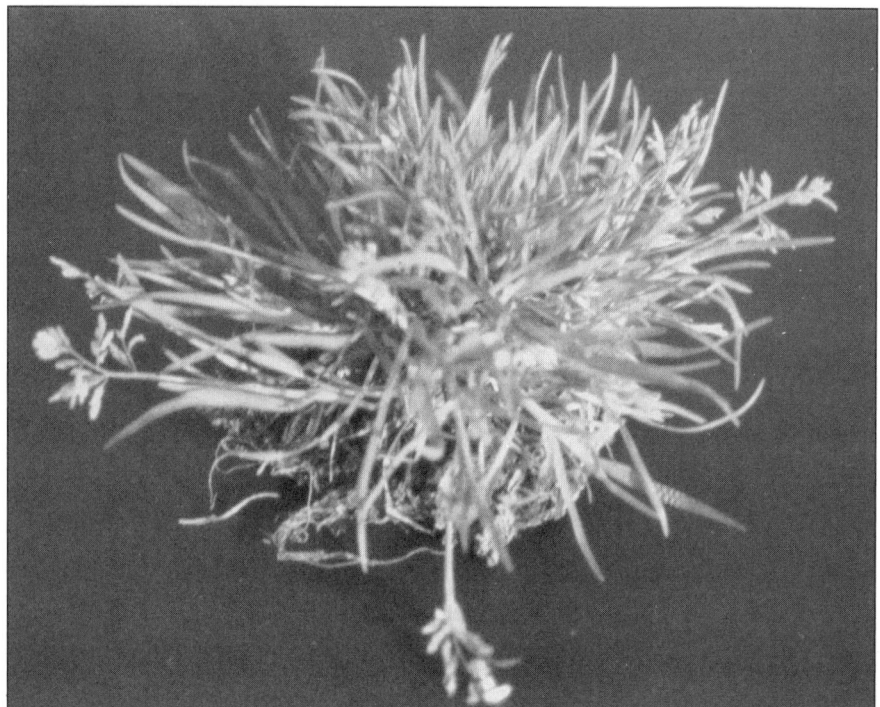

FIGURE 3.37 Annual bluegrass is managed as a turfgrass in many areas but is a troublesome weed for most turfgrass managers.

tends to be more common in turf and landscapes that occupy once rural areas. Tufted hardgrass is a winter annual most prevalent in the central regions of the US (Figure 3.39). It is often confused for annual bluegrass in the juvenile stage of growth. As the plant matures, it develops an open-faced and prostrate growth habit that is similar to goosegrass. While tufted hardgrass is thus sometimes prone to be mistaken for goosegrass, its compact seedhead and spring flowering period help differentiate the two species. Another potential cause for confusion between tufted hardgrass and goosegrass is that they both thrive on compacted soils. Hardgrass was discovered as a problem in trafficked areas of athletic field turf by being mistaken for early-season goosegrass by practitioners who were unfamiliar with the species.

Weeds Found in Turf and Landscapes **147**

FIGURE 3.38 Little barley is a wild cousin of the cultivated winter cereal crop.

Perennial grassy weeds present the biggest challenges to turfgrass practitioners. This challenge is rooted in the fact that practitioners must control an unwanted perennial grass species within turfgrasses

FIGURE 3.39 Tufted hardgrass is a growing problem on athletic fields with higher levels of soil compaction.

that are themselves perennial. This close biological similarity usually translates into few viable options for effective control. Some of the most significant turfgrass renovations or species changes at a facility are based upon this very problem. Figures 3.40 through 3.44 illustrate some of the more prevalent perennial grasses that pose major challenges for turfgrass managers. These same weeds can also be very difficult in landscape settings but more herbicide options exist in these areas with greater numbers of broadleaf plantings.

Some of the perennial grasses listed in Table 3.3 and/or pictured may be more problematic in your area than others. Let's briefly cover some of the environments and/or areas where they are most common. Bermudagrass represents one of the most ongoing love/hate relationships that exist in turfgrass management:

- *Bermudagrass is the most desired and most commonly used turfgrass species.*
- *Tolerance to low mowing*
- *Superior heat tolerance*
- *Drought tolerance*
- *Excellent recuperative ability*
- *Ideal option for southern athletic fields, golf courses, and lawn-type turf.*
- *An annoying weed problem where other turfgrasses are being grown (Figure 3.40).*
- *An aggressive network of both rhizomes and stolons make unwanted bermudagrass both robust and invasive.*

Nimblewill is a perennial grass that can sometimes be mistaken for bermudagrass, due to similar leaf structure. However, nimblewill is less invasive and is more prevalent in shaded areas that the sun-loving bermudagrass might not occupy.

Orchardgrass is a robust perennial species that is not invasive but forms thick clumps that are unsightly in most turfgrass settings (Figure 3.41). The species is a common forage crop and is usually introduced to turfgrass via seed contamination. Many regional seed production facilities, particularly those that produce the Kentucky 31 (or K-31) variety of tall fescue, have large populations of orchardgrass. Both tall fescue and orchardgrass can be and are used as forages so the existence of both species at these seed production facilities is common. Similar seed size between orchardgrass and tall fescue makes it difficult to remove orchardgrass from commercially available seed and results in the weed's introduction to new turf areas. Few chemical options are available for control of orchardgrass so it can be difficult to get rid of once it has become established. Tall fescue is recognized as an excellent turfgrass species for lawn turf, particularly in the transition zone.

FIGURE 3.40 Bermudagrass can be a managed turfgrass or a highly invasive weed species, shown here encroaching into zoysiagrass.

FIGURE 3.41 Orchardgrass is a common forage grass but can be a persistent weed species in managed turf.

PESTPOINTER

Tall fescue is recognized as an excellent turfgrass species for lawn turf, particularly in the transition zone. However, it can also form unsightly clumps when it appears as a weed in other turfgrass species.

However, it can also form unsightly clumps when it appears as a weed in other turfgrass species.

Perennial weedy grasses that are more prevalent in southern regions of the US include dallisgrass, bahiagrass, and torpedograss. Dallisgrass is a robust perennial that has very thick rhizomes (Figure 3.42). It is very commonly found on roadsides in the southern region but can also be a contaminant in close-mowed turf. Under mowed conditions, dallisgrass can form thick clumps and can very difficult to control. Bahiagrass is sometimes used as a turfgrass species but is perhaps most prevalent as a roadside species that is an effective ground stabilizer. It has a very robust rhizome system, which can make it difficult to control when it appears as an undesirable weed species. The characteristic V-shaped seedhead is a clas-

PESTPOINTER

Bahiagrass is sometimes used as a turfgrass species but is perhaps most prevalent as a roadside species that is an effective ground stabilizer.

FIGURE 3.42 Dallisgrass is a very robust perennial weed species in the Southern USA.

sic indicator of this plant's existence during its late summer flowering period.

Torpedograss is a perennial weed species that only exists in subtropical to tropical regions that experience little to no cold weather. Its name is derived from its pointed rhizomes that facilitate its invasive tendencies. This species can be an aggressive invader of any managed turfgrass systems, including aggressive species like bermudagrass. Torpedograss can tolerate a variety of mowing heights and thus can be a problem even on golf course putting greens.

Perennial weedy grasses that present greater challenges for northern practitioners include quackgrass and rough bluegrass. Quackgrass is a species that can spread via rhizomes and can thus be fairly invasive (Figure 3.43). It has some visual similarity to common turfgrass species like tall fescue or perennial ryegrass but can be distinguished by its clasping auricles, which are white, hook-like appendages that circle around the stem at the point where leaves develop. Quackgrass naturally prefers environments where it can grow taller but it can tolerate lower mowing heights, even those found on golf courses. Rough bluegrass, often referred to by its scientific name of Poa trivialis, can be difficult to differentiate from annual bluegrass. It is not as aggressive, in terms of seed production, as its annual cousin but also is used in many areas as a desired turfgrass.

Many overseeded golf course putting greens in states like Arizona or Florida use rough bluegrass, since it can adequately tolerate the close mowing that is necessary. In similar fashion to annual bluegrass,

---

PESTPOINTER

Torpedograss can tolerate a variety of mowing heights and thus can be a problem even on golf course putting greens.

FIGURE 3.43 Quackgrass is a spreading perennial that can be a stiff competitor for cool-season turfgrasses.

poor tolerance to heat limits the use of rough bluegrass as a desired turf. Its prevalence as a weed species is similar to that for orchardgrass in that it is a common contaminant of commercial turfgrass seed, particularly seed produced in western Oregon. Rough bluegrass

> **PESTPOINTER**
>
> Many overseeded golf course putting greens in states like Arizona or Florida use rough bluegrass, since it can adequately tolerate the close mowing that is necessary.

spreads via stolons and can form large patches within other turfgrass species. Its appearance is not always objectionable but its texture differs considerably from species like tall fescue and its poor heat tolerance can result in declining or dead patches in desired turf during summer. Another perennial species that can grow throughout much of the country is johnsongrass. It does not tolerate close mowing but can be a robust contaminant of ornamental areas or in unmowed native grass areas found at many facilities (Figure 3.44).

Sedges and sedge-like species all fall into the same monocot family of plants that also includes grasses. However, this group of plants can present unique challenges for turfgrass managers because many grow at a rate much faster than most turfgrasses. While some of these species tolerate close mowing better than others, their faster growth rate allows them to remain competitive in many managed environments. Figures 3.45 through 3.49 depict some of the common monocot weed species, other than grasses, which are common in turf and landscape systems.

The most common types of non-grass monocot weeds are sedges. Sedges are biologically more similar to rapid-growing species like rushes and horsetails than they are to grasses. One of the rush species, slender rush, is actually fairly common as a weed in turf and landscapes. Numerous sedge species exist as weeds, especially in turfgrass. The most common of these is yellow nutsedge, which thrives in many parts of the country (Figure 3.45). Common also in many field crops,

FIGURE 3.44 Johnsongrass is a tall-growing species that can aggressively invade low maintenance turf areas or native landscape areas.

yellow nutsedge is a perennial that can aggressively compete for available resources. The species can reproduce via seed but more commonly from underground tubers or nutlets, which serve as the sources for new plants each year. Generation of new plants from nutlets can increase the density of yellow nutsedge populations but the plants don't otherwise spread laterally. Other annual or perennial sedge species, including annual sedge, Texas sedge, and globe sedge, reproduce primarily through seed production.

FIGURE 3.45 Yellow nutsedge is one of the most widespread sedge species and can impact both field crops and the green industry.

Some sedge species are more invasive than the ones mentioned above. Purple nutsedge is a perennial species that looks similar to yellow nutsedge (Figure 3.46). It also relies upon tuber or nutlet production but the rhizomes on which the nutlets are formed can develop into long chains, making this species much more aggressive than yellow nutsedge. Purple nutsedge is more common in the southern US but many agricultural and turfgrass practitioners consider this weed to be one of the most difficult they have to contend with. Another group of sedges that is more common in turf than in production agriculture is the kyllingas. Several kyllinga species have been identified but the most common of these is green kyllinga (Figure 3.47). Its darker green foliage and propensity to form dense patches can make it more difficult to distinguish from desirable turf species and, although it is most commonly found in southern areas, it may be more widespread than scientists and practitioners realize. Green kyllinga develops from seed but also rhizomes that allow the formation of dense patches. The species prefers moist conditions such as low spots where moisture may tend to accumulate.

Other non-grassy monocot weeds fall outside the sedge family. Examples include the summer annual spreading dayflower and the perennials wild garlic, wild onion, and Star-of Bethlehem. Spreading dayflower can form into low-growing large patches in moist areas and may resemble broadleaf species like Virginia buttonweed that have a similar growth habit. However, close inspection of spread-

PESTPOINTER

Purple nutsedge is more common in the southern US but many agricultural and turfgrass practitioners consider this weed to be one of the most difficult they have to contend with.

FIGURE 3.46 Purple nutsedge is a highly invasive sedge species and can be very difficult to eradicate.

ing dayflower will reveal the parallel leaf vein formation common to all monocots. Wild garlic and wild onion are very similar to one another and both are wild cousins of the cultivated herbs that we may be more familiar with. Wild garlic can grow much taller than turf it infests and can be further identified by its hollow stems and characteristic garlicky odor (Figure 3.48). Wild onion is similar in

160 *Turfgrass Chemicals and Pesticides: A Practitioner's Guide*

FIGURE 3.47 Green kyllinga can be easily mistaken for a grass, due to its low growth habit and narrow leaf width.

appearance but has flat, rather than hollow, stems. Both plants produce small bulbs underground that can give rise to future generations of plants. Star-of-Bethlehem is commonly sold as a flowering

---

PESTPOINTER

Star-of-Bethlehem is commonly sold as a flowering ornamental bulb species but is prone to escape managed ornamental beds and can develop into patches in turf.

Weeds Found in Turf and Landscapes 161

FIGURE 3.48 Wild garlic is a smaller cousin of cultivated garlic but the smell tells you they are related.

ornamental bulb species but is prone to escape managed ornamental beds and can develop into patches in turf (Figure 3.49). While the flowers may be showy, Star-of-Bethlehem can continue to proliferate in turf if uncontrolled and is poisonous if it is consumed.

FIGURE 3.49 Star-of-Bethlehem is often sold as an ornamental plant but can be troublesome in undesired locations.

WEED ECOLOGY AND CULTURAL WAYS OF CONTROLLING WEEDS

There's an old adage that states the best defense against weeds is a good healthy stand of turf. As practitioners, we must remember that weed management is founded upon solid turf management. Insects and diseases directly feed upon or damage turf, and thus seem to prefer turf that is healthy or well managed. Weeds exact a more indirect effect upon either turfgrass systems or landscape plantings. Their role as pests is founded upon their ability to compete with desirable plants for common resources like moisture, sunlight, and nutrients. When turf is disadvantaged or poorly maintained, weeds are usually the first pest problems that result. Is that a coincidence? As was stated earlier, weeds are simply defined as plants out of place. When we provide them with space to occupy, they are very adept at using that opportunity to their advantage. Over the next few paragraphs, we will discuss/review some of the basics of weed ecology and turfgrass management, and how breakdowns in management can result in weed infestations.

Turfgrasses and landscapes represent a departure from what one could call a natural setting. Species are selected for these areas based upon reasonable adaptation to the local climate. Does this mean they are well suited for the area in which they are planted? This is not necessarily the case. Table 3.5 lists some of the major turfgrass species grown in the US and where they originated.

PESTPOINTER

Insects and diseases directly feed upon or damage turf, and thus seem to prefer turf that is healthy or well managed.

TABLE 3.5 Local origins for common turfgrass species.

Species	Warm or Cool-Season	Origin
Perennial ryegrass	Cool-season	Northern Europe
Creeping bentgrass	Cool-season	Northern Europe
Kentucky bluegrass	Cool-season	Northern Europe
Tall fescue	Cool-season	Northern Europe
Fine fescues	Cool-season	Northern Europe
Bermudagrass	Warm-season	Africa
Zoysiagrass	Warm-season	Eastern Asia
St. Augustinegrass	Warm-season	Africa/Caribbean
Centipedegrass	Warm-season	Southeast Asia
Buffalograss	Warm-season	Central Great Plains

As you can see, buffalograss is the only common turfgrass species that can be considered indigenous or native to the United States. What does this mean to the practitioner who manages this or the assortment of imports they must choose from? From an ecological standpoint, most turfgrasses are out of place where they are planted. The same could be said for many of our ornamental landscape plantings. Stated another way, if nature was the sole driving force behind the plant composition of an area, turfgrasses and landscape plantings would likely not exist. We, as practitioners, force the issue and

PESTPOINTER

When turf is disadvantaged or poorly maintained, weeds are usually the first pest problems that result insist upon species native to other areas to be the prevalent plants at a managed location. This so-called defiance of nature and the laws of ecology should not be seen as a bad thing. However, this concept is presented so that we may better understand why turfgrasses and landscapes are often difficult to manage. Climatic stress and pest problems represent nature holding us accountable for growing species beyond where they originated. As such, we now have entire scientific fields that are devoted to optimizing turfgrass and landscape management, due to these ever-present challenges.

The role of weeds in our managed turfgrasses and landscapes has a lot to do with this concept of ecology. We know weeds are opportunistic. This attribute boils down to the fact that most of our common weeds are better adapted to the area in which they are found than our desired plantings. Weeds rely upon reproduction and proliferation of their species to ensure survival. This gets to the core of what the laws of nature dictate. Table 3.6 explores a hypothetical setting in the Southeast US.

What we see in this example is the gradual tendency for an area that is maintained for agriculture to revert back to what was likely there before humans arrived. This transition will of course vary considerably in duration and the ultimate species composition across areas but the idea is clear. The occurrence of annual and perennial weeds in our managed turfgrass and landscape systems is a sign that nature is trying to shift back to a more natural ecosystem. Practices like turfgrass maintenance and weed control practices help us avoid these

TABLE 3.6 Species composition of an undisturbed area in the Southeast US over a 100 year period.

Time Frame	Primary Species
Years 0-1	Commercial corn (field then abandoned)
Years 1-10	Wide array of annual and perennial weeds
Years 10-50	Pine forest, annual species phase out
Years 50-100	Hardwood forest, pines become less prominent

PESTPOINTER

From an ecological standpoint, most turfgrasses are out of place where they are planted.

natural tendencies. As it stands, nature is a powerful foe and we must focus on the management tools that allow us to maintain and grow turf or landscape plantings where we want them.

How do turfgrass management practices affect the occurrence and proliferation of weeds? Our four primary cultural practices are mowing, fertilization, irrigation, and aeration. Each can affect the development of weeds and are collectively the cornerstone of a solid weed management program. Mowing has a direct impact on the types of weeds found in turf. Not coincidentally, species that are more prostrate in growth succeed more in turf than in taller crops like corn, where they would be shaded out and not able to compete for available resources. Some taller weeds can succeed in turf and landscapes but they typically are found in less maintained areas like native grass stands. Shorter mowing heights, such as those on golf courses, can further limit the types of weeds found in turf but can also predispose adaptable weeds to be more competitive. Examples are summer annual grasses like crabgrass or goosegrass that, once established, are generally more competitive for available resources than the turfgrasses they inhabit. Shorter mowed turf has a shallower root system but tends to be denser. Weeds thus have a more difficult time getting started but, once they do, they can be more vigorous.

The challenge facing turfgrass managers is to optimize mowing heights, such that turf is as healthy and competitive against weeds as possible. Proper turfgrass species selection is a key component of this strategy. For example, many varieties of Kentucky bluegrass are adapted to be mowed at fairway heights of cut and can provide a dense, competitive stand. By contrast, tall fescue is better adapted to taller mowing heights and, when it is mowed too short, it weakens and is more susceptible to weed invasion. Placing a species or variety under conditions they are not well suited to will diminish turfgrass competitiveness and create additional weed management challenges.

Cultural practices like irrigation and fertilization are critical to the development and maintenance of quality turfgrass stands. They also can present challenges to weed management in that these practices provide resources that can be used by both turf and weeds. It is thus critical to focus these inputs on times when the turf can best utilize them. For example, early spring fertilization of warm-season grasses or summer fertilization of cool-season grasses is usually not recommended. These mistimed applications can create an unwanted stimulation of growth when temperatures do not support active growth. While such circumstances can be potentially detrimental to turf, they can also increase the incidence of weeds. Summer fertilization of cool-season lawn turf like tall fescue may not benefit the stand and will undoubtedly provide a source of nutrition for opportunistic summer weeds like crabgrass. Similarly, water-

PESTPOINTER

The challenge facing turfgrass managers is to optimize mowing heights, such that turf is as healthy and competitive against weeds as possible.

ing too often or in too much volume for what turf needs can be just what moisture-craving weeds like yellow nutsedge need to develop. Focus these key cultural practices on periods of active turf growth to achieve optimum benefit to the turf stand and avoid unintentionally supporting weeds.

The last key cultural practice in turf is aeration. Aeration is a critical soil conditioning practice that is unique to turf because it is a perennial ground cover. Aside from new establishment projects, we don't have the option of annually tilling soil to loosen it and prepare it for planting. Aeration serves to relieve soil compaction, can improve water infiltration through soil, and is an excellent tool for managing thatch development. All of these benefits are critical to maintaining soils that turf can effectively grow in. Lack of regular aeration can result in excessive thatch levels in turf, which can weaken the stand and open doors for more competitive weeds. Compacted soils, common to athletic fields and some golf course or lawn areas, discourage the development of most turfgrasses and can result in proliferation of weeds that thrive in compacted areas. Example species are goosegrass, tufted hardgrass, prostrate knotweed, and spotted or prostrate spurge. Controlling the weeds becomes a necessity in these cases but loosening the soil is a better long-term solution to problems related to compacted soils.

Recognition of the relationship between certain weeds and an adverse environment for turfgrass can be critical to developing a solid weed management program. Weeds sometimes occur in

PESTPOINTER

Watering too often or in too much volume for what turf needs can be just what moisture-craving weeds like yellow nutsedge need to develop.

> **PESTPOINTER**
>
> Lack of regular aeration can result in excessive thatch levels in turf, which can weaken the stand and open doors for more competitive weeds.

healthy, vigorous turf stands but are especially prevalent in compromised areas where turf growth is poor. Weeds can thus serve as a tool to help identify problems that exist and that usually can be corrected. Consider the example species from the previous paragraph:

- *A persistent and aggressive outbreak of goosegrass may be a coincidence but will usually point to a soil compaction problem that can be remedied with aeration.*
- *Prolific nutsedge outbreaks often point to overwatering or an irrigation leak as the culprits, which are again fixable problems.*
- *Annual bluegrass can grow in many areas but is often more concentrated in areas that are heavily watered and fertilized.*
- *Weeds that succeed in areas where turf performs poorly are often called indicator weeds.*

Knowing what weeds fit into this category and what problems they point to can help you develop management strategies that both promote better turf growth and help reduce populations of the weed species.

The next chapter discusses in greater detail the theory behind and principles of chemical weed control. While herbicides are an essential component of most weed management programs, it can become easy to rely upon chemical control as the principal means of eliminating weeds. When herbicides become a crutch, rather than a

> **PESTPOINTER**
>
> Weeds can serve as a tool to help identify problems that exist and that usually can be corrected.

component of a total turf management program, the basics of turf management inevitably become compromised and weed problems can actually worsen. Adherence to the basics of weed control: proper weed species and life cycle identification, site remediation, and proper turf management techniques, will optimize the performance of herbicides that do need to be used and result in a better overall approach to weed control at your facility.

SUMMARY QUESTIONS

- *What weeds pose the biggest challenges to you at your facility? Are they annuals or perennials? Are they grasses, broadleaf weeds, or other like sedges?*

- *How would you rank your ability to properly identify weed species and their life cycles? Do you see need for improvement in this area for either you or your staff?*

- *What factors most influence your decision to control weeds? (Example answers might include aesthetics, playability, or cost of control.)*

- *To what extent do clients, patrons, or supervisors influence your perception of what weeds should or should not be controlled?*

Chapter 4

Herbicides Used in Turf and Landscape

Chemical control of weeds is the most direct form of control but is only one component of a comprehensive weed management program. We cannot forget that weed management extends beyond the scope of herbicide use. The steps to take for a total weed management program are as follows:

1. Scout for weeds and confirm their presence and location.
2. Properly identify the weed species present and their life cycles.
3. Correct or remediate site conditions that may be promoting weed development.
4. Select herbicides for control of future or existing weeds.
5. Apply herbicides at recommended times and rates of application.
6. Monitor the success of the program, both shortly following herbicide applications and in the long-term, to gauge if employed strategies are reducing weed populations.

As you can see, herbicides are mentioned in only two of the six steps outlined above. Proper identification of weeds and life cycles is a crucial component of weed management, along with making areas at your facility suitable for growing turf. Following these early steps and using resources you have available to you beats guessing and can save you money in the long run. Thousands or even millions of dollars are spent each year on pesticide applications that are either unnecessary or improper. Knowing your pests can help avoid this pitfall. Proper scouting for pests and also monitoring the success of your applications can also be valuable time and money savers. Many weeds do not occupy the entirety of your facility. Scouting where they are located can lead to targeted applications in these areas only, decreasing the amount of both herbicide used and the time required for application. Success monitoring can help avoid unnecessary applications to areas where weed problems have been previously eliminated.

The steps outlined and discussed above are not meant to disparage the use of herbicides but rather to highlight them as tools within a larger program. Herbicides are indeed a fundamental part of any weed management program as they, like no other tool available to us, offer the fastest and most complete means of controlling problem weeds. We can manage weeds by maintaining quality turf but routine maintenance only takes us so far. Weeds will inevitably appear and, when they do, we must have a plan of action in place so they can be controlled safely, effectively, and in a timely manner. Herbicides can be distinguished from each other by several means. They are selective or nonselective, contact or systemic, and preemergence or postemergence. Differences between preemergence and postemergence, selective and nonselective, and contact and systemic herbicides will be discussed later in this chapter.

PESTPOINTER

Herbicides can be distinguished from each other by several means. They are selective or nonselective, contact or systemic, and preemergence or postemergence.

HERBICIDE FAMILIES

Herbicides can be classified into families, just as the weeds they target. Herbicide families are not necessarily something that must be committed to memory to successfully control weeds. However, it is important to recognize that the families differ significantly in their chemistries and can thus dramatically affect the kinds of weeds they target. Herbicide classification is based upon the mode of action, which is the means by which that family of herbicides targets susceptible weeds. As will be discussed later, similar modes of action among herbicides in a particular family are why herbicide rotations must deviate from one family to another to be effective. What herbicide families are there and how do they function? Table 4.1 lists the most common herbicide families, along with their modes of action.

Now that the herbicide families have been listed, you're probably thinking, "I can't even pronounce some of these. Where are you

TABLE 4.1 Herbicide families, listed with their modes of action and an example herbicide material.

Herbicide family	Example	Mode of action
Arsenical	MSMA	Undefined
Aryloxyphenoxypropionate	Fexoxaprop-ethyl	ACCase inhibition
Benzamide	Pronamide	Cell division inhibition
Dinitroaniline	Prodiamine	Cell division inhibition
Phenylurea	Siduron	Cell division inhibition
Benzoic	Dicamba	Auxin disruption
Phenoxy	2,4-D	Auxin disruption
Pyridine	Triclopyr	Auxin disruption
Benzothiadiazole	Bentazon	Photosynthesis disruption
Triazine	Simazine	Photosynthesis disruption
Bipyridilium	Diquat	Cell membrane disruption
Imidazolinone	Imazaquin	Amino acid disruption
Organophosphate	Glyphosate	Amino acid disruption
Sulfonylurea	Halosulfuron	Amino acid disruption

going with this?" Whether you're thinking it or not, it's a very fair question. The way to approach herbicide families is to focus less on the names, since they reflect the complex chemistry behind many herbicides. Instead, focus on the mode of action and the specific herbicides that have certain modes of action. These details will improve your understanding of herbicide activity. Let's discuss these different modes of action and what they mean to the practitioner.

Arsenic-based herbicides are descendants of some of the original materials used for control of weeds dating back nearly 100 years. These materials have lost their niche in most production agriculture because of the risk of arsenic appearing in food crops. However, for inedible crops like turf, these herbicides can still be very effective. MSMA is the most common example of a herbicide from this family. Its mode of action is somewhat unclear. However, it does cause chlorosis (yellowing) in the leaves of susceptible weeds, followed by death. Some warm-season turfgrasses have reasonable to good tolerance to herbicides like MSMA. They may exhibit the yellowing symptoms for a brief period but will recover. Summer weeds like crabgrass and sedges are common targets for these types of herbicides.

Aryloxyphenoxypropionate herbicides have clearly one of the longest and most confusing family names but they can be best remembered as the grass-killing herbicide family. The modes of action for these herbicides can vary but they are used specifically to target grassy species. Due to biological differences, most

PESTPOINTER

Arsenic-based herbicides are descendants of some of the original materials used for control of weeds dating back nearly 100 years.

PESTPOINTER

The grass-killing family of herbicides is not used as commonly in turfgrass as in many field crops because of the sensitivity of many turfgrasses.

broadleaf weeds are not sensitive to these types of herbicides. Sometimes, herbicides in this family are referred to as "fops" and "dims". This is because of the names of some of the more common herbicides in the family, such as fluazifop, diclofop, sethoxydim, and clethodim. The grass-killing family of herbicides is not used as commonly in turfgrass as in many field crops because of the sensitivity of many turfgrasses. However, they are very common for grassy weed control in ornamentals and landscapes. They do also have some niche uses in turfgrass systems. Diclofop is a herbicide that is used for postemergence control of summer annual grasses like goosegrass in warm-season turf. Sethoxydim is occasionally used at very low rates to stunt the growth of low maintenance or roadside turf. Fenoxaprop-ethyl is probably the most widely used herbicide from this family for turfgrass applications. Many cool-season turfgrasses have good tolerance to it so it has a variety of applications, from killing existing crabgrass to suppressing invasive bermudagrass.

Many of the most common turfgrass herbicides are cell division inhibitors:

- *Benzamide family*
- *Dinitroaniline family*
- *Phenylurea family*
- *Unique materials like dithiopyr*

PESTPOINTER

Most cell division inhibitors are used as preemergence herbicides, preventing many summer annual weeds from becoming established when used properly.

Most cell division inhibitors are used as preemergence herbicides, preventing many summer annual weeds from becoming established when used properly. Susceptible plants absorb the herbicide as they attempt to emerge from the ground. Subsequent growth that occurs via cell division is then prevented to the point where the plant can no longer sustain itself and it dies. Most of these herbicides are not chemically suited to target grasses better than broadleaf weeds or vice-versa. However, the usual spring timing of application and the large demand for crabgrass control herbicides have traditionally made these materials more consistently suited for annual grass control.

Auxin disrupting herbicides include some of the oldest that are still commonly used. The best example is 2,4-D, a member of the phenoxy herbicide family that has roots back to the World War II era. Auxin disrupters are so called because their mode of action prevents the normal action of the hormone auxin in susceptible plants. Auxin is present in all plants but is much more active in broadleaf plants and controls their normal growth and branching patterns. Disrupting auxin, using these types of herbicides, will characteristically cause leaves to curl up and stems to grow in convoluted directions, resulting in the plant essentially contorting itself to death (Figure 4.1). These herbicides have become the recognized standards for control of a wide variety of broadleaf weeds in turfgrass systems.

Photosynthesis is one of the most important processes in plants that permit them to grow and develop. Disrupting photosynthesis is

Herbicides Used in Turf and Landscape 177

FIGURE 4.1 Phenoxy herbicide damage to flowering stalks of dandelions.

thus a very good way to eliminate certain weeds. The mode of action for these herbicides is to block the normal transfer of energy during photosynthesis. This energy builds up in the leaves of susceptible plants

PESTPOINTER

Disrupting auxin, using these types of herbicides, will characteristically cause leaves to curl up and stems to grow in convoluted directions, resulting in the plant essentially contorting itself to death. These herbicides have become the recognized standards for control of a wide variety of broadleaf weeds in turfgrass systems.

to the point where it causes fatal damage to leaf tissues. Two primary families of herbicides feature this mode of action. One is the benzothiadiazole family, from which only bentazon is used at all in turf. The other is the triazine family, which includes a number of common herbicides. Atrazine is perhaps the most well known triazine herbicide but others, including simazine and metribuzin, are also common in turfgrass systems. Atrazine revolutionized weed control in corn production by being one of the first selective herbicides that controlled a large number of weeds in this crop. Triazine herbicides are safe to any warm-season grass including crops like corn or sorghum and warm-season turfgrasses. Safety is rendered by the ability of warm-season grasses to break down or metabolize the herbicide before its mode of action can kick in. Most or all cool-season grasses and broadleaf plants are susceptible to triazine herbicides (Figure 4.2) so they have become very commonplace for winter weed control in warm-season turf.

Other types of herbicides may indirectly affect photosynthesis. Usually they cause bleaching or loss of leaf pigments that are critical for photosynthesis (Figure 4.3). Examples of herbicides with this indirect mode of action are oxadiazon and carfentrazone. Oxadiazon is usually a preemergence herbicide while carfentrazone is a herbicide that is sometimes added to mixture products for broadleaf weed control. The preferred activity of oxadiazon on leaf tissues has created a unique niche for it among preemergence herbicides because it has a lesser effect on sensitive turfgrass roots.

PESTPOINTER

Most or all cool-season grasses and broadleaf plants are susceptible to triazine herbicides, so they have become very commonplace for winter weed control in warm-season turf.

Herbicides Used in Turf and Landscape   179

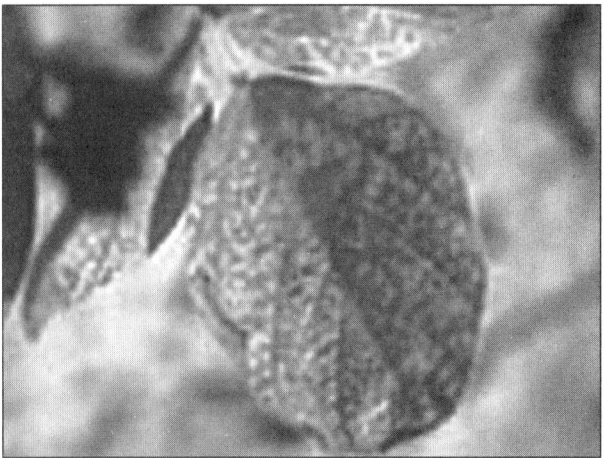

FIGURE 4.2   Triazine herbicide injury, showing yellowing between leaf veins.

FIGURE 4.3   Effects of a pigment disrupting herbicide.

A particularly nasty group of herbicides is the bipyridilium family. The best known of these herbicides is paraquat, although its cousin diquat is more common for turfgrass applications. When I refer to these as nasty, it is in reference to their particularly lethal mode of action. These herbicides result in rapid energy buildup in plant cells, which ruptures the cells and immediately kills them (Figure 4.4). This

180 Turfgrass Chemicals and Pesticides: A Practitioner's Guide

FIGURE 4.4   Paraquat effects on the leaves of a grassy plant.

effect can take place in 24 hours or less when temperatures are warm, so the herbicides waste no time in doing their job. Paraquat and diquat are highly nonselective, meaning they will kill any type of plant, and they only kill tissues they come into direct contact with. The highly lethal mode of action and limited internal movement of these herbicides can make them useful for edging and for rapid kill of undesired vegetation. However, care must be exercised to avoid drift as they can unwittingly damage other plants.

The final group of herbicide families is the amino acid disruption group. This is a diverse group that includes three primary families:

- The imidazolinones
- The organophosphates
- The sulfonylureas

Imidazolinones are more common in field crops but their representatives in turf include imazaquin and imazapic. Organophosphates are best represented by the most popular herbicide in the world, glyphosate, which has a number of potential turf applications. Sulfonylureas are similar to imidazolinones in that they are more commonly used in field crops. However, the need for niche herbicides in turf has resulted in a recent influx of these materials into the turf market. Examples include halosulfuron, rimsulfuron, chlorsulfuron, and metsulfuron. All of these families have a similar mode of action in that they prevent the development of key amino acids in susceptible plants. The result is basically a starvation process that, depending upon the specific herbicide, can either be fairly rapid or cause a slow death. Glyphosate is a nonselective herbicide but has become widespread in field crops with the development of crop plants that resist its effects. Most other amino acid disrupters are selective herbicides but they target a broad range of weeds.

Even though the names of herbicide families and even the common names of some herbicides can be fairly complex, knowing the mode of action does not have to be as difficult. Your past experiences may have led you to rely more upon trade names and label directions for your product recognition. While this strategy may be effective, I challenge you as a practitioner to delve a bit deeper and really venture to understand how these materials work. Your overall understanding of weed control will improve, as will your selection and handling of herbicides. Let's now discuss some of the ways herbicides can be differentiated from the standpoint of the applicator, rather than the chemistry.

Preemergence and Postemergence Herbicides

Preemergence herbicides are applied with the intent of stopping weeds before they can develop and ever become problems. These kinds of herbicides therefore target annual weeds that rely upon seed germination every year as their means of development and survival. These types of herbicides offer little in the way of controlling perennial weeds that have existing storage structures but can sometimes limit the spread of invasive perennials by retarding root development of new rhizomes or stolons. The species of annual weeds that preemergence herbicides target can be numerous but they are usually applied to target grasses. Crabgrass is usually the most sought after victim for these herbicides but annual bluegrass, goosegrass, and sandbur are other common targets.

The principle of preemergence weed control is simple. Herbicides must be in place prior to the period of active germination for the target weed species. The focus of effective preemergence weed control is therefore on proper timing of application. This timing is of course based upon the characteristics of the target weed species. For summer annuals, the proper timing of application can range from late winter to late spring, depending upon the area you live in. The corresponding range of application dates for winter annual weed control can be anywhere from August to October. Because these timing ranges can be so variable, there must be some easier way to determine the proper timing for your area. The common thread that we seek for this issue is soil temperature. Regardless of where you live, soil temperature has the most influence on when annual weeds germinate. Table 4.2 lists some requisite soil temperatures for germination of common annual weeds.

PESTPOINTER

The focus of effective preemergence weed control is on proper timing of application. This timing is based upon the characteristics of the target weed species.

TABLE 4.2 Soil temperature requirements for common annual weeds.

Weed Species	Life Cycle	Required Soil Temperature for Germination
Prostrate knotweed	Summer annual	45°F
Spotted spurge	Summer annual	50°F
Crabgrass	Summer annual	55°F
Goosegrass	Summer annual	60°F
Sandbur	Summer annual	60°F
Annual bluegrass	Winter annual	55°F
Henbit	Winter annual	50°F
Chickweed	Winter annual	50°F

Knowing the characteristics of the annual weeds you must contend with will help you plan for their control. For example, in my home state of Kansas, the 55°F benchmark for crabgrass germination usually means preemergence applications must be made around early April. Environmental variables, like blooming patterns for certain flowering trees, can be used to assist with the proper timing for your area but soil temperature is the most consistent and is easy to measure. Emphasis on soil temperatures can be especially critical if you have variable environments at your facility. For example, south-facing slopes will tend to warm up faster in spring so applications may need to be timed for these areas to avoid escaped weeds.

PESTPOINTER

Regardless of where you live, soil temperature has the most influence on when annual weeds germinate.

PESTPOINTER

Most turfgrass managers would identify annual grasses as bigger problems than annual broadleaf weeds, so control strategies usually focus on these weeds.

With a range of germination temperatures for different weeds, which ones do you focus on? Broadleaf annuals tend to have a lower temperature requirement for germination but does that mean applications should be timed for their control? The answers to these questions must be sought on a case-by-case basis. Most turfgrass managers would identify annual grasses as bigger problems than annual broadleaf weeds so control strategies usually focus on these weeds. Identification of what weeds are the biggest priorities at your facility will assist you in determining when preemergence herbicides need to be applied.

How do preemergence herbicides work? As was discussed in the previous section, most act by inhibiting cell division. For this mode of action to be most effective, herbicides must target susceptible weeds very early in their life cycle. Preemergence herbicides are applied, at appropriate times, as a broadcast application to areas where control is desired. The result is a thin layer of herbicide at or beneath the soil surface. When germinating weeds encounter this layer, they absorb the herbicide and the mode of action prevents the weeds from growing any further. A common misnomer with preemergence herbicides is that they actually prevent germination. Although this is not the case, it is easy to see how some might think so. Weeds controlled by preemergence herbicides are not seen so it might make sense that the herbicide stopped the germination process. Television advertisements also allude to this untruth. The reality is that we want controlled weeds to germinate before they are killed.

Germination is an irreversible process so, once the plant starts to grow, it cannot turn back. If weeds could not germinate because of pre-emergence herbicides, they would simply lie in wait until conditions were more suitable for their development.

Selective and Nonselective Herbicides

Herbicide selectivity is one of the most important issues to consider when planning a weed management program. The basis of what we call selectivity is a variable response between target weeds and desired plants to a particular herbicide. Herbicides that result in this variable response are called selective while those that affect all plants equally are called nonselective. Nonselective herbicides are less common, in terms of available products for turf use. Some materials have nonselective properties but are not viable herbicides. Examples would include most petroleum-based materials. Any of us who have witnessed a gasoline spill or a hydraulic leak on turf could attest to this. True nonselective herbicides include:

- Diquat
- Glyphosate
- Glufosinate
- Pelargonic acid.

Of these, diquat is a contact herbicide while the other two are systemic. Advantages and disadvantages of these attributes will be discussed in the next section.

Traditionally, nonselective herbicides have had limited utility because of their toxic effects on desirable turf or landscape plants. Common uses have been for edging manicured areas, total vegetation control in areas slated for renovation, and winter weed control in dormant warm-season turfgrasses. The latter use represents an effective and cost efficient option for warm-season turfgrass managers. When warm-season turfgrasses enter full dormancy, they

> **PESTPOINTER**
>
> Partially dormant turf can be affected by these applications, so making sure the turf is devoid of green growth is critical to avoiding turfgrass injury.

are inactive enough to be unaffected by nonselective herbicides. Partially dormant turf can be affected by these applications, so making sure the turf is devoid of green growth is critical to avoiding turfgrass injury.

Recent technological advances have created new uses for nonselective herbicides. Genetic alterations of common crops like corn, soybean, and cotton have rendered certain varieties resistant to the nonselective herbicide glyphosate. This 'Roundup-Ready' technology has changed the face of weed management in these cropping systems and steps are being taken to bring this technology to turfgrass managers. The first species in development for glyphosate resistance is creeping bentgrass. While the technology has clear appeal from the standpoint of weed management, there are reasons to approach it cautiously. Questions to ask would include:

- *Are the varieties (cultivars) being developed with glyphosate resistance suitable for my climate and situation?*
- *How do I control the enhanced plant if it escapes its target area?*
- *Is the enhanced plant as strong and vigorous as current turfgrass varieties?*
- *Is the value of the technology enough to justify the costs associated with complete renovation of target turfgrass areas at my facility?*

In fairness, developers of this new turfgrass technology have accounted for many of these issues but there are still considerations you should heed as a practitioner. Public fear over genetically modified organism (GMO) technology has focused on its impact on world food crops. We don't consume turf but there is prevailing concern that this technology may be relied upon too exclusively and diminish the importance of non-herbicide related weed management principles.

Selective herbicides have been the traditional cornerstones of chemical weed control. One of the first turf herbicides, 2,4-D, was accidentally discovered because of its preference for killing broadleaf plants. Since then, the quest for finding herbicides that control weeds but do not target crops has been a key focus for any product manufacturer. In fact, if a newly discovered material is not adequately selective, its further development is often halted. What makes a herbicide lethal to some plants and safe on others is called the basis for selectivity. For example, herbicides like 2,4-D are safe to most turfgrasses because the hormone they disrupt is not as critical to the development of grasses. In this case, the herbicide's mode of action is the basis for selectivity. Another common basis for selectivity is herbicide metabolism. Selectivity with triazine herbicides is founded upon the ability of tolerant plants to break down or metabolize the herbicide before its mode of action can harm the plant. Selectivity that is based upon either herbicide mode of action or metabolism is usually a very reliable way to plan for selective weed control.

PESTPOINTER

Selectivity that is based upon either herbicide mode of action or metabolism is usually a very reliable way to plan for selective weed control.

PESTPOINTER

Labels will only recommend uses that have been verified with sound research, so you can trust what they are telling you. If in doubt, test a particular material in a less visible or nursery area to verify what you can expect at your own facility.

Selectivity is not always a black and white process. For example, some grass-killing herbicides will have a more potent effect on certain grass species than others will. The desirable species in such a case may not be completely tolerant but, if the application is made at the right time and at the right rate, some selectivity can be achieved. A great example of this is the use of grass-killing herbicides like fluazifop or fenoxaprop for control of volunteer bermudagrass in cool-season turf. These herbicides normally might harm cool-season grasses but, when they are applied during summer, the desired species are less actively growing and less susceptible to the herbicide. By contrast, actively growing summer weeds like crabgrass or volunteer bermudagrass will tend to be more vulnerable to these herbicides. Taking advantage of variable growth rates among grasses to achieve selectivity can be more risky, so following label instructions can be highly critical to achieve success. The take-home message with herbicide selectivity is to rely upon your experiences and, more importantly, the content of herbicide labels. Labels will only recommend uses that have been verified with sound research so you can trust what they are telling you. If in doubt, test a particular material in a less visible or nursery area to verify what you can expect at your own facility.

Contact and Systemic Herbicides

The distinction between contact and systemic herbicides comes after the material has been applied and after key weed management decisions have already been made. However, knowing if a material is contact or systemic can influence where it is used and for what purposes. Contact herbicides are so named because they only execute their mode of action at points where herbicide droplets directly encounter the target plant. By this reasoning, contact materials can only be those that are applied to the leaves. Herbicides applied to the soil and taken up by plant roots cannot be contact in nature. The primary advantage of contact materials is that they are more predictable, relying less upon plant activity to be effective. Unfortunately, contact herbicides have several disadvantages:

- *They have a very low margin for error. Inaccurate application patterns, resulting in poor coverage on target leaves, can diminish activity.*

- *They only target tissues they come directly in contact with, limiting the spectrum of weeds they control. Annual weeds may be great targets, considering most of their growth is above ground and resources are invested in the leaves these herbicides best target. Perennial weeds pose a challenge in that they survive with the help of below ground storage structures.*

- *Contact materials may destroy existing leaves but they will not reach storage organs. Perennial weeds can usually survive the effects of a contact herbicide application.*

- *Lastly, few herbicides are of the contact variety, creating a limited number of choices if a practitioner desired such an option.*

PESTPOINTER

Contact materials can only be those that are applied to the leaves. Herbicides applied to the soil and taken up by plant roots cannot be contact in nature.

Only two herbicide families feature contact type herbicide materials. These are the bipyridilium family and the benzothiadiazole families. This limits the number of contact herbicides to just two that would commonly be used in turf: diquat and bentazon. As was discussed earlier, diquat is usually only used as a nonselective herbicide and can be useful for edging or for winter weed control in dormant warm-season turf. Diquat is commonly marketed under the trade name Reward®. Bentazon is a herbicide that selectively controls certain weeds in many turfgrasses. It is most commonly used for control of sedges and is more suitable for use in warm-season turf. Bentazon is most commonly sold under the trade name Basagran® but it is also found with alternative trade names.

The alternative to a contact herbicide is a systemic herbicide. Most herbicides used in turf or landscapes are inherently systemic, meaning that they translocate or move within the plant once roots or leaves absorb them. Where a herbicide material is absorbed and where it moves to in the plant depends upon the herbicide's mode of action. For example, herbicides that inhibit cell division will tend to move to growing points of the plant. The crown of grass plants and the buds of broadleaf plants, where new leaves begin, are the most common growing points for turfgrass and landscape weeds. Alternatively, herbicides that inhibit either amino acid production or photosynthesis will move to the leaves, where these key plant processes occur. How do herbicides know where to go? They often have chemical simi-

PESTPOINTER

Most herbicides used in turf or landscapes are inherently systemic, meaning that they translocate or move within the plant once roots or leaves absorb them.

> **PESTPOINTER**
>
> The chemical properties of an herbicide dictate where the plant absorbs it. Some materials are naturally suited to enter the pores on plant leaves while others are better suited to enter through the roots. Either way, the plant will send the herbicide to where it needs to go.

larity to or are able to bind to certain plant enzymes so, once they enter the target plant, they are automatically sent to where the plant thinks they belong. The chemical properties of a herbicide dictate where the plant absorbs it. Some materials are naturally suited to enter the pores on plant leaves while others are better suited to enter through the roots. Either way, the plant will send the herbicide to where it needs to go.

Systemic herbicides may not act as quickly as contact herbicides because time is required for them to move to where they need to be and to exert their mode of action. However, they tend to be more complete in how well they control weeds and they are better for controlling perennial weeds because the herbicide is able to move into perennial storage structures. Since most herbicides are systemic, weed control recommendations usually are made with this in mind. For example, recommending that a herbicide be applied when target weeds are actively growing is so that a herbicide can be effectively absorbed and translocated. Weeds under stress may not absorb herbicides as well and movement throughout the plant might be impaired, diminishing the success of weed control. It is also recommended that weeds, especially annuals, be controlled when they are young. This is because vital plant processes that herbicides target are most active during this stage of growth, rendering the herbicides more effective.

INTERPRETING AND FOLLOWING HERBICIDE RECOMMENDATIONS

So far, the issue of herbicide use has been approached from a very conceptual standpoint. It is time to get to the heart of the matter and provide you with some of the tools you will need to develop an effective weed management program that includes herbicides. Table 4.3 lists some of the most common herbicides used in turf.

Selecting a herbicide product will depend upon a number of issues. One is, of course, the cost which can vary among distributors and formulations. Formulation can be an important consideration of its own. Some herbicide products are only sold as liquids but many preemergence and broadleaf herbicides are available in both liquid and granular forms. As discussed in Chapter 2, liquids have the advantage of providing more even coverage on leaves if leaf uptake is the primary objective. Liquids are also more readily available for uptake by target weeds and evidence of their application is not as noticeable. However, granules tend to be more resilient and are not as subject to the degrading effects of wind, moisture, and sunlight. If you have a choice among formulations for a particular type of herbicide, consider the following things:

- *Is the application being made during a period of heavy rainfall in your area? If so, liquids are less susceptible to runoff.*

- *Is irrigation available at the time of application? Some liquids need to be moved into the soil to be effective. Granules with the same requirement will last longer without this necessary moisture.*

- *What types of application equipment do you have? Many larger facilities have large liquid spray units to accommodate big jobs while smaller ones may not.*

- *Who are your clientele? The sight of a liquid sprayer can provoke fear and uncertainty in many people, especially homeowners.*

- *How much time do you have for herbicide applications? Many granular herbicides also contain fertilizers so multiple tasks can be achieved at once.*

TABLE 4.3 Common herbicides used in turfgrass systems.

Common Name	Family	Common Trade Name(s)
Atrazine	Triazine	Aatrex
Benefin	Dinitroaniline	Balan, Team*
Bentazon	Benzothiadiazole	Basagran, Lescogran
Carfentrazone	N/A	Speed Zone*, Power Zone*
Chlorsulfuron	Sulfonylurea	Corsair, TFC, Telar
Clethodim	Aryloxyphenoxypropionate	Envoy
Clopyralid	Pyridine	Lontrel, Confront, Millennium*
Dicamba	Benzoic	Banvel, various broadleaf herbicide combinations
Diclofop	Aryloxyphenoxypropionate	Illoxan
Diquat	Bipyridilium	Reward
Dithiopyr	N/A	Dimension
Ethofumesate	N/A	Prograss
Fenoxaprop-ethyl	Aryloxyphenoxypropionate	Acclaim Extra
Fluazifop-butyl	Aryloxyphenoxypropionate	Fusilade
Glufosinate	N/A	Finale
Glyphosate	Organophosphate	Roundup Pro, Touchdown Pro, others
Halosulfuron	Sulfonylurea	Manage
Imazapic	Imidazolinone	Plateau
Imazaquin	Imidazolinone	Image
Isoxaben	N/A	Gallery
MCPA	Phenoxy	Various broadleaf herbicide combinations
MCPP	Phenoxy	Various broadleaf herbicide combinations
Metolachlor	N/A	Pennant
Metribuzin	Triazine	Sencor
Metsulfuron	Sulfonylurea	Manor
Oxadiazon	N/A	Ronstar

Continued on next page

TABLE 4.3 *(continued)* Common herbicides used in turfgrass systems.

Common Name	Family	Common Trade Name(s)
Pelargonic acid	N/A	Scythe
Pendimethalin	Dinitroaniline	Pre-M, Pendulum
Prodiamine	Dinitroaniline	Barricade
Quinclorac	N/A	Drive
Rimsulfuron	Sulfonylurea	TranXit
Sethoxydim	Aryloxyphenoxypropionate	Vantage
Siduron	Phenylurea	Tupersan
Simazine	Triazine	Princep
2,4-D	Phenoxy	Various broadleaf herbicide combinations
Triclopyr	Pyridine	Turflon Ester, Confront*
Trifluralin	Dinitroaniline	Treflan, Team*

* Indicates those that also contain other herbicides

Selecting a herbicide for a particular task is the ultimate goal to achieve before the application is made. We've discussed some of the necessary precursors to this selection process, such as proper weed identification and site remediation. Another important step to make is to develop an annual plan of attack. Some herbicide applications are made based on "spur of the moment" decisions for an unexpected outbreak of weeds. However, most are planned well in advance, based upon past experiences and the usual complement of weeds you see in your area. Plans of attack can vary considerably across geographic regions, in terms of timing, types of weeds targeted, and total necessary numbers of applications. Tables 4.4 and 4.5 show some possible weed management plans for two different geographic areas. Use these as foundations for developing programs that specifically meet the needs of your area and your facility.

Naturally, these scenarios may not include all the weed problems in your area but they give you an idea of what may be customary for different regions. Southern turfgrass managers typically must make more applications each year because their climate sup-

TABLE 4.4 A sample herbicide plan for a northern cool-season turfgrass system.

Timing	Example weeds targeted	Type of herbicide
Early/mid-spring	Dandelion, clover	Hormone disrupter, postemergence
Mid-spring	Crabgrass	Cell division inhibitor, preemergence
Mid-summer	Crabgrass	Grass-killer, postemergence
Fall	Dandelion, clover	Hormone disrupter, postemergence

TABLE 4.5 A sample herbicide plan for a Southern warm-season turfgrass system.

Timing	Weeds targeted	Type of herbicide
Winter	Annual bluegrass, henbit	Nonselective, postemergence
Late winter	Crabgrass, goosegrass	Cell division inhibitor, preemergence
Early/mid-spring	Dandelion, clover	Hormone disrupter, postemergence
Mid-spring	Crabgrass, goosegrass	Cell division inhibitor, preemergence, repeat application
Mid-summer	Crabgrass, goosegrass	Grass-killer, postemergence
Early fall	Annual bluegrass	Cell division inhibitor, preemergence
Fall	Dandelion, clover	Hormone disrupter, postemergence
Late fall	Annual bluegrass, henbit	Photosynthesis inhibitor, pre/postemergence

ports a broader spectrum of weeds over a greater portion of any calendar year. I encourage you to take these basic scenarios and mold them into ones that best fit your area. Consult with your product representatives and local extension resources to best determine which specific products will best fit your needs and the weed species you need to control. Of course, the final step to take, once you have developed your plan and selected the products you will use, is to make the applications properly. Details of how this is best achieved are covered later in this text.

HERBICIDE USE AND TURF ESTABLISHMENT

One of the most common issues that can get in the way of a standard weed management plan is turfgrass establishment. Sometimes establishment can be a planned process that expands the scope of managed turfgrass facility or can result in new clients. However, establishment can also be an unplanned event that becomes necessary as a result of damage cause by pests or environmental stress. In either case, understanding how herbicide use impacts the establishment process can be a valuable tool to avoid unwanted turfgrass injury. The key component to understanding this process is to have a cautious approach to any herbicide application. This is because young turfgrass plants may not have the tolerance to herbicides that older plants will. In the following paragraphs, I will outline some of the most common instances where herbicides may conflict with establishment and how you can best avoid such conflicts.

Probably the most common types of herbicides that can impair establishment are preemergence herbicides. If you recall, most of these herbicides inhibit cell division and can stunt leaf and root growth of susceptible weeds to the point where the weeds can no longer sur-

PESTPOINTER

Probably the most common types of herbicides that can impair establishment are preemergence herbicides.

vive. Turfgrasses can also be susceptible to this mode of action. The reason why most turfgrasses are tolerant to these herbicides is that their roots lie beneath the thin layer of herbicide that is created at the soil surface. Young turfgrass plants or those that are just germinating do not have this level of root development and can thus be exposed to the herbicide just like a weedy grass species (Figure 4.5). Results in this case can severely diminish the successful establishment of turf. The key to avoiding this issue with turfgrass establishment is timing. Preemergence herbicides are most active immediately after they are applied and diminish in potency over time. Planting turf when the herbicides have 25 percent or more of their full potency will detrimentally affect establishment.

Strategies to avoid preemergence herbicide damage to new turf can be numerous:

- *The best one is to plant in an alternate season to the one when the herbicide is applied. Fall planting of cool-season turfgrasses is common in many areas and can allow for successful establishment well before preemergence herbicides are applied in the spring.*

FIGURE 4.5   Preemergence herbicide injury on turfgrass roots.

- *Avoid the use of preemergence herbicides for a season to allow turf to become established. This strategy will result in weed infestations but we must focus on successful turf first and worry about the weeds later.*
- *Simply apply the preemergence herbicide later than normal to accommodate turf establishment. This will also likely result in more weeds than may be desired but it is better than allowing weeds to develop for the entire season. This last strategy may be most useful in northern climates where spring planting may be more common.*

The general rule of thumb to follow in all cases is to allow new turf enough time to warrant three or four mowings before applying preemergence herbicides. By this time, the root system in the newly planted area will be deep enough to be tolerant of the applied herbicides.

While planting cool-season turf in fall can be an excellent means of avoiding conflicts with preemergence herbicide programs, there are still challenges that can appear. These challenges are almost always associated with more persistent herbicides like prodiamine or dithiopyr. These herbicides are marketed for their ability to last longer than other preemergence products and they live up to

PESTPOINTER

The general rule of thumb to follow in all cases is to allow new turf enough time to warrant three or four mowings before applying preemergence herbicides. By this time, the root system in the newly planted area will be deep enough to be tolerant of the applied herbicides.

> **PESTPOINTER**
>
> If droughty conditions are prevalent in your area, consider a less potent preemergence herbicide or rotations between more potent and less potent materials to help avoid this challenge.

these claims. However, when summer conditions are more dry than normal, these herbicides may persist into early fall and still be present with enough potency to affect new plantings. There is no reliable way to predict this type of challenge since we rarely know what seasonal weather patterns will bring. However, if droughty conditions are prevalent in your area, consider a less potent preemergence herbicide or rotations between more potent and less potent materials to help avoid this challenge.

Another confounding issue with preemergence herbicide use is the uncertainty of whether or not new plantings will be necessary. In sensitive areas like the transition zone, spring renovations of warm-season turfgrasses are often necessary, as are fall renovations of cool-season turfgrass stands. Weather usually dictates the need for these types of plantings so they often are not planned in advance. Evidence of winter damaged warm-season turf is often not apparent until after preemergence herbicides have already been applied. This can cause a problem because many preemergence herbicides can significantly delay the renovative establishment of these species. Solutions to this challenge might be to use a less potent herbicide if winter conditions in your area have been harsh or to use a preemergence herbicide like oxadiazon that uniquely does not impair root growth. Fall renovations of cool-season grasses may be similarly susceptible to more potent preemergence herbicides so herbicide selection can be critical in areas where these types of renovations are common.

Another challenge that faces successful turfgrass establishment is the conventional use of postemergence herbicides. Practitioners would certainly not use grass-killing herbicides on young turf but they might be tempted to use broadleaf herbicides because turfgrasses are usually tolerant to these materials. However, discretion must also be used here, as turfgrasses may be unusually vulnerable to broadleaf herbicides when they are immature. Young turfgrass plants have not only a weaker root system, but also leaves that have a thinner cuticle, a protective waxy layer surrounding them. In this immature form, turfgrasses are rapidly developing and can be damaged by herbicides we wouldn't suspect would harm them. As for the example above, the safest strategy to use is to allow new turf to be mowed three or four times before using broadleaf herbicides. This will ensure the turf is sufficiently developed to have acquired tolerance to these herbicides. It should be stated that broadleaf herbicides would likely not kill young turfgrass plants like preemergence herbicides can. However, stunted development can slow down the establishment process and perhaps make fall-planted turf more susceptible to other challenges like winter damage. Keeping these types of challenges in mind as you prepare for any form of turfgrass establishment can help you avoid unsavory consequences and maintain solid weed management programs.

PESTPOINTER

The safest strategy to use is to allow new turf to be mowed three or four times before using broadleaf herbicides. This will ensure the turf is sufficiently developed to have acquired tolerance to these herbicides.

HERBICIDE RESISTANCE

The pursuit of weed control in a wide variety of cropping systems has created expectations that now vastly exceed those held by our predecessors. Fifty years ago, it was considered unreasonable and it was probably not feasible to achieve weed-free turf. Times have changed dramatically, haven't they? It still may not be possible to have completely weed-free turf but expectations dictate that it can and often should be so. Improvements in herbicide technology have fueled this shift in public opinion and, while there is prevalent and vocal opposition to the use of herbicides, the majority of the public would view weed-free turf as highly desirable. This component of the public includes clients of or financial decision-makers for the practitioners to whom this text is targeted. How do we, as practitioners, manage this demand for high quality and pest-free turf?

The easiest way to handle this challenge has been to pursue a variety of herbicide strategies. Herbicides are a tangible way to illustrate weed control and, due to the competitive nature of weeds, are often easily justifiable as a management cost. We must remember, however, that herbicides are simply a tool that we have available to us as a part of a total weed control program. Herbicides that we have access to today are very effective when used properly and control weeds that practitioners even 20 years ago may have thought to be uncontrollable. It can therefore be very easy to place the majority of the burden of weed control on chemicals. Logic tells us that if it

---

PESTPOINTER

We must remember that herbicides are simply a tool that we have available to us as a part of a total weed control program.

PESTPOINTER

History now tells us that heavy reliance upon a small spectrum of herbicides can result in a unique ecological problem: herbicide-resistant weeds.

is working, why change? The problem is that when herbicides become the focus, many practitioners start relying upon a particular type of herbicide product to meet their needs. History now tells us that heavy reliance upon a small spectrum of herbicides can result in a unique ecological problem: herbicide-resistant weeds.

Herbicide resistance is an issue that first appeared in a variety of field crops. The first documented case of resistance to herbicides occurred roughly 25 years ago and the problem has become more widespread since then. What exactly is herbicide resistance? The best way to address this question is to review some basic relationships between herbicides and target crops or weeds. In most settings, crops (including turf) will be either susceptible to or tolerant of a particular herbicide. A susceptible plant is one that, under normal circumstances, will succumb to the effects of a herbicide. By contrast, a tolerant plant is one that, under normal circumstances, will be unaffected by a herbicide. Earlier in this chapter, we discussed some of the different ways tolerant plants are able to withstand the effects of applied herbicides. Resistance is when a normally susceptible plant is not affected by a particular herbicide and should thus not be confused with herbicide tolerance.

Let's use a documented case of herbicide resistance in turf as an example. A common method of control for annual bluegrass in dormant warm-season grasses like bermudagrass is the use of herbicides like simazine or atrazine. Practitioners across the southeast recognize

PESTPOINTER

Resistance is when a normally susceptible plant is not affected by a particular herbicide and should thus not be confused with herbicide tolerance.

that this method is highly effective and cost efficient. Now what happens when a herbicide like simazine seems to no longer affect annual bluegrass? Is it a mistake by the applicator or a more severe problem like resistance? In either case, a weed that is normally controlled by such an application now is left unharmed, posing a unique challenge for the practitioner.

Herbicide resistance is not a result of a modification or mutation in plants that were once susceptible to a particular herbicide. It may be easy to think that this is the case since the problem tends to evolve slowly over time. The means by which herbicide resistance occurs is through natural selection. Most of the weeds that have shown documented resistance to herbicides are annuals. Annual weeds produce large quantities of seed every year. Plants that grow from these seeds reflect the characteristics of their parents but they are genetically unique. Occasionally, a new plant has a trait that gives it the ability to resist herbicides that would normally kill other members of an existing population. Over time, the appearance of resistance in a population is usually the result of very few weeds that, by virtue of being resistant, produce future generations of resistant plants.

We typically don't see resistance in as many perennial species because they are not as dependent upon seed production for their survival. Invasive perennials may occupy large amounts of space but the plant material may be all genetically identical if it spread from an original source. The occurrence of herbicide resistance requires a lot of variation in the genetic makeup of a population of weeds so annuals that produce a lot of seed tend to be the most likely culprits.

To date, nearly 300 different herbicide resistant weed biotypes have been identified in the US alone. These weeds come from numerous plant families and most have been identified in field crop environments, where herbicide use has historically been more aggressive. While very few of the reported cases of herbicide resistance in this country pertain directly to the weed problems encountered by turfgrass practitioners, this is not a problem that we can disregard. Example species for which resistance has been reported include crabgrass, goosegrass, the aforementioned annual bluegrass, and several species of ryegrass. Herbicide resistance may or may not be a problem you have had to contend with at your facility, but it is important to recognize which species are most likely to develop resistance and how to properly prevent and/or manage it.

Management of herbicide resistance can include both prevention measures and measures taken to eliminate an existing problem. Judicious herbicide selection is the best way to manage this potential problem. The great majority of circumstances in which herbicide- resistant weeds have been identified have been those where herbicide or crop rotation practices were not implemented. For

PESTPOINTER

Herbicide resistance is not a result of a modification or mutation in plants that were once susceptible to a particular herbicide. The means by which herbicide resistance occurs is through natural selection.

PESTPOINTER

To date, nearly 300 different herbicide resistant weed biotypes have been identified in the US alone.

example, the most common family of herbicides that weeds are resistant to is the triazine family. This family includes common herbicides like atrazine, simazine, and metribuzin. Tolerance of warm-season grass crops like corn to these herbicides revolutionized weed control in corn production but also led to very regular use of these herbicides. Years of heavy corn production and also triazine herbicide use in many parts of the country have resulted in many of the resistant weeds that we now see reported. Crop rotation and herbicide rotation are two of the most common practices that are used to manage herbicide resistance in these areas.

Crop rotation is a useful and reasonable strategy for field crops since most crops are planted as annuals. Switching from corn production to soybeans changes the spectrum of herbicides that are used to control weeds and can thus help eliminate any weeds that may be resistant. Rotation of the primary turf species is not a practical solution for the turfgrass practitioner so we must rely upon herbicide rotation to manage resistance. Herbicide rotation is not as simple as just changing the product that is used for control of a particular weed species. Rotation means also changing the family of herbicides that is used.

Let's return to the example of annual bluegrass that is resistant to a herbicide like simazine. Switching from simazine to another triazine herbicide like atrazine or metribuzin is not an effective rotation because of what is termed cross-resistance. Cross-resistance is when a resistant weed is resistant to an entire family of herbicides

> **PESTPOINTER**
>
> Crop rotation and herbicide rotation are two of the most common practices that are used to manage herbicide resistance.

that have a similar mode of action. Solving the problem of resistant annual bluegrass in the southeast required using herbicides like pronamide or glyphosate to get resistant weeds under control.

Some biotypes of crabgrass and goosegrass have shown resistance to dinitroaniline (DNA) herbicides. Many of the preemergence herbicides we have to choose from fall into this category and include prodiamine, trifluralin, benefin, pendimethalin, and oryzalin. DNA resistant summer annual grasses will likely be resistant to any of these herbicides so an effective rotation would require switching to a preemergence product with a different mode of action, such as either dithiopyr or oxadiazon. These products may represent higher costs but using them once every three to five years can help you avoid the occurrence of resistance at your facility. Understanding the issue of herbicide resistance involves knowing what weeds are most sensitive and also knowing your herbicides, so that effective rotation programs can be implemented.

As you have noticed, there are many issues that go into developing a solid weed management program. I've endeavored to cover some of the key basics that will give you the tools to develop such a program in your area. Weed management issues do differ considerably across regions and it is important to consider these differences so your program is as successful as possible. It cannot be emphasized enough that local resources available to you are a critical complement to the contents of this chapter. Certain herbicides may or may not be registered for use in your locality.

PESTPOINTER

Understanding the issue of herbicide resistance involves knowing what weeds are most sensitive and also knowing your herbicides, so that effective rotation programs can be implemented.

Timings of applications may vary considerably from state to state and even across geographical regions of a state. Your peers, extension personnel, and sales representatives can help you mold these basic concepts into programs that reflect your local climates, the products you have to choose from, and the weeds you need to manage. Herbicide labels are also valuable in making decisions that reflect the characteristics of your region. Be proactive in using these resources and they will most effectively augment what has been covered in this text. The next chapters will discuss diseases and insects, which also represent significant challenges to turfgrass and landscape managers.

SUMMARY QUESTIONS

- *What current herbicides do you use at your facility? What weeds or types of weeds do they target?*

- *What criteria do you use in selecting herbicides? Example answers might be spectrum of weeds controlled, peer recommendations, cost, and formulation/ease of application.*

- *What types of herbicides would you find most useful at your facility, in terms of those you currently don't use or that may not yet exist?*

- *How do herbicide costs compare to costs of other pesticides for your facility? Does the cost breakdown reflect the relative importance that weeds have to you?*
- *Is herbicide resistance a consideration you make when choosing a herbicide program? Do you rotate herbicides and, if so, how often?*

Chapter 5

Disease Problems and Fungicides Used in Turfgrass

Turfgrass diseases are some of the most feared pests in the industry because of their potential for dramatic visual impact and rapid development, and their penchant for attacking turf when it is most vulnerable. This combination of attributes has made disease control a premium concern in fine turf and an entire discipline of study. The importance of diseases in turfgrass management has increased with the level of maintenance expectation that turfgrass is subjected to. However, concurrent improvements in technology and management practices have contributed to the ever-changing face of disease control. University researchers and professors, chemical manufacturers, and turf industry professionals have answered the call by discovering new and improved methods to provide improved disease management practices and better turf conditions.

How best to approach disease management at your facility is the subject of this chapter. Modern turf management experts have adopted the concept known as Integrated Pest management or IPM. Theoretically, IPM is a sound practice. However, familiarity with the location you are managing is important, as is establishment of how much disease damage is tolerable for your situation. More will be discussed on strategies for developing disease management programs

> **PESTPOINTER**
>
> Modern turf management experts have adopted the concept known as Integrated Pest Management or IPM.

later in the chapter. However, as this topic is introduced, remember that management of diseases involves many of the fundamental concepts that have already been or will be discussed in this text.

Knowledge of the pests, understanding the turf you are managing, recognition of diverse control options, and proper communication of your control programs will help ensure a safe, responsible, and effective program. With these things in mind, let's get started!

DISEASE PROBLEMS IN TURF

Before discussing control options and strategies for any pest, it is important to understand the characteristics of the pests themselves. For diseases, this key precursory step means proper diagnosis of disease problems, recognition of the key disease threats in your area, and comprehension of where these problems are most likely to occur. Disease identification can be difficult or simple, but understanding the types and the patterns of disease development will make prognosis clearer. Disease pathogens are naturally occurring microorganism entities in the soil. When turf is grown in areas where a pathogen exists, the potential for disease outbreaks is created. In addition to a susceptible turfgrass host and the presence of a pathogen, there is a third criterion that must be in place for disease to occur: a proper environment for disease development. These three criteria are collectively referred to as the disease triangle because, much as the absence of one

PESTPOINTER

Environment affects disease development and explains why diseases seem to appear at the same time each year.

side of a triangle disrupts its structure, the absence of any of these three criteria prohibits disease development. Usually, environment is the wild card in this "triangular" relationship. Environment affects disease development more than the other two factors and explains why diseases seem to appear at the same time each year. It is likely that both the pathogen and the turfgrass are there the entire time, regardless of the conditions around them.

There are three main types of disease-causing organisms:

- *Fungi*
- *Viruses*
- *Bacteria*

Parasitic nematodes are often included as a fourth member of this group, since their activity may cause similar symptoms. Pathogens attack plants in different ways and require various techniques to identify, control, and prevent them. Diseases can be diagnosed in two ways: signs and symptoms. Disease signs are physical features of a pathogen that can be found on infected turfgrass plants:

- *Mushrooms*
- *Spore bodies*
- *Fungal hyphae*
- *These are often visible to either the naked eye or with a hand lens.*

Disease symptoms are more easily identified. They are the tangible effects of a pathogen on infected turfgrass:

- *Lesions on turfgrass leaves*
- *Rotted crowns*
- *Death*

Being able to recognize disease symptoms is crucial to proper diagnosis and timely treatment of a disease. The following paragraphs discuss some of the most common troubling disease problems a turf manager can be faced with at any given time.

Anthracnose is a fungal disease that readily attacks annual bluegrass but can spread from infected annual bluegrass to desirable species like creeping bentgrass, perennial ryegrass, Kentucky bluegrass and bermudagrass. Typically this disease occurs under warm and humid summer conditions and is especially prevalent with night temperatures above 68°F and 80 percent or more humidity. It can appear and survive with much cooler temperatures. If the disease is allowed to exist and thrive untreated on the leaf tissue, it will persist and infect the crown of the plant. Black fruiting bodies are visible to the naked eye when examining leaf tissue, as well as an orange appearance of shoots in infected plants (Figure 5.1). At this stage, it is the point of no return and permanent damage can occur. Early applications of

PESTPOINTER

Early applications of a systemic fungicide labeled for anthracnose control is most effective. Members of the strobularin family of fungicides have shown excellent control results, especially when accompanied by a balanced nutritional program.

Disease Problems and Fungicides Used in Turfgrass 213

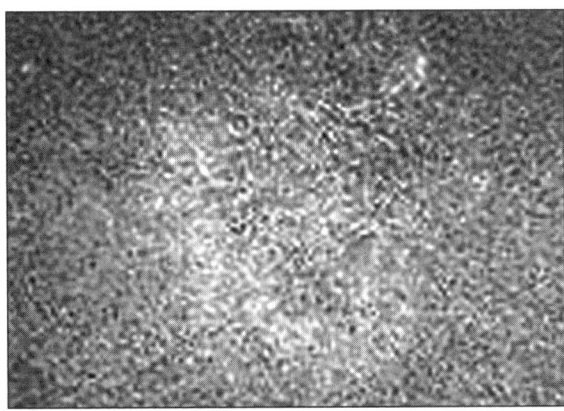

FIGURE 5.1 Symptoms of anthracnose infection are most evident in the crowns of turfgrass plants.

a systemic fungicide labeled for anthracnose control will be most effective. Members of the strobularin family of fungicides have shown excellent control results, especially when accompanied by a balanced nutritional program.

Brown patch, or *Rhizoctonia solani*, has a smoke ring appearance and has been detected in all types of turfgrasses (Figure 5.2). Closer inspection of infected leaves will reveal clear lesions caused by the pathogen (Figure 5.3). Moist morning conditions can also present opportunities to identify this disease by the presence of cottony mycelia (Figure 5.4). It is most severe at low mowing heights, specifically those maintained below $1/2$ inch. It appears in hot, humid weather and on all turfgrass types, especially in conditions with limited or no air movement and high use of nitrogen-based fertilizers. Constant, repetitive mowing will increase the spread of brown patch. In locations that are ripe for brown patch development, the infection can be stubborn and can persist for extended periods until weather conditions are favorable to promote uninfected turfgrass growth. A cousin of brown patch, sometimes referred to as cool-season brown patch or yellow patch, also appears in many turf stands during cool damp periods (Figure 5.5). Although it is less prone to cause permanent damage to turf, the presence of the yellow patch pathogen can weaken desirable turf to the

FIGURE 5.2 Brown patch causes a classic smoke ring appearance in infected turf.

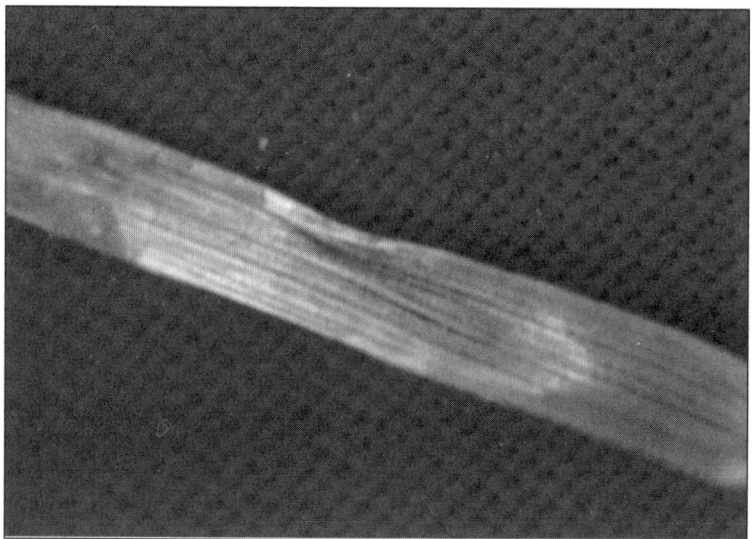

FIGURE 5.3 Brown patch lesion on an infected turfgrass leaf.

Disease Problems and Fungicides Used in Turfgrass 215

FIGURE 5.4 Brown patch mycelia are most visible during cool morning hours.

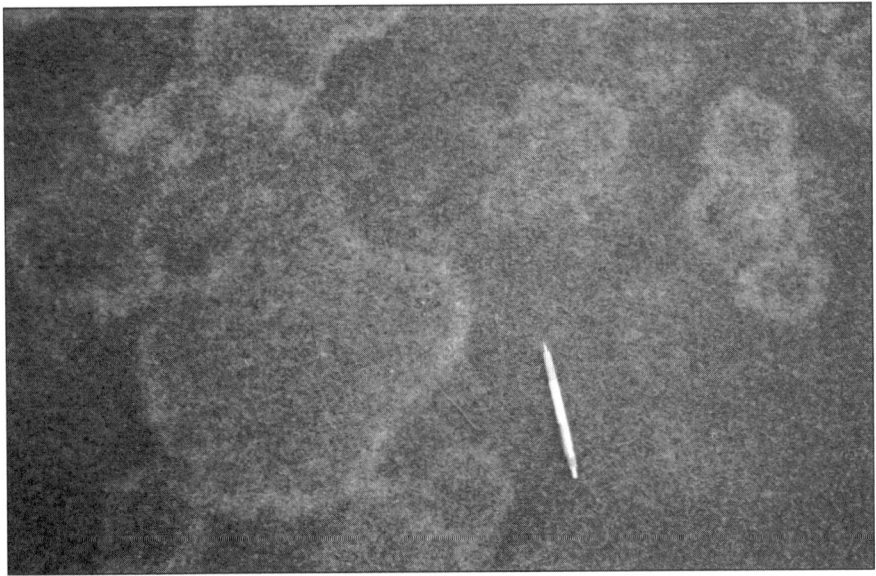

FIGURE 5.5 Cool-season brown patch, commonly called yellow patch, is more prevalent under cooler conditions and can be visually disruptive, even if it doesn't cause permanent damage.

PESTPOINTER

Contact fungicides tend to work the best on brown patch, although follow up applications of a systemic fungicide may be necessary if disease conditions persist.

point of invasion by a secondary pest. Contact fungicides tend to work the best on brown patch, although follow up applications of a systemic fungicide may be necessary if disease conditions persist.

While several diseases are known by a colorful name, two stand out in this regard. Red thread is a disease known to attack perennial ryegrass, fescues, Kentucky bluegrass, and bentgrasses. It appears under nutrient deficiencies and when turf is usually growing slowly, typically early and late season. It is a disease that a turf manager who

FIGURE 5.6 Red thread disease gets its name from the thread-like structures seen here among turfgrass leaves.

seasonally overseeds should be wary of. Red thread infection can be recognized by the characteristic leaf tip die-back and presence of reddish sclerotia, thread-like structures that permit survival of the pathogen in a turfgrass canopy (Figure 5.6). Recognition of these features can help differentiate red thread from other diseases. Because red thread is usually present where turf is nutrient deficient, a turf manager may consider adjusting fertilizer practices to remedy this disease. Another "red" disease is rust, which appears in many turfgrasses, although perennial ryegrass and bluegrasses are most commonly affected. Rust can be easily recognized by the rusty-colored spore structures that are easily visible and also rub off on your shoes when walking on infected turf (Figure 5.7). A severe outbreak of this disease could also be an indication of nutrition problems.

FIGURE 5.7 Rust spore structures, seen here on turfgrass leaves, give the disease its name.

PESTPOINTER

Because red thread is usually present where turf is nutrient deficient, a turf manager may consider adjusting fertilizer practices to remedy this disease.

Pythium blight is a highly destructive disease that can wipe out a turf stand in very little time. Often called damping off in new turf stands, pythium will cause germinating seedlings to wither and die. The moist, almost wet environment that is necessary for seedling establishment is ideal for the destructive damping off action of pythium. The pythium disease family is broad based and attacks in a variety of ways, making it one of the most difficult diseases to pinpoint and understand. Proper identification can make it easier to manage because it is a disease that will spread rapidly from mowing or other mechanical equipment to nearby uninfected turf. Strains of pythium diseases can be present in many different settings, seasons, and turfgrasses. All turfgrasses are susceptible to pythium outbreaks, but cool-season grasses are most affected. The pathogen easily overwinters and infects plants during cool, wet periods, although it seldom displays signs and symptoms at this time.

PESTPOINTER

During a seeding project, granular pythium-specific fungicides should be applied to the seed bed. Protecting seedlings is vital on highly visible and intensely managed turf.

The most recognizable and severe damage to foliage appears during hot and humid weather conditions when a turfgrass stand could be obliterated in less than one day. Infected turf can be covered with a thick, white mycelium web or can appear greasy and lucid under severe infection (Figure 5.8).

Pythium cannot be taken lightly by turf managers, due to its rapid and destructive behavior. During a seeding project, granular pythium-specific fungicides should be applied to the seed bed. Protecting seedlings is vital on highly visible and intensely managed turf. On areas where pythium has historically appeared, a preventative application should be made to the rootzone to protect plants from further infection.

Snow molds are diseases whose names can be a bit misleading because snow is not an absolute prerequisite for infection. Pink snow

FIGURE 5.8 A close-up picture of pythium infestation in tall fescue turf.

mold or *Microdochium nivale* is most destructive to annual bluegrass but it infects many turfgrass species under a wide range of environmental conditions. However, it is not active when air temperatures exceed 70°F. It is active in turf areas with:

- *High levels of nitrogen*
- *High humidity*
- *Rain*
- *Snow*
- *Shade*
- *Little or no air movement*

Snow falling on unfrozen ground is a common precursor for this copper colored, leaf-infecting pathogen. It tends to develop best under snow cover, producing a pink halo of damaged turf that is covered by a sparse, cottony mycelium (Figures 5.9 and 5.10).

Gray snow mold or *Typhula incarnata* does have a true snow requirement, needing 40 days or more of snow cover for infection. Gray snow mold can take advantage of weakened turf under snow cover, and can be more damaging than its pink cousin. In advanced stages, matted water soaked leaves, in a distinctive circular pattern, are clear indications of gray snow mold (Figure 5.11). Post infection control of gray snow mold is difficult. Preventative measures are most

PESTPOINTER

Post infection control of gray snow mold is difficult. Preventative measures are most effective when a late fall fungicide application for snow mold protection is made.

FIGURE 5.9 Mid stage of pink snow mold development with dense cottony mycelia.

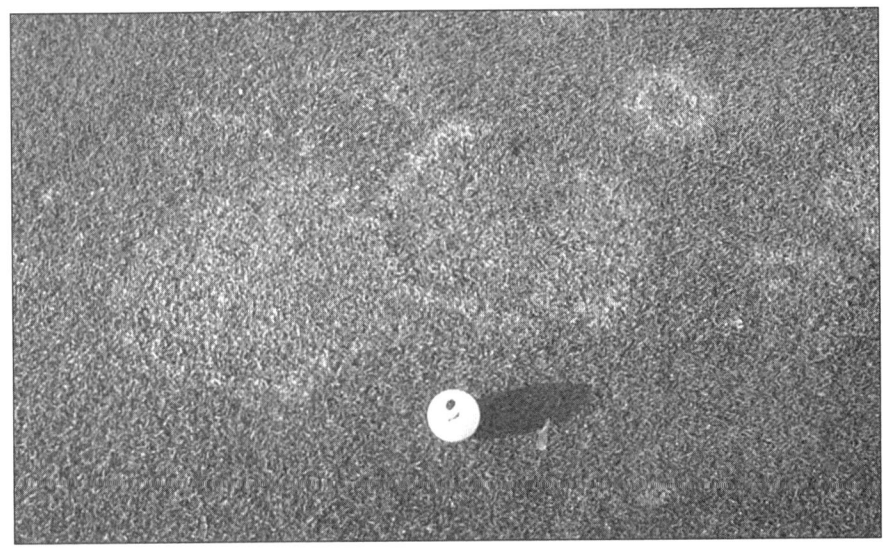

FIGURE 5.10 Fully developed pink snow mold infection on a golf course putting green.

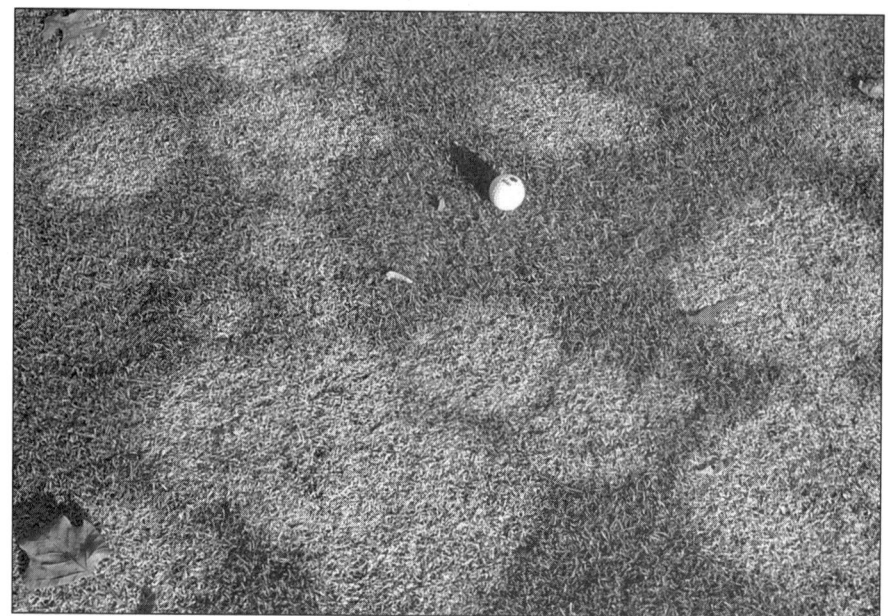

FIGURE 5.11   Advanced stages of a gray snow mold infection.

effective when a late fall fungicide application for snow mold protection is made. A curative approach and turf replacement should commence when growing conditions are favorable and active. At that time, a systemic fungicide should be applied when the turf is able to absorb the fungicide. Curative applications of fungicides are ill-advised before the turf breaks winter dormancy.

Summer patch is a root disease that is prevalent in hot humid seasons. It appears most commonly on heavily irrigated and fertilized turf maintained at low mowing heights. Summer patch is most commonly found to infect Kentucky bluegrass, annual bluegrass, and creeping bentgrass. It is compounded by concentrated and heavy traffic from machinery and mowing equipment. Infection from summer patch can be devastating, since afflicted species are often unable to grow successfully at the time damage from the pathogen occurs (Figure 5.12). On intensely managed turf, a preventative application of a systemic fungicide should be made before the onset of severe summer conditions.

Disease Problems and Fungicides Used in Turfgrass 223

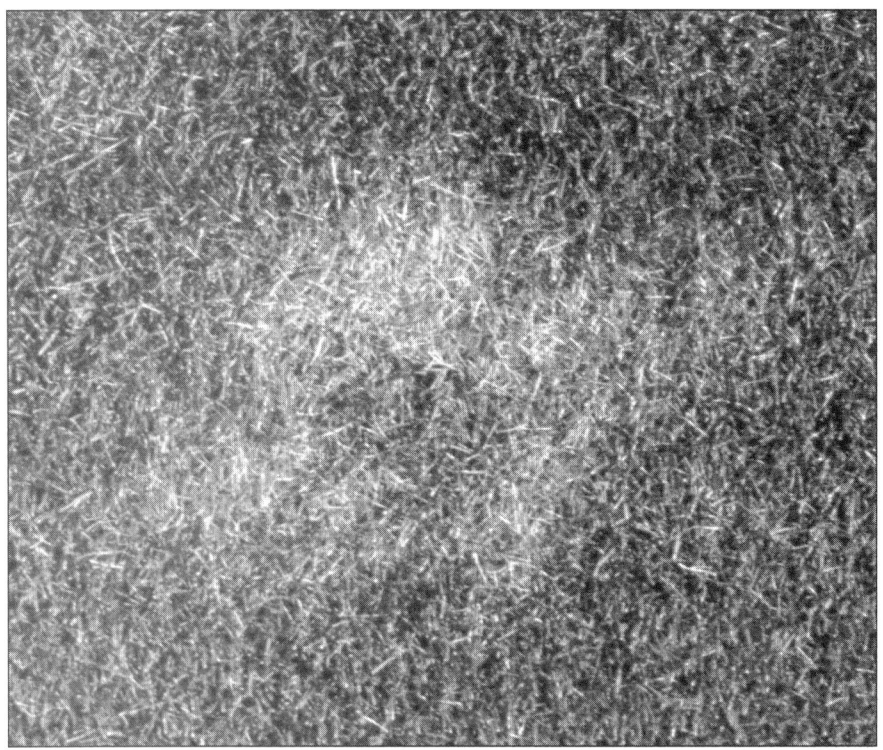

FIGURE 5.12 Summer patch can have injurious effects on susceptible species like Kentucky bluegrass.

Take all patch or *Gaeumannomyces graminis* is a disease that is predisposed by cool wet years and most definitely by the pH of the soil. Adjustments to lower a high soil pH can make a significant difference in the development of this disease. This pathogen wreaks havoc with young bentgrass, as infected plants die in irregular circular patches or rings. Southern turf is less prone but not immune to take all patch. It has been detected on intensely managed bermudagrass and should be monitored on overseeded greens that are sown to cool-season species. Sand-based greens with high soil pH values are most susceptible and a fungicide application of fenarimol drenched into the rootzone is an effective means of control. Take all patch can be difficult to correctly diagnose. It is a common diagnosis for turf

> **PESTPOINTER**
>
> Sand-based greens with high soil pH values are most susceptible to take all patch and a fungicide application of fenarimol drenched into the rootzone is an effective means of control.

samples with unknown problems, prompting it to be referred to sometimes as "catch all" patch.

Dollar spot is aptly named not for just its appearance of silver dollar size clusters of infected turf but for the great deal of money spent on fungicides, fertilizers, and cultural practices to combat this disease. *Sclerotinia homoeocarpa* is the most common pathogen that causes dollar spot, although other dollar spot strains are common among all turfgrasses in various settings. Cool, damp nights that produce heavy dews, coupled with warm days, provide ripe conditions for dollar spot, especially where turf is nutrient deficient. On closely mowed turf where clippings are collected, infected leaves coalesce and the disease infection forms sunken pockets that produce uneven and unsightly ruts in the thatch. On turf that is maintained above two inches, the acceptable threshold for dollar spot can be set much higher but the disease can still be easily identified in this setting. Close examination of the leaves will show tan lesions with brown margins (Figure 5.13). Because of nutrient deficiencies and stressful conditions, late season infestations are common and can be very destructive until the first heavy frost (Figure 5.14). To successfully combat dollar spot infection, adequate fertilizer must be applied to promote turf growth. When infection does occur, early intervention with fungicides will clear the disease up fairly easily. Many fungicides are effective against this pathogen, but alternating systemic fungicides with various modes of action work the best. Contact fungicides such as chlorothalonil have also proven to be effective. Be cautious when

Disease Problems and Fungicides Used in Turfgrass 225

FIGURE 5.13 Dollar spot lesion on a Kentucky bluegrass leaf.

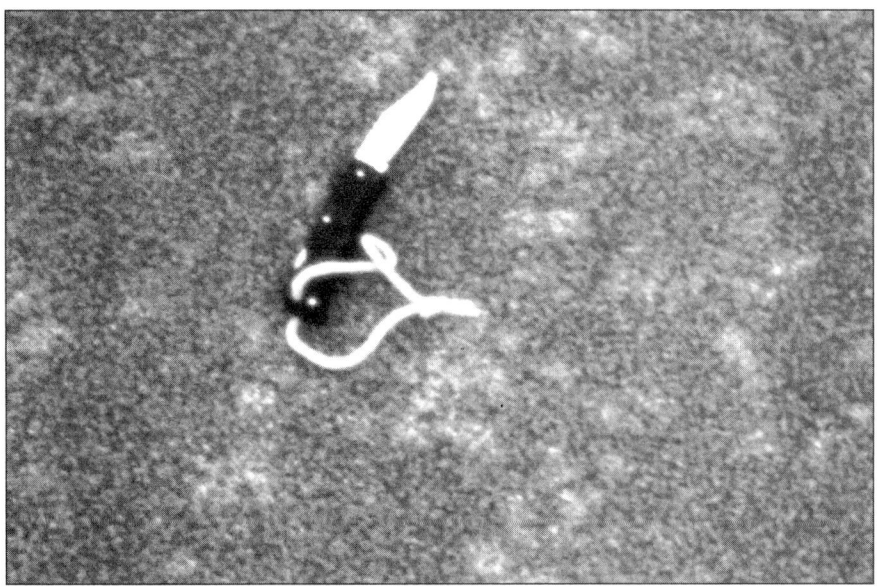

FIGURE 5.14 A heavy infestation of dollar spot in creeping bentgrass turf.

> **PESTPOINTER**
>
> To successfully combat dollar spot infection, adequate fertilizer must be applied to promote turf growth.

applying strobularin fungicides to target other diseases because, in areas where dollar spot has been detected, they may actually increase dollar spot incidence.

So aptly named "blast," gray leaf spot exploded onto the scene to ravage perennial ryegrass in the mid-Atlantic, the Northeast, and the Midwest of the United States in the 1990s (Figure 5.15). Previous to recent outbreaks in perennial ryegrass, the pathogen had only been reported as common in St. Augustinegrass. Gray leaf spot weakens the plant from the top down. Leaves begin to blight and become twisted as the plant weakens. The pathogen then moves to the crown and on into the root system. Detection and diagnosis of gray leaf spot may not be definite at the outset, due to some similarities to the bacterial pathogen *Helminthosporium*. During warm and humid periods, symptoms are capable of destroying grass plants quickly. Gray leaf spot thrives in poorly drained areas, with long periods of leaf wetness and with daytime temperatures around 85°F. It can spread via equipment traffic and drainage patterns, much like pythium diseases. It is at this stage that careful diagnosis is necessary because the two diseases may be confused. Above 90°F, gray leaf spot is less infectious, but the pathogen may still be present.

Spring dead spot is a fungal disease that is exclusive in turf to bermudagrass. Spring dead spot establishes itself in the fall, colonizes during the winter, and manifests itself in the spring as large, brown patches of turf that fail to green up (Figure 5.16). Re-growth of dormant turf is slow in the infected and damaged areas and can appear year after year in the same locations. Spring dead spot is a

Disease Problems and Fungicides Used in Turfgrass 227

FIGURE 5.15 Gray leaf spot or "blast" can cause massive damage in perennial ryegrass turf.

more prevalent disease in mature turfs that are intensely managed. Frequently, patches coalesce and the damage is nonuniform, thus appearing similar to some types of winter injury. The disease is less severe or absent on turfgrasses maintained at low fertility levels. Heavy late fall applications of nitrogen fertilizers should be avoided to lessen the severity of spring dead spot. Late fall applications of a systemic fungicide that are drenched into the rootzone

PESTPOINTER

Late fall applications of a systemic fungicide that are drenched into the rootzone may be an effective means of control for spring dead spot.

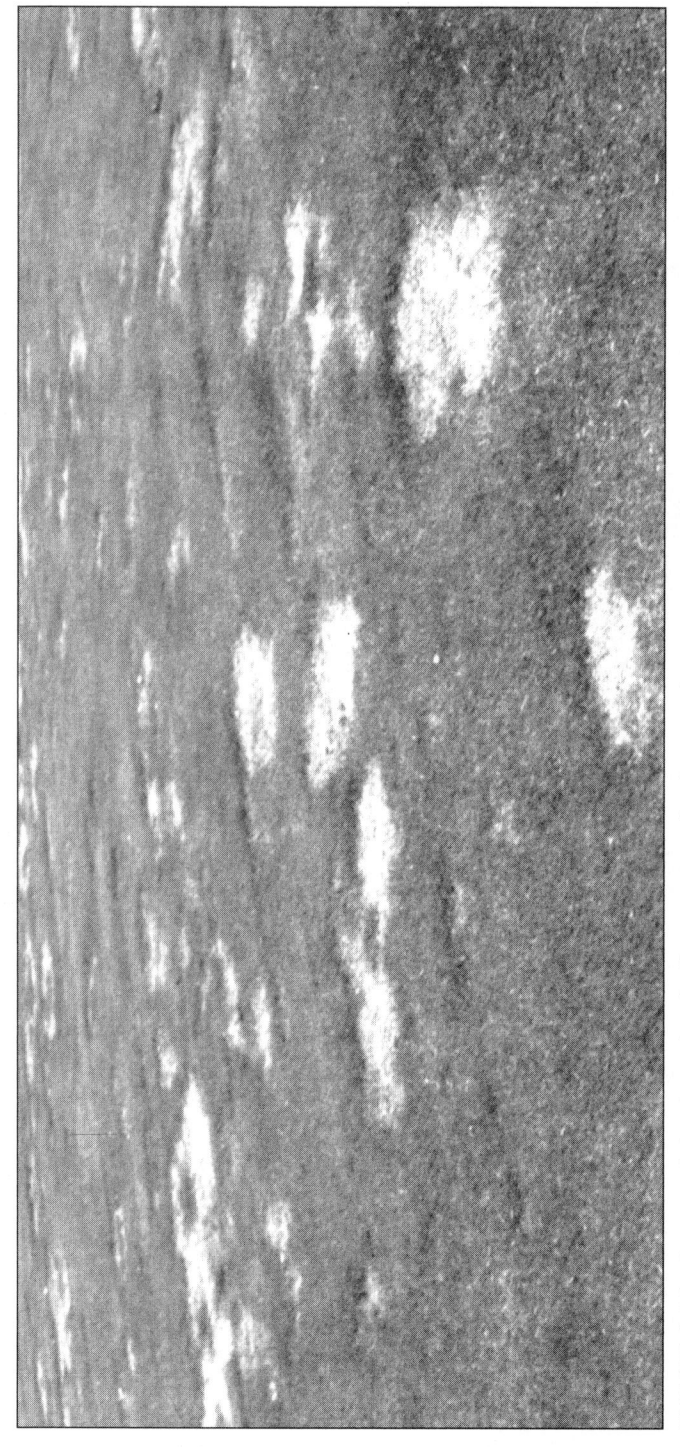

FIGURE 5.16 Scattered dead patches indicate spring dead spot infestations in bermudagrass.

Disease Problems and Fungicides Used in Turfgrass 229

may be an effective means of control for spring dead spot. Large patch is a similar disease to spring dead spot but occurs primarily in zoysiagrass turf. Like spring dead spot, large patch is characterized by infected areas that fail to green up in spring (Figure 5.17). However, the naturally slower growth rate of zoysiagrass can make recovery from this disease a challenging problem.

St. Augustine decline (SAD) is unusual among turfgrass diseases in that it is caused by a virus, rather than by the fungi that are responsible for all the diseases discussed up to this point. SAD is caused by a strain of *Panicum* mosaic virus. SAD results in St. Augustinegrass turf that is weak and shows small yellow spots on the leaves (Figure 5.18). Entire lawns can succumb to SAD because of the turf's weakened and vulnerable state, allowing other pests to attack the weakened plants. SAD is a slow moving viral disease that can resemble mite damage or numerous other diseases such as downy mildew. It can be spread when grass is mowed wet. The disease is one to watch out for because of the difficulty in effectively controlling it.

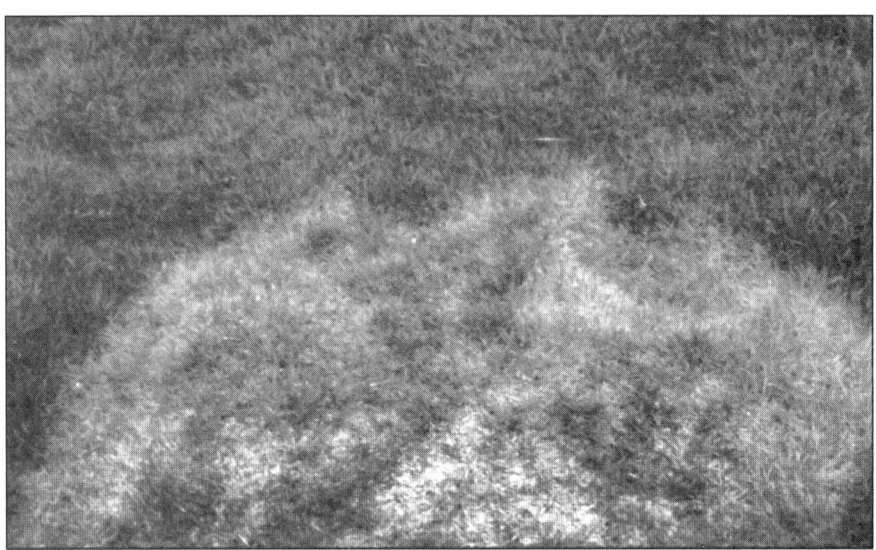

FIGURE 5.17 Large patch is observable in zoysiagrass during the green up phase in spring.

230 Turfgrass Chemicals and Pesticides: A Practitioner's Guide

FIGURE 5.18 Small yellow spots characterize St. Augustinegrass decline (SAD) infection.

Helminthosporium leaf spot or melting out is a bacterial disease, caused by several bacterial pathogens, that infects a variety of turfgrass species. The name *Helminthosporium* reflects an older taxonomic group for these pathogens that has since been split into more specific groups. Many avenues of infection are present with this pathogen. Leaf, crown, and root infections are all possible with *Helminthosporium* pathogens. All turfgrasses are susceptible to forms of this disease. In the spring when cool, damp conditions persist, the leaf spot variety is common and, when the pathogen is allowed to be active, severe infection can occur. If unchecked into summer, the infection migrates to the crown. The severity of the damage increases and will be fatal if not treated before it reaches the root zone.

Some keys to recognizing the symptoms of *Helminthosporium* are the blighting of the leaf tips that appear red to brown in color

Disease Problems and Fungicides Used in Turfgrass 231

PESTPOINTER

Clipping removal, adequate irrigation, and balanced nutrition are measures to combat whichever form of the *Helminthosporium* family of diseases is present. A systemic fungicide provides an effective means of control, especially if allowed to migrate to the rootzone for absorption, thus providing excellent protection against crown and root damage.

(Figure 5.19). During major infections, the crowns of the plant display a spotty, black, greasy algae-like appearance. During warm wet weather, the disease multiplies rapidly, particularly if grass clippings are left to decompose. At this point, if another pathogen is present, the two will work together and devour a turf stand. When *Helminthosporium* reaches the root infection stage, it is nearly fatal and very difficult to control. Turf managers in all regions should not let *Helminthosporium* linger. If conditions are favorable for

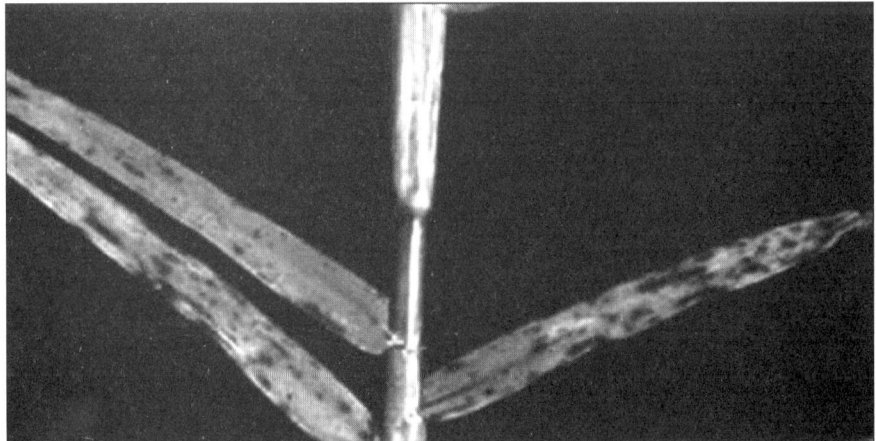

FIGURE 5.19 Symptoms of *Helminthosporium* leaf spot on turfgrass leaves.

development, take measures to control and implement appropriate cultural practices. Clipping removal, adequate irrigation, and balanced nutrition are measures to combat whichever form of the *Helminthosporium* family of diseases is present. A systemic fungicide provides an effective means of control, especially if allowed to migrate to the rootzone for absorption, thus providing excellent protection against crown and root damage.

One very different fungal disease that is a recurring topic of conversation with southern turf managers is fairy ring. Fairy rings uniquely result in three distinct types of signs or symptoms. At different points in the life cycle, fairy ring can result in mushrooms, dark or deep green rings, and rings of dying turf (Figure 5.20). Fairy ring spores find a home in a desirable location and then begin to infiltrate and infect turfgrass plants. Fairy rings develop because of concentrated sections of organic matter within the soil environment. It could be soil amendments such as peat moss or composted materials, or even concentrations of decomposing leaves or buried tree stumps. High levels of organic matter can result in such a high concentration of fairy ring fungi that they become toxic to turfgrass plants. Decomposing organic matter is a perfect home for fairy rings and the thatch layer of turf grass is the perfect launching pad for specimen fairy rings.

FIGURE 5.20 Fairy ring is often identified by the presence of mushrooms.

A good recipe for control of fairy ring is a regular maintenance regime of core cultivation and application of wetting agents to discourage water repellency in the soil. In addition, a fungicide application of azoxystrobularin that is drenched into the root zone may be an effective means of control. The impervious layer of fungal activity is very difficult to control and may appear year after year in the same location, especially when turf appears nutrient deficient. Applications of balanced fertilizers and micronutrients to mask the symptoms of fairy rings can also be an effective control measure.

Nematodes can cause disease-like symptoms but are unlike other pathogens. Nematodes resemble worms but are only visible under a microscope. Many species of nematodes exist but only a few are harmful to turf and other cultivated plants. Nematode infestations are much more widespread and damaging in southern turf, causing damage primarily to bermudagrass, zoysiagrass, and St. Augustinegrass. Nematodes, although present in cooler regions, are less severe in northern climates and do not produce the amount of turf damage as in southern turf. Direct nematode damage via feeding is usually to the roots of turfgrass plants. Nematodes can also bore into and occupy root conducting tissue like xylem, thus restricting plant water use and resulting in symptoms of drought or heat stress. Pinpoint diagnosis is very difficult when dealing with parasitic nematodes so it is wise for a turf manager to enlist the assistance of

PESTPOINTER

A good recipe for control of fairy ring is a regular maintenance regime of core cultivation and application of wetting agents to discourage water repellency in the soil. In addition, a fungicide application of azoxystrobularin that is drenched into the root zone may be an effective means of control.

PESTPOINTER

Pinpoint diagnosis is very difficult when dealing with parasitic nematodes, so it is wise for a turf manager to enlist the assistance of a plant pathologist to accurately identify the type of nematode present and suitable treatment options.

a plant pathologist to accurately identify the type of nematode present and suitable treatment options. When turf loss is evident, an application of a nematicide would be advised because fungicides have minimal effect on nematode populations.

FUNGICIDE FAMILIES AND MODES OF ACTION

By definition, a fungicide is a chemical that is used to kill fungi in a short period of time. This distinction helps explain why control is often difficult for bacterial, viral, or nematode infections because these organisms require unique control measures. Fortunately, most turfgrass diseases are caused by fungi so fungicides are reliable tools for most problems you may encounter. Fungicides are classified in groups according to their biochemical and topical modes of action. There are two general classifications. Contact fungicides are those fungicides that are effective only on fungi present on the external tissues of the plant. This contact mode of action can help protect the plant from potential infection, especially if fungicides remain on the plant surface for longer periods of time. Examples of contact fungicides are chlorothalonil, etradiazole, mancozeb, quintozene, and thiram. Contact fungicides kill spores and mycelia by direct engagement with the spore-producing pathogen. Contact fungicides possess the trait of providing a protective coating to the plant creating a barrier of protection from further infection. The following factors will rapidly decrease the fungicide efficacy of contact materials by physically removing them from plant surfaces:

- *Plant growth*
- *Precipitation*
- *Volatilization*
- *Mowing*

Penetrant fungicides are another group of fungicides that are described under the general heading of systemic. These compounds are absorbed into the plant tissue and kill the target pathogen. Because they are most effective when absorbed into plant tissues, systemic fungicides should not be applied to dormant turf or severely injured turf that is unable to effectively absorb the fungicide. Systemic fungicides can function in two ways in that they can kill pathogens on the tissue surface as well as inside the plant. Systemic fungicides tend to offer a longer period of protection against infection, because they are able to reach interior portions of the plant. Table 5.1 lists some common turfgrass fungicides, their chemical families, and their contact/systemic mode of action.

Within the penetrant or systemic group are three sub groups. First, there is the localized penetrant group, which enters the tissue and stays there to control pathogens near the site of fungicide absorption. Examples of localized penetrants are:

- *Vinclozolin*
- *Iprodione*
- *Propamocarb*

Second, there is the acropetal penetrant group. These compounds only move in the water conducting tissue (xylem) of the plant. Examples of acropetal penetrants are:

- *Fenarimol*
- *Flutalinol*

TABLE 5.1 Common Fungicides Used in Turf Management.

Fungicides	Trade names	Fungicide Families	Contact/Systemic
Azoxistrobularin	Heritage	Strobularin	Systemic
Chlorothalonil	Daconil, Echo, Concorde	Substituted Aromatic	Contact
Cyproconizole	Sentinel	Triazole	Systemic
Etradiazole	Koban	Thiadiazole	Contact
Fenarimol	Rubigan	Pyrimidine/Sterol Inhibiter	Systemic
Fludioxonil	Medallion	Phenylpyrrole	Systemic
Flutalonil	Pro Star	Benzinilide	Systemic
Forsetyl-Al	Signature Alliette	Phosphonate	Systemic
Iprodione	26GT, Fungicide X	Dicarboximide	Systemic
Mancozeb	Fore	Dithiocarbamate	Contact
Mefenoxam/Metalaxyl	Subdue Maxx	Phenylamide	Systemic
Myclobutanil	Eagle	Triazole	Systemic
PCNB	Turfcide, Terraclor	Substituted Aromatic	Contact
Polyoxin D	Endorse	Chitin synthase inhibiter	Systemic
Propamocarb	Banol	Carbamate	Systemic
Propaconazole	Banner Maxx	Triazole / Sterol Inhibiter	Systemic
Pyraclostrobin	Insignia	Strobularin	Systemic
Thiophanate-Methyl	Cleary's 3336	Benzimidizole	Systemic
Triadimefon	Bayleton	Triazole/Sterol Inhibiter	Systemic
Trifloxystrobin	Compass	Strobularin	Systemic
Vinclozolin	Curalan	Dicarboximide	Systemic

- Triademefon
- Thiophanates
- Propiconazole

The third group of penetrant fungicides are systemic penetrants. These fungicides are getting the most attention in research for new product development. Systemic penetrants can move in both the xylem and the phloem, meaning they can move in any direction within the plant. Example fungicides are:

- Ethylphosphonate
- Asoxystrobularin
- Polyoxin D zinc salt

These materials are exciting because of both their newness and because most common pathogens have not been previously exposed to this mode of action.

Deciding between a contact and a systemic fungicide usually is not an either/or case. In that light, perhaps the topic of discussion should be contact and systemic rather than contact versus systemic. The two types of fungicides are partners in turf management and should be used in conjunction with one another. Rotation between and use of these fungicides together are effective means for combating disease development. It is always a wise practice, for turf managers planning a disease control strategy, to alternate between fungicide families, rotate between modes of action, and tank mix contact and systemic fungicides for enhanced control. Rotation of fungicides follows the same logic that was described in the previous chapter on herbicides.

Disease pathogens, like other pests, can become resistant to fungicide applications. University research and field experience have clearly shown this to be true. Through use of fungicide rotation and tank-mixes of different products, resistance can be avoided and control can actually be enhanced. The theory is that contact and systemic

> ## PESTPOINTER
>
> The two types of fungicides are partners in turf management and should be used in conjunction with one another. Rotation between and use of these fungicides together are effective means for combating disease development.

fungicides, applied together, can be collectively more effective than if each were applied separately. It is believed that the contact fungicide makes the target disease-causing organism more vulnerable to the systemic fungicide. This approach is particularly effective for turfgrass disease control programs under a high degree of maintenance, such as on golf courses, where disease control is at a premium.

Fungicide formulations can be numerous. They come packaged as:

- *Dispersible granules*
- *Emulsifiable concentrates*
- *Wettable powders*
- *Water dispersible bags*
- *Flowables*
- *Granular products on inert carriers*

Deciding among these formulations will depend upon your situation. Many of the products you have to choose from are designed for liquid sprays so that even coverage on turfgrass surfaces can be achieved. This ensures that the contact activity of most fungicides will take effect. For systemic fungicides, placement of the fungicide can affect its performance. This decision involves understanding the characteristics of the disease you are targeting. For example, pythium-specific fungicides such as carbamates, acylalanines, and ethyl phosphonates are systemic fungicides that are typically absorbed by the plant either through the leaf crown or the roots. However, to maximize control, it is best that they be irrigated into the soil to enter through the root system. In a case such as this, tank mixing a contact and a systemic fungicide may not be recommended since the necessary irrigation that moves the systemic material where it needs to go would impact the effectiveness of the contact fungicide.

Follow label instructions and think of such issues before any fungicide application is made. Although most fungicide products are manufactured for sprayable functions, many fungicide formulations are also available to turf managers in granular or particle-based products. These products are excellent tools to have around because they are easy to apply for spot treatments, treatments during winter thaws, or during emergency situations. For example, when overseeding small but critical turf areas, a granular pythium fungicide product is a very useful tool.

Lastly, recognize that certain turf fungicides can have detrimental side effects on desirable grasses. Although fungicides aren't designed to kill plants like many herbicides, some fungicides have growth retarding properties. This effect may be more pronounced on certain turf species and cultivars and can be even more detrimental if materials are applied when the plant is under stress. Awareness of a fungicide's potential for this effect and of the sensitivity of a particular turfgrass species can help you avoid unwanted

PESTPOINTER

Tank mixing a contact and a systemic fungicide may not be recommended since the necessary irrigation that moves the systemic material where it needs to go would impact the effectiveness of the contact fungicide.

> **PESTPOINTER**
>
> Although most fungicide products are manufactured for sprayable functions, many fungicide formulations are also available to turf managers in granular or particle-based products. These products are excellent tools to have around because they are easy to apply for spot treatments, treatments during winter thaws, or during emergency situations.

difficulties. Fungicide formulations can also affect turfgrass health. Contact materials are designed to coat the surface of treated turfgrass leaves, invariably reducing the water-conducting abilities of the plants for a period of time. Applied under very hot conditions, these fungicides can impair turfgrass transpiration and cooling, resulting in unwanted environmental stress. Although somewhat rare, this type of detrimental circumstance can be avoided by using contact materials with caution, particularly in extreme heat and on sensitive turf areas like putting greens.

> **PESTPOINTER**
>
> Contact materials are designed to coat the surface of treated turfgrass leaves, invariably reducing the water-conducting abilities of the plants for a period of time. Applied under very hot conditions, these fungicides can impair turfgrass transpiration and cooling, resulting in unwanted environmental stress.

CONTROL STRATEGIES FOR TURFGRASS DISEASES

For a turf manager, there are many fungicide options and alternatives when planning a disease management program. Accurate disease identification is the most important factor when selecting a fungicide to apply for correcting a disease problem on turfgrass. Furthermore, focus on establishing a threshold for disease presence before intervention and what is an acceptable level of appearance to customers, property owners, club members, and officials. Be sure to clearly and openly discuss your turf management program with the main goal being education and communication of these threshold factors. It is important to understand the concept of economic outcome and effects as it relates to threshold or how much disease we can live with. If disease pressure is present to the point of significant turf loss and damage, where replacement is above and beyond routine maintenance, this could be considered economic loss. When a turf manager must apply fungicides at curative rates to correct a situation, and then absorb the costs of turfgrass renovation, this can be considered economic loss.

If healthy, aesthetically pleasing turf is desired, a turf manager must implement some form of control strategy to elude pathogen infestation. However, cost-based approaches to fungicide use can vary. A golf course superintendent would take a different approach than a highway superintendent regarding turf management. Fine turf managers must place more emphasis on turf aesthetics, and disease control programs will mirror this objective. Aggressive cultural practices and a preventative fungicide program are common components of fine turf management, where little or no turf loss can be tolerated. Lower maintenance turf can usually tolerate lesser emphasis on disease control, saving these measures for extreme cases of need.

Understanding the three main components of the disease triangle is an important concept to grasp before implementing a disease

> **PESTPOINTER**
>
> Accurate disease identification is the most important factor when selecting a fungicide to apply for correcting a disease problem on turfgrass. Furthermore, focus on establishing a threshold for disease presence before intervention and what is an acceptable level of appearance to customers, property owners, club members, and officials.

management program. There are three segments that must each be in place for disease infection to occur:

- A susceptible host plant
- An active pathogenic entity
- A favorable environment for disease development

Understanding the triangular relationship of turfgrass diseases means first understanding the diseases and their life cycle patterns. Pathogenic spores naturally occur in the soil. If left alone and unchecked, these spores can perpetuate the disease cycle. Spores invade plant cells through multiple ports of entry, infecting the plants with detrimental pathogens. Symptoms appear on infected plants. The desirable host will essentially get sick and die from disease. If left unchecked and untreated, the pathogen will coalesce to involve larger areas of turf. Disease spores overwinter in the soil and fester in the thatch/soil interface, feeding on dead and dying organic matter. Understanding this phase of disease development will assist in developing a plan for well-timed control strategies.

The second key component of the disease triangle is the susceptible host. In this regard, positive identification of the turfgrass species you are dealing with is a must. It is important because some turfgrass species or cultivars are more susceptible to pathogens than others and some diseases are species specific. Annual bluegrass is a common turfgrass species that turf managers are in a constant struggle to maintain. This plant at any mowing height and almost any degree of maintenance is succulent, weak, and can be susceptible to many diseases such as anthracnose. Avoiding excessive maintenance to annual bluegrass can help reduce disease potential. Perennial ryegrass is also at particular risk from either gray leaf spot or pink snowmold. Dollar spot causes high levels of damage to bentgrasses on golf courses, while spring dead spot exclusively damages bermudagrass in southern regions. It is wise to design a fungicide program that suits the budget, aesthetics, and health requirements of the turfgrass stand being maintained. Choose the correct grass cultivar for the site. Seed producers, sod growers, and researchers are continually producing and introducing grass cultivars that are bred to withstand pests or are more tolerant to the sorts of pests that love to feed on turfgrass plants.

The last component of the disease triangle is environment. Weather significantly affects the presence of disease-causing organisms, as many of them are seasonal in nature. Humidity, soil temperature, sunlight, ambient temperatures, and air movement are important factors for disease development. Developing a weather-based approach to disease control is best based upon past experience. Historical documentation of growth habits and past weather conditions is also helpful, as is being able to react to weather changes that may induce disease development. To give you an idea of how the three components of the disease triangle work together, let's look at some example situations.

Example 1. With overnight temperatures of 68°F, 80 percent humidity, and no breeze, a practitioner should deem this representative of a prime climate for many diseases. Strains of pythium, dollar spot, brown patch, and anthracnose should be anticipated, control measures should be initiated, and sites that are prone to infection should be monitored.

Example 2. In the Great Lakes Region, it is not uncommon for pink snowmold to appear with air temperatures as high as 65°F under overcast skies and damp conditions. However, when air temperatures climb above 70°F with low humidity and adequate air movement, this disease activity will cease. By monitoring weather factors, there may be no need for the turf manager to intervene with a fungicide application. Conversely, gray snowmold is a disease where extended periods of snow cover are needed. If extended periods of snow cover are common to your area, the proper control strategy in this case would be a preventative fungicide application if economic loss of turf is to be avoided.

Example 3. On the Fourth of July, a thunder storm drops 2.5 inches of rain on already saturated soil. In this case, fine turf managers should be prepared to make fungicide applications immediately to avoid threats of diseases like brown patch, summer patch, or pythium.

Example 4. Overseeding of warm-season turfgrasses during periods of dormancy presents a different set of circumstances for the manager of fine turf. Germinating seedlings and immature turf are susceptible to many diseases. Newer varieties of perennial ryegrasses are constantly being introduced for disease resistance but pythium blight and gray leaf spot should never be taken lightly in these circumstances. Overseeded putting greens in warm-season turfgrass areas have the propensity to become infected with seedling diseases. *Poa trivialis* and bentgrasses have been known to contract take all patch, pythium, and *Helminthosporium* strains of pathogens. Carefully prescribed and timed applications of fungicides are necessary because the threshold on these turf areas is very low. There is a substantial financial investment at stake and plant protection from disease infection and potential loss could be potentially devastating. It is critical to keep this overseeded turf healthy.

Example 5. Many disease symptoms resemble one another. For example, pythium infection can resemble dollar spot incidence. The choice of control strategy in this case is critical because of the similarity of symptoms at the early stages. A turf manager could be applying fungicides for dollar spot when in reality, pythium infection is present. In this case the pythium could become more severe and, as a result, time and money could be wasted because of an improper diagnosis. Comparing observed symptoms to ambient weather conditions in this case can help determine the true identity of a disease.

Turfgrass cultural practices like mowing, irrigation, and fertilization can be contributing factors to disease infestations. Sufficiently irrigated home lawns that are fertilized adequately and mowed above two inches in height are less likely to develop disease problems. Lower mowing heights require higher degrees of turfgrass maintenance and are at greater risk for diseases. Raising the mowing height can be an effective disease control strategy that a turf manager can utilize but may be difficult or unacceptable in some circumstances. For example a three-inch mowing height may not be acceptable for a football field at the professional sports level but is satisfactory for recreational usage or a home lawn. Golf course putting greens require more intense mowing to achieve desired results, and are thus are more likely to develop disease problems. However, raising mowing heights above ¼ inch on golf course greens would be not acceptable. Equipment care is as important a part of mowing as the mowing height, when it comes to diseases. Sharp and clean mowing equipment not only is a sign of sound and well maintained machinery, but is vital to curb spread of disease. A fungicide application should be scheduled in conjunction with invasive maintenance practices.

PESTPOINTER

Turfgrass cultural practices like mowing, irrigation, and fertilization can be contributing factors to disease infestations.

Excessive irrigation is usually a recipe for disease, especially during extreme hot, humid conditions. Moderate irrigation practices that do not result in moist turf for extended periods of time can help reduce this risk. Overfertilization can sometimes promote diseases by producing a succulent plant that may be less hardy than those receiving fewer nutrients. However, some diseases thrive in areas that receive low fertility (e.g. dollar spot) so provision of fertilizer at times of high disease pressure can help minimize the threat. Understanding the nature of the pests you encounter will help you make proper fertility decisions.

Other cultural issues to consider when designing a disease management program are core cultivation, verticutting, top dressing, and soil conditioning procedures that are invasive and injurious to turfgrass plants. The wounds produced during these aggressive practices could weaken the grass plants and provide points of entry for pathogens or other destructive pests. Turf experts often recommend to golf course superintendents to avoid topdressing or other aggressive activities during periods of high stress because the sand granules and consequent brushing could damage leaf tissue and accelerate disease activity. Timing of these abrasive practices has to be judicious during periods of high disease pressure.

When selecting fungicides for a particular problem, always know what is registered for use in your region or state. A majority

PESTPOINTER

Turf experts often recommend to golf course superintendents to avoid topdressing or other aggressive activities during periods of high stress because the sand granules and consequent brushing could damage leaf tissue and accelerate disease activity.

of fungicides are restricted use and should only be handled by certified pesticide applicators. Judicious care and handling by professionally trained applicators utilizing personal protection equipment is a wise and necessary practice (Figure 5.21). Reliable and properly calibrated spraying equipment is always important, whether it is a sprayer or a spreader, in order to make the task more efficient. Also pay attention to the specific spraying instructions for fungicides. Contact materials need to be applied in at least 40 gallons of water per acre to properly coat the plant and to protect it from disease. Some systemic fungicides need to be irrigated into the root zone to maximize plant uptake. The newer forms of strobularin fungicides require tank-mixing with a fungicide that has a different mode of action because repeated use could promote other diseases like dollar spot. Pay attention to these sorts of details in order to get the most out of your applications.

FIGURE 5.21 A sprayer used for a fungicide application, being operated by an applicator with proper protective equipment.

When it comes time for a turf manager to implement a reliable disease management plan, the best defense is a good offense when it comes to fungicide selection and application in turf management. Table 5.2 lists some common fungicide products and the diseases they control. The primary control strategy is a good, healthy turf achieved by means of sound cultural practices, balanced nutrition, and consistent monitoring. When your acceptable disease threshold is threatened by pathogen activity or weather conditions, the control phase of disease management must be introduced.

In most cases, it is best not to delay disease control action until visual symptoms of disease are present. By that time, it may be too late to reverse the pathogenic damage. Early implementation of a fungicide application can halt or even reverse the trend of a destructive disease infection. Be sure to document and record the site conditions,

TABLE 5.2 Common Turf Fungicides and Key Diseases They Control.

Fungicide Trade Name	Diseases Controlled
Heritage	Anthracnose, brown patch, fairy rings, gray leaf spot
Daconil, Echo, Concorde	Dollar spot, brown patch, fusarium blight
Koban	Pythium
Rubigan	Take all patch, spring dead spot
Forsetyl-Al	Pythium, summer stress
26GT	Brown patch, dollar spot, helminthosporium
Subdue	Pythium, seedling damping off
Turfcide 400	Snow mold, pink and gray
Endorse	Anthracnose, brown patch, zoysia patch
Banol	Root pythium
Banner MAXX	Brown patch, dollarspot, summer patch
Bayleton	Anthracnose, brown patch, dollar spot,
Compass	Anthracnose, brown patch, summer patch
Cleary's 3336	Dollarspot, fusarium blight, red thread
Curalan	Brown patch, dollar spot, pink snow mold

the environmental conditions, and the materials applied, because concise record keeping, continued monitoring, and success evaluation will assist in better decision making for future practices, especially if a particular disease is a persistent problem. In summary, a manager of fine turf can use the following five-step plan when designing and implementing a fungicide program:

1. Early detection. Through scouting and monitoring, a turf manager can establish indicator areas to watch for pockets of disease development. Poor sites should be constantly observed.

2. Early diagnosis. Positive identification, by utilizing the information from signs and symptoms that are present to catch the pathogen as early in development and infection as possible, will make control more effective. A diagnostic lab, a plant pathologist, or a turf expert would be helpful.

3. Early intervention. Applying a corrective fungicide at label rates with reliable equipment at the earliest stage of disease development provides the best control.

4. Evaluate. Follow up monitoring of the infected site is important. Here a turf manager can observe if the desired results were achieved by the fungicide application and the pathogen activity has ceased.

5. Historical data. Record keeping is invaluable in designing and implementing future disease management practices. It will provide a map for future scouting and monitoring.

By following these key steps and committing to good comprehension of both diseases and their control methods, you will be able to make good judgements and optimize turfgrass disease control at your facility. Develop a sound disease management program through education. Be consistent, be proactive, and stick to what works. The next chapter takes this topic one step further to include disease problems and control strategies in landscape plants. Let's continue, shall we?

SUMMARY QUESTIONS

- What are three components that have to be in place for disease development?
- What diseases are the biggest threats to turfgrass at your facility?
- What areas at your facility are most prone to disease infection? Why do you think this is the case?
- What is the difference in mode of action between contact and systemic fungicides? Is there a clear advantage to either one?
- What fungicides, if any, have you used for disease control in the past? What made you choose these particular products?
- What is the most important factor in designing a disease management program? What secondary factors should also be considered and why?

Chapter 6

Disease Problems and Fungicide Use in the Landscape

Chemical controls are still the most common way for controlling the wide variety of diseases that occur on landscaping plants. However, this may not be the best way or the least expensive way to prevent, treat, and cure a disease problem. A balance of chemical usage in combination with IPM practices and other cultural control methods provides the most desirable results in the long run. I would like you to think about controlling diseases and using fungicides in a preventative way, only when it is absolutely necessary.

Numerous diseases occur in nature and every variety or species of plant has specific pathogens that will try to infect it. Disease is actually the visible results that are caused by the presence of a plant pathogen. However, pathogens are only one part of what makes up disease. A combination of pathogen, host, and the correct environment must occur together for disease to develop. This is also known as the disease triangle. Without any one component of this triangle, landscaping plants will not show actual signs of disease. It is very possible to have pathogens on plants but the plant may not be affected by the pathogen because it is not the correct host. It also possible for pathogens to be present but the environmental conditions are not suited for disease development. This very important concept to controlling diseases on landscaping plants is frequently misunderstood by many who work within the industry.

Since environment has such a key role in causing disease you should consider it as a primary strategy for control. It is more difficult to change environmental factors for landscaping plants than some other plant types but there are still some techniques that should be considered. Disease occurrence in a nursery can really be affected by limiting the environmental factors such as water and sun exposure. The factors of temperature, moisture, light, and humidity can all affect disease occurrence. In a situation where irrigation is used, the water can easily be manipulated to affect the occurrence of many diseases in turfgrass and the surrounding landscaping plants. The same is possible for a sunlight problem that might be correctable by removing other less valuable tree species in the area or thinning the existing cover. Temperature and humidity are hard to control but indirectly can be affected by changing the light and water issues already described. All environmental factors are very closely related and can be changed to produce conditions more desirable for plant success and less desirable for disease.

Changing environmental factors is the least expensive control measure that can be done to try to control a disease problem. Spraying chemicals is often cost prohibitive and can also be the most difficult control measure to use. In the past, it was almost impossi-

---

PESTPOINTER

Changing environmental factors is the least expensive control measure that can be done to try to control a disease problem.

ble to have the equipment available to spray large ornamentals and when this was done, the results were not always as planned. Ornamental plant diseases were thought to be hard to control and were looked at as a total loss. Replacement was often cheaper than control and certainly much easier. This is still the case for many ornamental diseases today but, with technology and the development of more systemic fungicides, the overall methods of disease control have improved considerably.

What once required a large high pressure sprayer can now be performed with a small injection point into the tree or even a root drip or soaking method (Figure 6.1). Fungicides are more effective and much safer to use. One of the biggest problems with using foliar

FIGURE 6.1 High pressure sprayers can be as small as 10 gallon up to 500 gallons. The pumps can be either electric or gas.

controls is the amount of drift and off site movement that is experienced with using a spray control. Spraying must be done so that coverage of a specimen tree is sufficient enough to cover all infected areas. If the coverage is lacking or the spray does not reach the top of the tree, reoccurrence of the disease could occur in a short period of time. In most cases, there is also the problem of the disease reoccurring the following season. Pathogens can often over-winter in the immediate area or can be reintroduced from other species or hosts in close proximity. Some diseases can spread over miles because the pathogens that cause the disease are very mobile and have many modes of transportation from specimen to specimen.

One very simple thing that can be done to reduce disease occurrences is to remove debris which can harbor pathogens or pests. Leaves, needles, and other tree debris are a good hideout for plant pathogens. Debris can keep the pathogen in the immediate area for infection of the species the next year. By doing thorough leaf clean ups and eliminating this debris every fall, the chances of a repeat infection of the plant is reduced. It takes more than mulching the leaf debris up. What actually needs to happen is complete removal and destruction of the debris by burning which will kill the pathogen. Mulching up the material with a mower reintroduces the pathogen and scatters it around where it will lay until the following season.

PESTPOINTER

One very simple thing that can be done to reduce disease occurrences is to remove debris that can harbor pathogens or pests. Leaves, needles, and other tree debris are good hideouts for plant pathogens.

> # PESTPOINTER
>
> Another way to limit the contact with a plant pathogen is by removing the primary or secondary host of the plant pathogen.

Another way to limit the contact with a plant pathogen is by removing the primary or secondary host of the plant pathogen. Many times plant pathogens will harbor in a secondary location or in a plant that is not affected by the pathogen's presence. Then, once environmental conditions are favorable for disease, the pathogen will spread to the plant that it will soon infect. Soon thereafter, disease will occur and the pathogen may spread mycelium or spores around the area for the next infection period. To successfully use this type of prevention you must learn a great deal about the pathogen, host, and environment that causes the disease. Keep in mind any possible hosts and remember that many ornamentals have closely related hosts that can occur in the wild. Also remember that pathogens can spread over miles in a matter of minutes by wind or other carriers. This makes elimination of all possible hosts very difficult.

Many people in the industry do not account for all of the possible hosts or the carriers of plant pathogens. These carriers should be examined for potential control strategies. For instance, control of a particular insect that is known to harbor a pathogen may be easier than trying to control the pathogen itself. There are other carriers that can spread diseases around, including humans. Any person, insect, animal, or even tool could be a potential carrier of a plant pathogen. The use of pruning equipment is probably one of the most common carriers of a pathogen.

As a person trims, saws, or prunes a specimen plant, the blades are entering the infected portions of the plant that is being worked on. Often the reason that pruning is being done in the first place is

> **PESTPOINTER**
>
> The use of pruning equipment is probably one of the most common carriers of a pathogen.

to remove some damaged or diseased portion of a plant. The pathogen may have spores or mycelium that can attach to the pruning equipment and stay there until the tools are later used on healthy specimens. Pruning tools or anything that comes into contact with plants can become a potential carrier for plant disease.

DISEASE CONTROL

The newest trend in disease treatment is the use of systemic injections directly into the tree tissues below the bark. These new chemicals can be systemically introduced into the plants and are very effective. With the injection method (Figure 6.2), there is little loss of the product being used for control. Most tree injectors are very easy to use and are now becoming readily available in the retail market for homeowners to use themselves. Tree injections have been used by arborist for quite a few years but the controls were not that effective and the use of injectors was limited to a very specialized trade. As the systemic chemicals and the methods for application have become more effective, the use of chemical controls is considerably better than it was even just a decade ago.

A tree injection can be done with several devices but the idea is very simple and easy to use. In most cases it involves the use of a small drill with a $1/8$ inch or $1/4$ inch drill bit. A series of several

Disease Problems and Fungicide Use in the Landscape 257

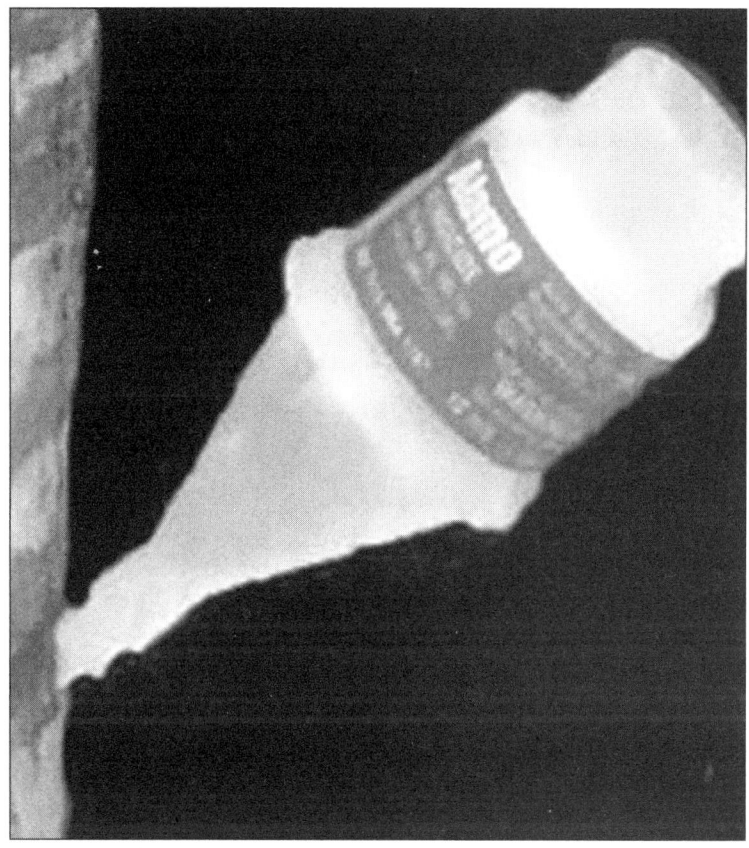

FIGURE 6.2   Tree injectors like the one above require little time and are very cost effective and efficient when controlling ornamental diseases.

holes needs to be drilled into the tree or plant around the base of the plant. In a larger plant, such as a 20 foot shade tree, these holes may be drilled every four to six inches all the way around the trunk of the tree, two to three feet off of the ground. In a shrub of a smaller ornamental plant it might only be necessary to drill one hole in the trunk of the plant which would treat the entire plant system. The idea is to make a hole in the trunk of the plant into the xylem and phloem tissues which are responsible for the transportation of fluids in a plant system. The chemical then is moved through the entire system of the plant from the root zone clear out to the leaf structure, giving you complete control over the pathogen that is invading the species. The injection process only takes a few minutes and there is little chance for exposure to the user of the product or chance of any off-site movement of the product. Clean up is easy and requires the disposal of some small injection tubes. This is now the most preferred method, for not only fungicide treatments in trees and shrubs, but also the most effective control for many insects as well.

The use of tree injections has created a much more cost effective way to be able to use chemicals for the control of ornamental diseases. The injections do vary in price and some can be quite expensive when treating a larger tree or shrub that is infected with some sort of a plant pathogen. The treatment methods and amounts of product used are based on the size of the plant and the trunk diameter in most cases. It is important that enough active ingredient is introduced into the plant to insure movement throughout the entire specimen.

A lot of the chemicals that were used for spraying in the past were contact fungicides which is why they had limited success for the treatment of ornamental plant diseases. The chemical would have had to come into contact with all of the infected areas to be a success

---

PESTPOINTER

The use of tree injections has created a much more cost-effective way to be able to use chemicals for the control of ornamental diseases.

and that was very hard to do especially when treating larger plant materials. There are still many contact fungicides in use today and spraying is still a very common control method. However the use of tree injectors for controlling disease is certainly the most preferred method. The only drawback to using an injection system is the small holes which are drilled into the tree for the use of injectors. There is the possibility that this could become an entry point for future occurrences of the pathogen or worse yet, an entry point for a new or different pathogen.

Liquid chemicals are used for almost all treatments of disease in ornamental plants. I cannot think of any instance where a dry product would be used for successful control of any type of disease in ornamental landscaping. There would be no way for the plant to take up the product and coverage of the plant would be impossible. Another way treatments are given to ornamentals is through the use of drip systems or root soakings. This is one way to keep from drilling into the plant trunk and causing a small point of injury.

Root soaking is done by mixing a volume of a systemic product in a certain amount of water. The product is mixed and then poured around both the base of the tree and also out at the root zone (Figure 6.3). The tree should then pick up the material through the root zone and carry the product throughout the rest of the infected plant specimen.

Spraying is still one of the most common methods used for controlling pests in ornamental trees and shrubs and can be an effective control measure. When using systemic chemicals, complete coverage of the plant is not always necessary to eliminate the disease. It is important to get as much coverage as possible. Sprays are more effective on smaller plants because drift and coverage can be controlled more easily. It might be more difficult to treat a small plant with an injection of some other type of fungicide treatment instead of using a fast foliar spray. I think the key to spraying successfully is to do it only when it is the most convenient and safe way to treat for disease. Safety should always be an issue with any pesticide

260 Turfgrass Chemicals and Pesticides: A Practitioner's Guide

FIGURE 6.3 Mixing up the chemical in a bucket and pouring it in the drip zone and trunk area can be an effective way to treat the entire tree.

PESTPOINTER

The key to spraying successfully is to do it only when it is the most convenient and safe way to treat for disease. Safety should always be an issue with any pesticide usage but especially when considering the use of high pressure sprayers or something that can cause off-site drift.

usage but especially when considering the use of high pressure sprayers or something that can cause off-site drift. Most sprays are contact-type fungicides which require efficient coverage of the infected plant specimen. This is another reason that smaller plants are much easier to treat than larger ones.

PLANT HEALTH

Proper plant health can be a good deterrent for any plant disease occurrence. As with all living things, it is truly survival of the fittest and this includes plant species. When plants are healthy and actively growing, they are less likely to be infected by a plant pathogen or anything else that would be detrimental to the overall survival of the plant. When there is a row of plants of all the same variety and size and one somewhere in the middle becomes infected with a pathogen and starts showing a disease problem, that is an indication of one of two things. It is possible that something caused only that individual to be exposed to the pathogen. For example, a bug carrying a virus happened to land only on that plant. The other possibility is that the plant was not as healthy as the other plants and therefore could not fight off the pathogen on its own. Plant systems are much like the human body in fighting off disease problems. If you are sick or not eating right, you are more likely to get an infection or catch a virus. Plants react the same way to poor health and nutrition. If a plant is under stress from some nutritional or environmental factor, then it is more likely to become infected or show a disease problem.

The way to prevent the possibility of disease is through proper care of all of your ornamental plants. This is Mother Nature's way to insure that survival of the fittest will pertain to the plants in your care. Having a good program to address the water and nutrition needs of the plant will be the most beneficial preventative measure. Plant health also will help to reduce misdiagnosis of plant health problems. It is hard to distinguish a plant pathogen problem from a nutrition problem on unhealthy landscaping plants. A major leaf discoloration or unusually light color may be from a lack of nitrogen and iron and not the result of some chronic disease plaguing the plant. This is another good reason to have healthy plants in your landscape.

The age of an infected plant can have important implications for the plant's recovery. A young, more vigorously growing plant will respond much better to treatment and be able to outgrow the disease in some cases. However, plants that are very young and ten-

PESTPOINTER

Plant systems are much like the human body in fighting off disease problems. If you are sick or not eating right, you are more likely to get an infection or catch a virus.

der are more prone to die. There is a point of establishment that must be present where the plant is past the vulnerable transplant stage. If the plant is very small or just recently transplanted, it is likely that the plant is already suffering from environmental stresses. Disease then becomes an even larger obstacle for plant survival. Mature plants that are very old also do not have the ability to recover from a disease problem. I would compare a plant's life to a human in response to the flu. If you are very young or very old, the common flu can be very hard on the human body. A plant's ability to recover also has a lot to do with the age of the plant and how healthy the plant was before it was infected.

Diagnosis of disease is difficult and there is often no surefire way to distinguish a nutritional deficiency from a plant disease with the human eye. The only way to guarantee diagnosis of a plant problem is through the use of a microscope to look for signs of a disease problem (Figure 6.4). There should not be any assumptions that a plant is suffering from disease, environmental stress, or nutrition deficiency. The only way to know for sure is to look for the actual pathogen. Many state universities and extension agencies provide testing of plant material for a low or nominal fee to help you determine what is causing unhealthy plants. Make sure and use these resources if you lack the knowledge and skills that are required for plant disease diagnosis. Even a plant pathologist with years of experience sometimes cannot say for sure that a pathogen exists without determination in a lab. Field diagnosis can be done with a

PESTPOINTER

A plant's ability to recover also has a lot to do with the age of the plant and how healthy the plant was before it was infected.

264 Turfgrass Chemicals and Pesticides: A Practitioner's Guide

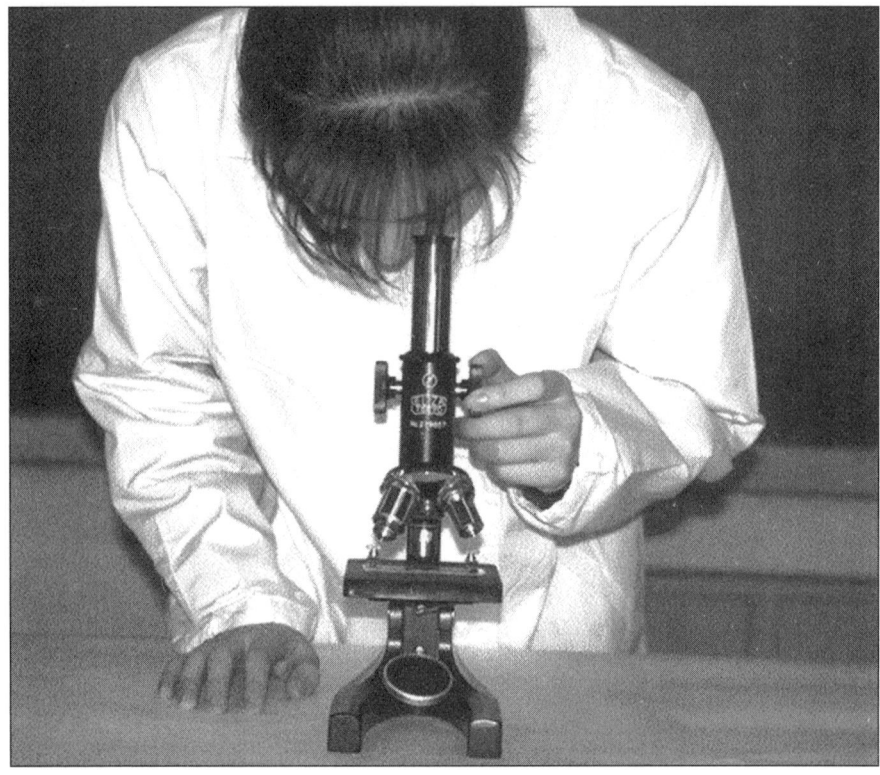

FIGURE 6.4 While diagnosis is possible in the field, many pathogens are not visible to the human eye. This is why a laboratory provides better disease diagnostic capabilities.

handheld lens or, in some cases, the naked eye but unless actual signs of the plant pathogen are seen, diagnosis is only a guess.

One way that is often overlooked for disease control is to simply use plants or varieties of plants that are not susceptible to disease. The point here is that a pathogen can be present but without a desirable host around, it will be impossible for disease to occur. There are new varieties of plants being developed every day for the purpose of disease resistance. This can also be an inexpensive way to prevent a problem over the entire lifespan of a plant even if it is a little more expensive to use the resistant cultivars to start with. With

PESTPOINTER

Many state universities and extension agencies provide testing of plant material for a low or nominal fee to help you determine what is causing unhealthy plants.

any plant species that is bred for disease resistance, it is very common to have to sacrifice one good characteristic to get a different one. Plants tend to give up one desirable trait in order to have a different desirable trait. This is not always true and plant breeding is at its highest level when it comes to the influence of technology. It may take a considerable amount of time for a plant breeder to develop a plant that has all of the features that are desired for that cultivar.

The best example of what I am talking about is the *Malus* species, otherwise known as crabapple trees. The crabapple tree is one of the most problematic plants that can be used in landscaping, but is also one of the most beautiful plants. Initially, there were only a few

PESTPOINTER

A pathogen can be present but without a desirable host around, it will be impossible for disease to occur. There are new varieties of plants being developed every day for the purpose of disease resistance.

select varieties of crabapples and many of these types were plagued by disease problems. There are disease problems such as scale, cedar apple rust, fire blight, leaf spot, and a variety of other less common diseases that all have crabapples on the list of suitable hosts. Over the years, resistant varieties of crabapple trees were developed that made its use more prevalent in the landscape once again. It is believed that there are currently over 800-plus varieties of *Malus* species being used in landscaping today. While some of the crabapples may not have the brilliant color of their predecessors, with so many choices it should be possible to find a crabapple that is exactly what you envisioned for the location in mind. Plant selection is as critical to limiting disease in plants as any good fungicide program.

MONOCULTURES

One of the leading causes of widespread disease in the landscape is from the overuse of one type of landscaping plant. The overuse usually comes about from the public buying one type of plant that they find a desirable characteristic in. It is up to us as lawn and landscape professionals to educate people on the uses of landscaping plants and reduce the chances for large monocultures in our landscaped areas. A monoculture is the overuse of a specific variety of tree or shrub in the landscape. When a major disease problem occurs in such plants, there are rampant outbreaks because there are so many host plants available.

The best example I can give for a history lesson in monoculture was the overplanting of the American Elm tree in the landscape some 40 to 50 years ago. After World War II and during the boom of suburbia in the United States, a pattern of plantings showed up in many residential settings. This pattern was not exclusive to any one part of the country because at the time, the elm tree was versatile enough that it could survive in a wide range of hardiness zones. The American Elm tree was planted in yards and especially along streets

PESTPOINTER

A monoculture is the overuse of a specific variety of tree or shrub in the landscape. When a major disease problem occurs in such plants, there are rampant outbreaks because there are so many host plants available.

from one side of the country to another. Many other trees were ignored because the neighbors had planted an elm tree so it was just accepted that everyone needed an elm tree in the yard. This tree became very standard in and around houses in large numbers until all of sudden, years later, the elm trees started to become sick.

The sickness in elms at first was unnoticed because it started in isolated areas and eventually progressed across the entire United States. Eventually, a near destruction of all of the mature elm tress in the whole country occurred over a very short period of years from what is known as Dutch Elm disease (Figure 6.5). The mature elm tree was literally wiped out and now is not recommended or used in the landscape at all. The reason this occurred is that once a pathogen started, there was no way to control the spread of the disease because there were so many host plants to travel on. The disease was easily able to overtake many trees and it quickly moved from one geographic location to the next because elm plantings existed everywhere. There were also no chemical controls for the pathogen that caused Dutch Elm disease so treatment was almost impossible. It was also later discovered that only mature trees were susceptible to the disease which was unknown for while. It took a period of time to realize that the disease was taking its destructive toll only on older trees. Unfortunately for a period of time after Dutch Elm was discovered, new trees were replanted that did not die for

FIGURE 6.5 Dutch Elm disease has killed many of the elm trees that existed in the United States.

several years. Younger trees had a tighter fitting bark that had no splits in it which prevented entry of the pathogen, but as the tree matured at 15 to 20 years of age, the trees all became infected and died. This disease will be the demise of all mature elms and still is present in the species today. You may see elms in fence rows or might know of some isolated trees that do not have the disease, but if they are American Elms, they will eventually end up getting the disease. Not all elm species are susceptible to Dutch Elm disease but many of the other elm types have undesirable characteristics that make them less suitable for use in the landscape.

The use of fungicides in this case was not only expensive but it was futile because exposure to this disease would have occurred at some point no matter what. There was no way to control the spread of the disease and the disease still exists on a widespread level today. Despite what happened to the elm trees, we are still are in a habit of creating new monocultures which someday may remind us of the painful results that occurred from Dutch Elm disease. Awareness of what has happened with the American Elm in the past should keep us from having the same thing happen in the future. However, I still am seeing many overplanted species in some regions. Some examples are the hybrid maples, specifically *Red Sunset* and *October Glory*. These are two of the most beautiful maple trees I have ever seen, but they need to be planted as specimens and not in multiples of 10 or 20. By limiting the use of some of these types of plants, you restrict the spread of disease. Another example of a tree that is over planted would be the Bradford pear (Figure 6.6). History has already proven that this tree which was once thought of as the perfect shade tree is actually far from that. This is one of the weakest trees used in the landscape for longevity and has problems similar to those of the silver maple. I know that this has nothing to do with disease but should there ever be a disease to attack either the Bradford pears or any of the hybrid maples, for instance, the results

PESTPOINTER

Despite what happened to the elm trees, we are still are in a habit of creating new monocultures which someday may remind us of the painful results that occurred from Dutch Elm disease.

FIGURE 6.6 It is possible that some of the new varieties of trees being planted in the United States could become widespread hosts in the future.

PESTPOINTER

If you are a landscape designer, you should limit the long term possibilities for disease occurrence by using a variety of plants in the landscaping projects you are doing.

would be catastrophic to the species. The reason being is the commonality that varieties like this have and the overplanting of the species that has already occurred.

If you are a landscape designer, you should limit the long term possibilities for disease occurrence by using a variety of plants in the landscaping projects you are doing. There can still be the use of the same plants from project to project, but the overuse can be avoided. There is no need for ten of the same type of tree or shrub in every single plan that goes through your office. If other, less common plants are intermixed with some of the more common ones, a major disease problem might seem less catastrophic.

SOME COMMON DISEASE PROBLEMS

There are too many types of pathogens that attack ornamental planting in the landscape to even begin to mention them all. I do want to tell you about the various types of pathogens there are and what signs of disease will result from infection. The causes for disease can vary and still result in a sick plant. As I mentioned before, it is also possible that the problem associated with the tree has no connection to a pathogen whatsoever.

Wilts or wilting problems are one very common sign of disease in ornamental plants. Wilts can be a result of a watering or heat

> **PESTPOINTER**
>
> Wilts can be a result of a watering or heat problem but often are associated with some type of plant pathogen.

problem but often are associated with some type of plant pathogen. The symptoms of a wilt are a drooping effect to the plant or curling of the leaves, appearing much like a moisture problem. Unfortunately, one of the things that can lead to further problems associated with wilt is to water a drooping plant that is believed to be suffering from moisture problem. Watering will only encourage disease activity which can lead to the death of the plant in many cases. There are some wilts, such as verticillium wilt, which have no cure and are impossible to prevent or control. Some wilts are treatable though with a multi purpose fungicide (Figure 6.7).

Blotches and blights are another type of disease problem associated with ornamental plants. The plants that suffer from these problems express a range of symptoms such as leaf browning, leaf curling, discoloration, or other altered appearance. Many blights are not curable once the plant is infected but some can be prevented or treated

> **PESTPOINTER**
>
> Many blights are not curable once the plant is infected, but some can be prevented or treated with a multi purpose fungicide such as Daconil® 2787.

Disease Problems and Fungicide Use in the Landscape 273

FIGURE 6.7 Verticillium wilt is often terminal, which makes fungicide use impractical in some cases.

with a multipurpose fungicide such as Daconil® 2787. As mentioned before, many plant pathogens create an untreatable condition in plants, that once infected, are damaged to a great extent.

Rusts are another disease that can be cured with a general multipurpose fungicide. As with blights and blotches, control measures should be done as early in the year as possible, usually just as buds and flowers begin to open on most species. If treatment of rusts is performed later in the year, the results are not as effective or may

> **PESTPOINTER**
>
> Rusts are another disease that can be cured with a general multipurpose fungicide. As with blights and blotches, control measures should be done as early in the year as possible, usually just as buds and flowers begin to open on most species.

not provide any control at all. Rusts show symptoms of discolored or speckled leaves and are often confused with some sort of insect or feeding insect problem. Hawthorn and crabapple trees are just a couple of the more common plants that can become infected with this disease.

Some diseases can be simply called leaf spots, but the causes for leaf spots can be numerous. Again, many leaf spots can be confused with insect problems and most are curable with a general purpose fungicide that can be systemic or contact. Captan is a common chemical used for treating leaf spots which can be found on many varieties of trees and shrubs. Orchards are often plagued with various leaf spot disease and treat for this problem throughout the growing season.

The last of the numerous diseases I want to mention are root rot and crown rot. These diseases are important because diagnosis of these problems often comes entirely too late and the plant may already be dying. Root rots occur underground so, until the tree starts showing signs that it is sick, there may not be an obvious problem at hand. These diseases are not always treatable and systemic chemical controls can provide some results. Always make sure to take into consideration that a sick plant could be suffering from something that is underground as well.

PESTPOINTER

Captan is a common chemical used for treating leaf spots which can be found on many varieties of trees and shrubs.

There are numerous other cankers, scorches, galls, and many other plant problems that are sometimes associated with disease. It would be impossible to cover the list of problems that exist and even more impossible to tell you the chemicals used to treat these problems. Plant diseases in the landscape should be treated more on a preventative process whenever possible because of the difficulties associated with diagnosis and treatment of plant pathogens. If the problem is curable, there are a wide variety of general purpose fungicides. Diagnosis is the most critical part of plant disease problems because sometimes treatment is not even a possibility.

PESTPOINTER

Plant diseases in the landscape should be treated more on a preventative process whenever possible because of the difficulties associated with diagnosis and treatment of plant pathogens. If the problem is curable, there are a wide variety of general purpose fungicides.

CONTROL STRATEGIES

Control strategies should start with the simple things like using prevention instead of treatment for successful disease control. The best way to control disease problems in the landscape is to take care of possible problems before anything occurs. After all, in comparison, wouldn't you rather avoid getting the flu than to suffer through it? Plants should be thought of along those same lines. It is best to avoid the problems and do things that will prevent ornamentals from ever developing a disease problem.

Start first with the concept of having more healthy plants in the lawn and landscape. This will mean fewer problems from any type of sickness whether it is caused by a pathogen or not. It certainly will provide you with a better looking landscape which is really the idea in the first place. Make sure to fertilize trees and shrubs on a yearly basis and if the weather brings about some drier periods, make sure to water the plants in the landscape along with the grass. It is a common misconception that established plants can survive maintenance free, which is not the case at all. All plants have the same needs of proper fertility and moisture, although established plants are able to tolerate the lack of either one better than younger plants without any root system.

Make sure to use a wide variety of trees and shrubs in the landscape and always use cultivars that are proven to be disease resistant and less likely to develop pest problems. In some cases, a certain

---

PESTPOINTER

It is a common misconception that established plants can survive maintenance free.

color or size might not be exactly what you want, but a sick plant will look a lot worse it you have a recurring problem with a disease in a particular area with the same old varieties. I know that not all species of plants have disease resistant counterparts but there are other plant species that can be substituted in most situations with similar size and characteristics. I have seen very few planting situations where there were not at least a half a dozen plants that would be suited for the same site location.

When a problem is noticed in an ornamental plant, make sure first to diagnose the problem before using a chemical control to insure that, once a control measure is put into place, it is the correct step for eliminating the problem. If you have the resources available in your area and the problem is not something that needs immediate attention, send your sample to a lab and have it diagnosed if you unsure of what you are dealing with. Once a diagnosis is made, establish what options are available to you for control or if there are even any control measures available. In the rare situation that there is no control for the specific problem, find out what should be done. Removal of the diseased species might be the best route to eliminate exposure to your landscapes or similar species in surrounding landscapes.

Once a problem is diagnosed as a positive disease occurrence, decide what chemical treatment will the most effective control for financial and aesthetic reasons. You should look at all of the possible chemicals available and assess their use for your particular problem.

PESTPOINTER

Removal of the diseased species might be the best route to eliminate exposure to your landscapes or similar species in surrounding landscapes.

PESTPOINTER

Contact chemicals work much more quickly and are generally less expensive.

If the specimen is quite large, a systemic treatment might be the only answer if there is even a control available for your particular disease problem. If the diseased specimen is small, a contact fungicide might provide a much faster and more desirable result. Contact chemicals work much more quickly and are generally less expensive.

By providing a healthy environment for all of the plants in your landscape and avoiding problems like excess watering or lack of a fertilization program, you can establish a good basis for healthy growth. Treat sick or unhealthy plants as needed or preventatively if there has been a reoccurring problem. It is possible to spray plants or treat plants in a preventative fashion if you are sure a pathogen is present or disease occurrence is likely. It is better however to only treat when absolutely necessary and use prevention as the key to success.

SUMMARY QUESTIONS

- *What preventative measures can be used to avoid the use of chemical fungicides to treat diseases of landscape plants and trees?*
- *What methods of applying chemicals are used at your facility?*
- *What is the surest way to diagnose plant disease?*
- *Why are monocultures so important in controlling disease?*

Chapter 7
Insect Problems and Insecticides Used in Turf

Insects are the most successful form of animal life on the planet. There are millions of known species that occupy each of the world's seven continents. Yes, that even includes Antarctica. Fortunately for agricultural and green industry practitioners, only a very small percentage of the known species of insects actually are damaging to cultivated plants. However, insects still represent a significant problem to managed plant systems, including turfgrass and landscapes.

Damage caused by insect pests is the primary impetus for controlling them. In ornamentals and turf, economically important insect damage can be classified as either direct or indirect. Direct damage is generally the result of feeding or from oviposition (laying of eggs) in plant tissues. Indirect damage is most often associated with anything affecting the aesthetics of the host plant. For example, sooty mold (associated with honeydew-producing insects) can make grasses/ornamentals look unsightly. Insects that cause feeding damage in turf will either possess chewing (jaw-like) or piercing-sucking (needle-like) mouthparts. These insects can be grouped or classified as surface feeders or as subterranean (below ground) feeders. Recognizing the types of damage or knowing the plant part that is being damaged can often times help in the identification of that particular pest.

To maintain healthy and aesthetically valuable turf, pests such as plant pathogens, vertebrates, and arthropods (primarily insects) must be controlled. This chapter focuses on some of the more common insect and related arthropod pests of turf. Although there are entire texts devoted to the issue of turfgrass insects, we are going to touch on the basics and offer the core information you will need to develop an insect control program at your facility. Since the types of insects that plague turf often differ from those that target landscape plants, we'll be discussing these issues in separate chapters. Certain insect and arthropod pests are going to be more problematic in some areas than in others but many of the principles and control strategies outlined herein will apply to any particular pest that plagues your area. Perhaps the best way to introduce insect problems and the appropriate control strategies is to provide some background on the pests themselves.

INSECT PROBLEMS IN TURF

Arthropods are very successful organisms and survive in virtually every known environment. They are characterized by having the presence of an exoskeleton (outer shell) and segmented bodies. Insects have three body segments (head, thorax, and abdomen), a pair of antennae, and wings (most of the time). Insect development is accomplished by a process termed metamorphosis, which can be holometabolous (complete) or paurometabolous (incomplete.) Complete metamorphosis has four distinct life cycle stages: egg, larva, pupa, and finally adult (Figure 7.1). Molting, the process of an insect shedding its old exoskeleton and forming a new, larger exoskeleton, allows insects to grow.

For holometabolous larvae, the growth stages between molts are called instars. Incomplete metamorphosis has three life cycle stages: egg, nymph (several nymphal stages usually present), and adult

Insect Problems and Insecticides Used in Turf   **281**

FIGURE 7.1   Stages of complete metamorphosis, including the larva, pupa, and adult.

(Figure 7.2). In contrast to holometabolous larvae, paurometabolous nymphs resemble miniature, wingless versions of the adult stage. For holometabolous insects infesting turf, the larval stage is usually the most destructive. Both the adult and nymphal stages for paurometabolous insects may damage turf. Common holometabolous insects found in turf include:

- *Coleoptera (beetles)*
- *Diptera (flies)*
- *Hymenoptera (ants, bees and wasps)*
- *Lepidoptera (butterflies,moths)*

The primary paurometabolous insects associated with turf are the Hemiptera, which include:

- *Chinch bugs*
- *Leafhoppers*

282 *Turfgrass Chemicals and Pesticides: A Practitioner's Guide*

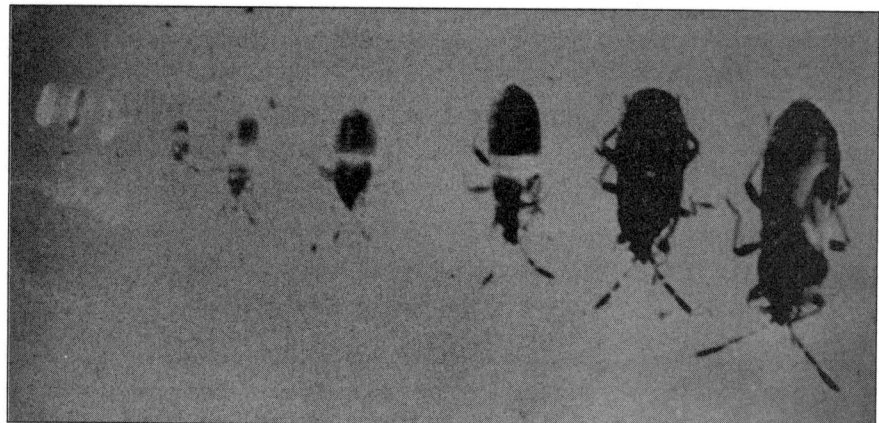

FIGURE 7.2 Incomplete metamorphosis stages, including egg, several nymphs, and the adult.

- *Ground pearls*
- *Mealybugs*
- *Spittlebugs*

Other non-insect arthropods that occasionally frequent turf include:

- *Spiders*
- *Mites*
- *Pillbugs*
- *Centipedes*
- *Scorpions*
- *Millipedes*

As mentioned earlier, insects possess an exoskeleton, have segmented bodies, and grow by molting. In contrast, non-insect arthropods are wingless, have more than six legs, and differ in the number of body segments. Spiders, scorpions, centipedes, and ticks don't directly damage turf but can be considered medically important nuisance pests. Both millipedes and pillbugs can be commonly found in or around turf but cause no direct damage. Generally, problems associated with millipedes and pillbugs deal primarily with home invasions. There are also many mite species found within the soil and some are considered turfgrass pests. Some of the most common turfgrass insect and non-insect arthropod pests are discussed below. Table 7.1 breaks down some of these key pest species by their means of metamorphosis. The discussion of each of these pests that will follow will give you some insight as to their habitats, turfgrass species they infest, physical traits of the insects, and when to expect them to be active.

TABLE 7.1 Key turfgrass insect pests and their means of metamorphosis.

Insect Type	Means of Metamorphosis	Where They Feed
Ground pearls	Incomplete	Roots of turf
Mealybugs	Incomplete	Leaves of turf
Leafhoppers	Incomplete	Leaves of turf
Spittlebugs	Incomplete	Leaves of turf
Chinch bugs	Incomplete	Stems and crowns of turf
Mole crickets	Complete	Roots and shoots of turf
White grubs	Complete	Roots of turf
Sod webworms	Complete	Leaves of turf
Armyworms	Complete	Leaves of turf
Cutworms	Complete	Roots of turf
Frit flies	Complete	Stems and crowns of turf
Crane flies	Complete	Roots and crowns of turf
Wireworms	Complete	Roots of turf
Billbugs	Complete	Stems and roots of turf
Fire ants	Complete	No feeding, nuisance pests
Mites	Molting (non-insect arthropod)	Leaves of turf

Ground Pearls

Ground pearls are members of a "primitive" family (Margarodidae) of scale insects that are considered to be potential serious pests of turf. Ground pearls most closely resemble mealybugs in body shape, color, and size (Figure 7.3). They are subterranean and feed primarily on the roots of grasses but have also been reported in other crops. They are most common as pests in Australia and the southern United States. The most commonly reported turfgrass hosts include bermudagrass, St. Augustinegrass, centipedegrass, and zoysia. Only two species of ground pearls are commonly associated with turf.

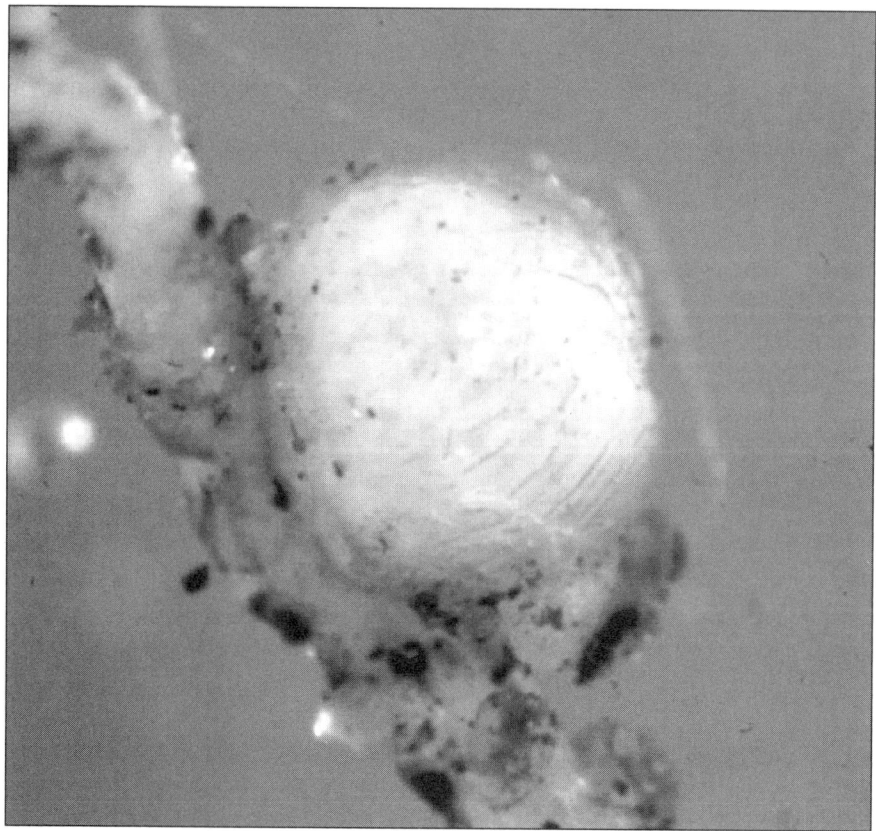

FIGURE 7.3 Ground pearls have a unique appearance in the mature form.

The adult females are approximately 0.1 inch long and generally pink in color with well-developed forelegs. They secrete a protective wax that surrounds their bodies. This wax closely resembles the waxes produced by mealybugs. Reproduction for the ground pearl can be sexual or asexual (parthenogenesis). When males do occur, they are small, brown to red gnat-like insects ranging from 0.1 to 0.4 inches in length. Eggs are generally pink in color and found in a white cottony mass full of eggs. The crawler stage (those that just hatched from the eggs) is very small, pink, six-legged, and mobile. The cyst form (immature female) is the stage generally referred to as the ground pearl. These individuals can be up to $1/8$ inch in diameter and do resemble "little golden pearls."

The overwintering life stage is the ground pearl. Females reach maturity in the spring and emerge from their cysts. The mature females locate a suitable feeding site and settle two to three inches deep in the soil. Once they settle, they begin secretion of a protective waxy coat. Females begin producing eggs (roughly 100 per female) and deposit these through the early summer. The crawlers emerge in midsummer and infest grass root tips. After the crawlers settle and begin to feed on the root tips, they develop into the "cyst" or ground pearl life stage. Generally, there is only one generation per year. If environmental conditions are not favorable, the encysted nymphs may remain in the ground pearl stage for several years. Control of ground

PESTPOINTER

In general, insecticides are not effective against ground pearls. The best control relies on good cultural practices such as watering and proper fertilization to help grasses recover from injury. Future controls will probably rely on timing of control methods to correspond with crawler flushes.

pearls can be difficult. In general, insecticides are not effective against ground pearls. The best control relies on good cultural practices such as watering and proper fertilization to help grasses recover from injury. Future controls will probably rely on timing of control methods to correspond with crawler flushes.

Mealybugs

The rhodesgrass mealybug is a grass and sedge-infesting mealybug that was first identified in the United States in 1942 by Harold Morrison. It is found in many regions of the world. In the United States, this mealybug is found from Florida through the Gulf States to California. Its name comes from its identified preference for feeding on the pasture species rhodesgrass (Figure 7.4). Since it was first discovered, rhodesgrass mealybug has been found to be a potential economic pest of rhodesgrass, johnsongrass, bermudagrass and St. Augustinegrass.

The adult females are generally are brown to purple, oval, and about $1/8$ inch long. Adults are legless and are often found within a

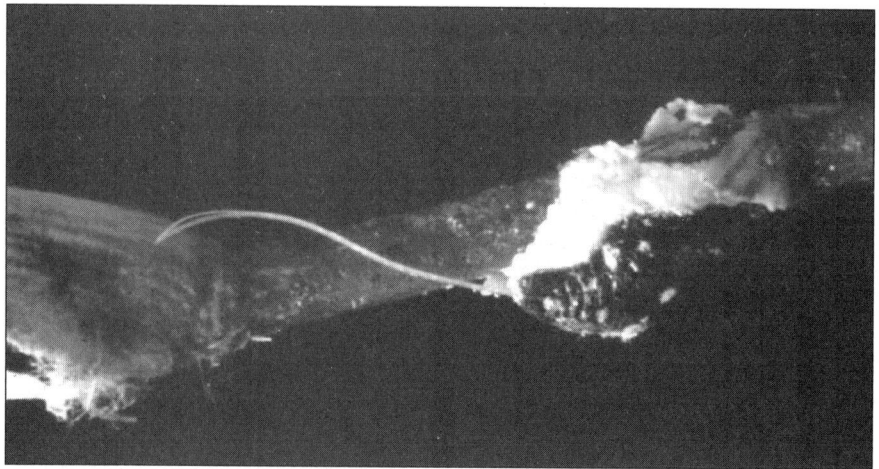

FIGURE 7.4 Rhodesgrass mealybug feeding upon its favorite host, the pasture species rhodesgrass.

felt-like sac found beneath the sheaths and nodal areas of grasses. The crawler stage is elongate-oval, yellow to cream colored, very small, and very active (Figure 7.5). This stage is the mobile life stage and still retains the presence of legs. Once the crawlers settle and molt into subsequent immature stages, the body is enclosed in the same (but smaller) felt-like sac that the females have. This felt-like sac is actually a secreted mass of wax that helps the insect avoid desiccation and offers some protection from natural enemies.

The rhodesgrass mealybug is parthenogenic (asexual reproduction) and reproduces live born young over approximately a 50 day interval. Once born, crawlers move about in search of a feeding site,

FIGURE 7.5 Arrows highlight crawler stage mealybugs on an infested grass leaf.

> **PESTPOINTER**
>
> The best option for control of mealybugs may lie in use of granular or soil drenches with systemic insecticides.

most frequently just beneath the leaf sheath at the crown or lower nodal areas. After the crawlers settle, they insert their mouthparts into the host plant and begin to feed. Molting into subsequent stages takes place and the mealybug eventually becomes sessile. The entire life cycle takes approximately 70 days with five or more generations occurring.

Cultural controls include collection and destruction of infested grass clippings. Unlike the "ground pearls," cultural control has not been proven to be practical. Likewise, biological control has offered little hope of attaining good control of this insect. The best option for control may lie in use of granular or soil drenches with systemic insecticides. For specific chemical controls, refer to the later portions of this chapter and consult agricultural extension service personnel.

Leafhoppers

Leafhoppers are common insects seen outdoors, occurring on almost all types of plants including garden crops, trees, shrubs, and grasses found throughout the United States. Leafhoppers feed primarily on the leaves and are usually host-specific. In turf, these insects are generally more of a nuisance pest, although heavy infestations can severely damage turf stands. Initial injury in turf appears as stippling on the leaf blades. Several leafhoppers also produce honeydew secretions that cause an increase in sooty mold on host plant surfaces.

There are quite a few species of leafhoppers that could potentially be encountered and body coloration ranges from tan to green to brown. Generally, adults are wedge-shaped and range from $1/8$ to $1/2$ inch in length (Figure 7.6). The immature leafhoppers are elongate, soft bodied and generally have a triangular or wedge-shaped head. Both adults and immature stages move sideways over leaf blades and stems when disturbed. Generally, leafhoppers have two to three generations per year depending on the species. Overwintering life stage varies from eggs deposited in leaf material to adults, depending on the species. Generally, highest population densities occur in mid-summer through the early fall. Leafhoppers often rely on strong wind currents for migration into new areas.

There are currently no economic threshold levels established for leafhoppers on turf. Control may be warranted when leafhoppers are found in relatively large numbers or if injury appears. Insecticide treatments are often effective, but multiple applications may be needed due to the possibility of re-infestation from other turf stands.

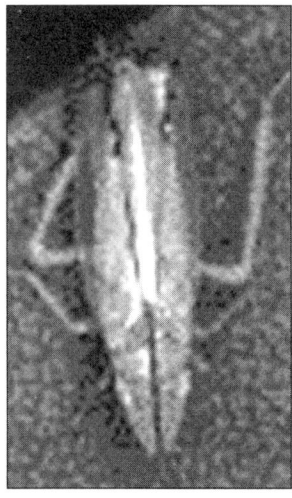

FIGURE 7.6 A mature leafhopper at rest on a plant leaf.

FIGURE 7.7 An adult two-lined spittlebug.

> **PESTPOINTER**
>
> Control may be warranted when leafhoppers are found in relatively large numbers or if injury appears. Insecticide treatments are often effective, but multiple applications may be needed due to the possibility of reinfestation from other turf stands.

Spittlebugs

The two-lined spittlebug has become an increasingly important pest to ornamentals and turf throughout the southeastern United States. It is distributed from Maine to Florida and westerly to Iowa, Kansas, and Oklahoma. It has been a sporadic pest of bermudagrass in the southeast since its discovery as a turfgrass pest. Since then, it has been associated with sporadic damage of several warm-season turfgrass species. Spittlebugs damage host plants by inserting their piercing-sucking mouthparts into plant tissue and sucking out plant nutrients. Damaged grass plants often become wilted, yellow, and then brown before death.

Adult two-lined spittlebugs are wedge-shaped, dark brown to black with two orange to red lines on the wings (Figure 7.7). They also have red eyes and legs. The abdominal color beneath the wings is red. Adults are generally $1/2$ inch long and wings are generally held in an inverted "V" shape. Immature stages are cream to ivory colored with brown heads and eyes. Immature stages produce the characteristic "spittle" mass that is associated with their name. The spittle acts as a protective barrier against natural enemies and, more importantly, against desiccation. Common turfgrasses damaged by two-lined spittlebugs include St. Augustinegrass, zoysiagrass, bermudagrass and centipedegrass. Centipedegrass tends to be the most severely damaged.

There are two generations, and possibly three, per year in the Southeast for the two-lined spittlebug, which overwinters as eggs in hollow stems behind leaf sheaths and in plant debris on the surface of the soil. Overwintering eggs hatch in March and April. After hatching, young nymphs move about and begin feeding in a humid, sheltered place. Nymphs become encompassed in a protective spittle mass during feeding. Nymphs will feed and molt through five nymphal instars with each instar lasting one to two months. Egg laying usually begins within the first two weeks following adult emergence, and the adult life span is approximately one month. Adult females generally lay between 40 to 50 eggs. The egg stage of development generally lasts no more than three weeks.

Infestations of two-lined spittlebugs are usually the result of an overdeveloped thatch layer. Management practices include de-thatching and topdressing turfgrasses, as well as scouting for flying adults and the cream colored immatures encased in spittle masses. Insecticide applications may be required if infestations are severe. Good coverage of insecticides is required and success is enhanced through mowing and proper irrigation prior to treatment. Insecticide treatments should be targeted for late in the day when the two-lined spittlebugs are on the higher portions of the grass. If possible, avoid using the combination of preferred hosts for adults (such as Japanese holly) and highly susceptible grasses such as centipedegrass.

PESTPOINTER

Infestations of two-lined spittlebugs are usually the result of an overdeveloped thatch layer. Management practices include de-thatching and topdressing turfgrasses, as well as scouting for flying adults and the cream colored immatures encased in spittle masses.

Chinch Bugs

Chinch bugs are best known for the damage they cause on sorghum and corn. They do feed on many different grasses and two species have been problematic on turfgrasses. The hairy chinch bug is primarily a pest of bluegrass, zoysiagrass, bentgrass, and fescue in the northern and upper midwestern United States. The southern chinch bug is found throughout the southeastern United States and can be a pest of many different turfgrasses but St. Augustinegrass is the primary host. Adult chinch bugs are roughly ⅛ inch long and keep their wings folded flat. Color of adult chinch bugs is generally dark brown (Figure 7.8). Wings are shiny white with triangular black markings on the outer edges and in the middle. All chinch bugs possess piercing-sucking mouthparts that they use to mechanically enter the host plant's phloem and xylem cells. The chinch bugs live in the thatch layer and readily feed on fluids from the crowns, stems, and stolons of the grasses. Mechanical damage and the insertion of toxic saliva can cause damaged grasses to yellow or even be killed. Chinch bugs often feed in small groups and damage from group feeding can cause dead patches to occur in the turf. Southern chinch bugs can frequently be found in sunny areas, areas under drought stress, and adjacent to sidewalks and driveways.

Adults overwinter in clumped grasses or in the thatch area of grasses. Adults begin to become active once daytime temperatures approach 70°F. Feeding and mating occur once chinch bugs are active, and adult females lay eggs between the grass sheaths or in soft soil surrounding the grass roots. Once eggs hatch, the reddish-colored nymphs begin feeding on the turf and undergo five nymphal instars. In the northern or Midwestern areas of the country, one to two generations of chinch bug will occur in a given year. However, if conditions are right (mild winter followed by dry spring and summer), three generations may occur in those regions. In the southern states, multiple generations can occur (between five and seven).

A key in management of chinch bugs is being able to detect them when they are present. There are a couple of methods avail-

Insect Problems and Insecticides Used in Turf **293**

FIGURE 7.8 Mature chinch bugs are dark brown while the nymphs are lighter in color.

able for detecting infestations of chinch bugs. One is to part the grass near chlorotic patches and examine the area near the base of the turf. Take more than one sample to insure that you get an accurate read to determine if the chinch bugs are present. A second method is to use what is termed the "flotation technique." For this method, take a can (plastic or metal) and cut out both ends. Push one end of this device three to four inches into the soil on the border between damaged and healthy areas of turf. Slowly fill with water to the surface level of the grass. After about five

> **PESTPOINTER**
>
> Culturally, prevention of drought stress by proper watering is helpful in dealing with the chinch bug. Reduction in the thatch layer will also reduce the population levels.

minutes, chinch bugs will float to the top if they are present. In areas that border sidewalks or pavement, chinch bugs can be seen migrating to new patches of grasses when populations are at high densities.

In the many areas, naturally occurring fungal pathogens will attack chinch bugs. Control can be good with the fungi but it is often times inconsistent because one area may have the right micro-climatic conditions for the fungi to persist while another area may not have a suitable environment. Culturally, prevention of drought stress by proper watering is helpful in dealing with this pest. Reduction in the thatch layer will also reduce the population levels of chinch bug. Reduction of the thatch layer can be accomplished by not over-watering or over-fertilizing the grass as well as proper mowing. Depending on the area of the country, resistant or tolerant varieties of turfgrass may exist. Chemical controls may be warranted if more than 25 chinch bugs are detected within one square foot of turf.

Mole Crickets

Mole crickets get their name from their burrowing and subterranean life habits. Mole crickets have emerged as the most damaging insect pest of turf in many southeastern states.

There are currently five species of mole crickets found in the United States:

- *Short-winged mole cricket*
- *Southern mole cricket*
- *Tawny mole cricket*
- *Northern mole cricket*
- *European mole cricket*

Of these five species, southern mole cricket, tawny mole cricket, and short winged mole crickets cause the most damage to turf and pasture grasses. The northern mole cricket is the only native species of the five and is rarely considered to be economically important.

All mole crickets have a similar appearance in being gray to brown in color, wings present, and digging front legs (Figure 7.9). The short winged mole cricket differs in appearance because of the short wings. The immature stages for all species resemble the adults, but are wingless and much smaller. Like other Orthopteran insects such as grasshoppers or cockroaches, mole crickets are omnivores. The southern mole cricket feeds mainly on organisms in the soil but can occasionally cause feeding damage on strawberries and tomatoes. The tawny mole cricket and short winged mole cricket feed primarily on vegetation from vegetable crops, weeds, and grasses. Turfgrasses that are most affected by tawny mole cricket include bahiagrass and bermudagrass and those most affected by short winged mole cricket are St. Augustinegrass and bermudagrass.

Mole crickets overwinter, as adults or large nymphs, deep in the soil. Overwintering nymphs complete their development to the adult life stage and then mate in the early spring as soil and air temperatures begin to warm. Males construct (except for the short winged species) special chambers and produce calls to attract mates. Once mated, females construct chambers beneath the soil and deposit 30 to 50 eggs. Eggs generally hatch within two to four

FIGURE 7.9 The adult mole cricket has powerful digging legs.

weeks. Soon after hatching, nymphs dig their way to the soil surface. Nymphs undergo many instars before becoming adults. Generally, there is one generation per year. Both nymphs and adults create networks of tunnels up to a foot beneath the surface of the soil. Most feeding occurs during the night hours. Both tawny and southern mole crickets are attracted to lights and can be seen in large numbers during the mating season when they are searching for mates.

Management of mole crickets requires the monitoring of populations and matching control efforts to the mole cricket's life cycle and demands of the turf. Different strategies may be required at different times of the year. The spring poses some difficulties in control because:

1. Most of the mole crickets are adults that cause both tunneling and feeding damage.

PESTPOINTER

Fall is a good time to make note of the locations of mole cricket populations in order to better target areas for treatment the following spring.

2. This is often the dispersal time when mole crickets are likely to spread.
3. Weather conditions are unpredictable and mole crickets can migrate deeper in the soil, rendering treatments ineffective. Insecticide treatments are also ineffective against the egg stage, which is prevalent after spring mating.

Treatments should only be applied in areas with severe damage and where cuttings will be made early in the season before the grass has the chance to recover. In areas where damage is not severe, practitioners should mark (or map) the area and plan for treatments after the eggs have hatched.

Mole cricket eggs hatch and nymphs emerge during late spring and early summer. Newly emerged nymphs are very small, and damage is difficult to detect. Small nymphs spend a majority of time near the surface of turf, and pesticide applications should focus on this vulnerable life stage. Adults are once again prevalent and comprise half the population during the fall. However, adult movement slows down at this time and damage is not as significant. Control during this time frame can be difficult, and only areas with critical damage should be treated. Fall is also a good time to make note of the locations of mole cricket populations in order to better target areas for treatment the following spring.

White Grubs

Some of the most serious turf pests can be found in the beetle family Scarabaeidae. There are over 1500 species of scarab beetles found in North America. They can be found in a variety of habitats, but many are associated with turf and pasturelands. The larval forms are referred to as white grubs. They possess a characteristic C-shaped body and three pairs of functional legs just behind the head capsule (Figure 7.10). The larval body color is generally white to cream-colored with a dark area near the posterior end of the abdomen. The adults are various sizes and colors, depending on the species.

Scarabs have a complete life cycle (eggs-larvae-pupae-adults) and the amount of time it takes from one generation to the next will vary from species to species. Some species may require three to four years to complete their life cycle but generally most complete their life cycles in one year. For many of the white grub species, eggs are laid one to two inches below the soil surface, usually in May or June. Eggs hatch July through August, and larvae begin feeding on turfgrass roots. The larvae will continue to feed on the roots of turf and molt two times prior to becoming dormant during the winter. As winter approaches, the grubs will burrow deeper into the soil to avoid the affects of cold temperatures. As temperatures begin to rise in the spring, the grubs will return to the root zone and begin feeding on the grass roots once again. Feeding will continue through much of the early portion of spring with pupation occurring in late spring. New adult beetles will emerge in May-June. For species taking more than one year to complete a life cycle, the extra time will be spent as grubs feeding throughout the season.

Unlike most white grubs, Ataenius beetles undergo two generations per year (Figure 7.11). Overwintering adults emerge from protective areas (ground covers) in March through April. Adults mate and lay eggs during late April through May. Larvae emerge in late May or early June, feed, and molt twice. In July, mature larvae cease feeding and burrow deeper into the soil to undergo pupation. New adults emerge in late July and into August. Soon after emergence, adults mate and begin laying eggs. The developmental process is

Insect Problems and Insecticides Used in Turf 299

FIGURE 7.10 Classic white grub shape is in a "C" form with the legs near the front.

FIGURE 7.11 Turfgrass Ataenius larva and mature beetle, shown with a penny for size comparison.

then repeated with a second generation of adults emerging in September-October and moving to overwintering sights.

Some white grubs associated with turf feed on the roots of the grasses while others may uproot turf and push up small mounds of soil. Feeding damage to the roots can cause large patches of turf to die. Grub damaged turf is easily pulled loose since most of the roots have been eaten away (Figure 7.12). The most prevalent feeding damage is seen with the larger and more mature grubs that are found during the summer and early fall months. Heavy infestations of grubs and severe damage can give the soil a loose and spongy texture. Grub infestations generally result in yellowing and wilting of the turfgrass, even when it is watered properly.

Like many of the other turf pests, inspection will play an important role in managing white grub pests. Grub inspection should be

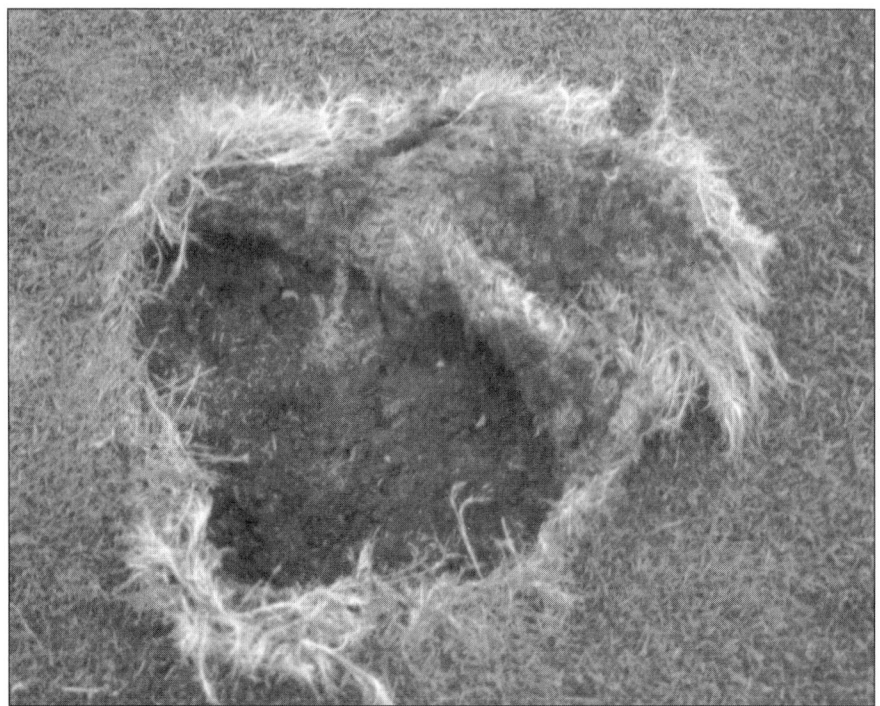

FIGURE 7.12 Grub damaged creeping bentgrass turf.

PESTPOINTER

Control through use of pesticides generally controls approximately 90 percent of the beetle grubs but better control can be realized when proper timing of applications is made.

done by using a spade to cut three sides of a one-foot section of turf, taking the spade and peeling back the cut section to inspect the roots for damage and presence of grubs. Many adult scarab beetles are attracted to night lights, so light trapping or monitoring what is present at night lights may forewarn growers of the types of beetles that are present.

Control of white grubs can be difficult with the subterranean habits of the larvae. Biological control using bacterium from Bacillus popillae, Bacillus thuringiensis (Bt), and parasitic nematodes does exist and can be successful. Both Bacillus bacteria can work at controlling beetle grubs but may require extended periods of time before control is realized. Control through use of pesticides generally controls approximately 90 percent of the beetle grubs but better control can be realized when proper timing of applications is made.

Sod Webworms

Several turfgrass moth pests are referred to as sod webworms. The sod webworms are native to North America, and generally restrict their feeding to grasses. The majority of the sod webworms are distributed throughout the United States. Several species of sod webworm exist. The silver striped webworm is widespread throughout the eastern United States. The bluegrass webworm ranges from Massachusetts west to Colorado and south to Texas and Florida. The bluegrass region of Kentucky and Tennessee is especially vulnerable to the bluegrass webworm. More abundant populations of the larger sod webworm occur in the northern United States. Even though the distribution of the striped sod webworm is widespread, it is most prevalent in the region including Illinois, Tennessee, and Pennsylvania. The corn root webworm is more problematic as a turf pest in the north central and northeastern United States, but can range further south. The subterranean webworm occurs throughout the northern United States, but is particularly a pest in the Pacific Northwest. The tropical sod webworm is distributed throughout the Caribbean and the southeastern United States.

The primary food sources for sod webworms are generally grasses. Except for the tropical sod webworm, sod webworms have been most commonly reported as pests of Kentucky bluegrass, perennial ryegrass, fescues, and creeping bentgrass. Even though turfgrass is generally a preferred host, most sod webworms will also feed on corn, wheat, rye, and oats. The corn root webworm will also feed on seedlings of corn and tobacco. In addition to feeding on the crown and roots of grasses grown for seed production, subterranean webworms destroy cranberry plants by girdling their roots. The tropical sod webworm feeds on bermudagrass, centipedegrass, St. Augustinegrass, zoysiagrass, and bahiagrass.

Adult moths of sod webworms have a snoutlike projection (Figure 7.13). Moths generally have a light gray to tan coloration with various patterns of silver, gold, yellow, brown, and black on their forewings. The body size of adult moths ranges from 0.5 to 0.7 inches with a wingspan of 0.5 to 1 inch. Eggs are very small and coloration may range from a white or beige to orange. Most larval sod webworms may be green, brown, or gray. The tropical sod webworm is light green. Except for the corn root webworm and the subterranean webworm, dark spots are scattered over the bodies of most sod webworms (Figure 7.14). Mature larval head capsules are usually light brown, but color varies from light brown to black. The number of larval instars varies from six to ten depending on species. Mature larvae of some species can be over an inch long. Pupae are located

Insect Problems and Insecticides Used in Turf 303

FIGURE 7.13   The adult sod webworm moth is tan and non-showy with a long snout.

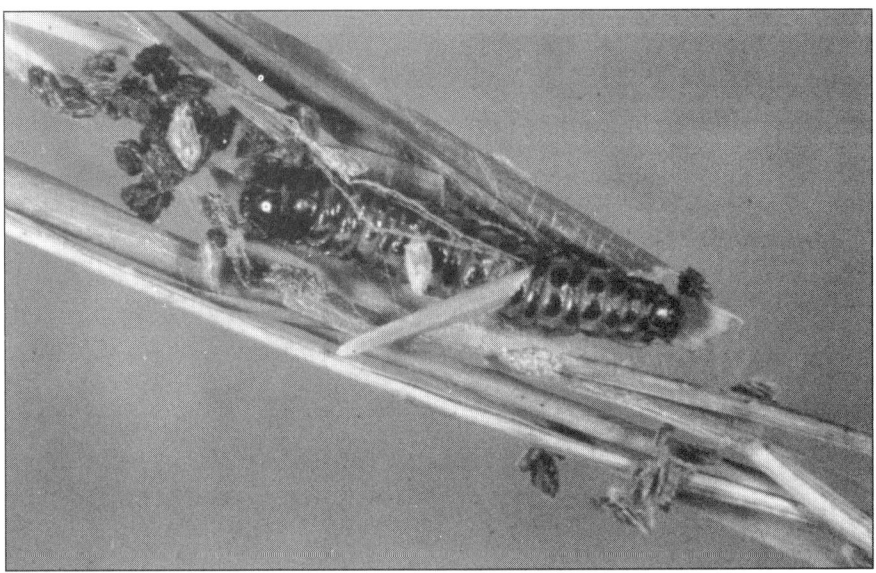

FIGURE 7.14   The usually spotted sod webworm larva will chew on grass leaf tissue and leave behind frass, or ground up plant tissue.

in the soil and are usually encased in cocoons of various debris, soil particles, and fecal pellets. Pupae mature from a pale yellow color to a dark brown prior to adult emergence. Sod webworm pupae are generally about ½ inches long.

Female sod webworm adults deposit eggs on turf at night. Eggs hatch within seven to ten days. The total time from egg laying to adult emergence for most sod webworm species is six weeks. Larval activity for sod webworms is also generally nocturnal, but occasionally larvae may be observed feeding on a cloudy day. Larvae construct silken tubes for feeding and as protection from natural enemies and desiccation. First instar larvae focus their feeding on the surface tissue of leaves and stems. Older larvae will devour all of the leaves and stems. The feeding habits of the tropical sod webworm differ slightly from the other sod webworm species. Tropical sod webworm larvae chew notches in the leaves of their host. Feeding by sod webworms will first be evident by small patches of yellow to brown leaves in an otherwise healthy lawn. Initial signs of infestation are frequently confused with disease.

Sod webworms overwinter as larvae in the soil. The subterranean webworm and the corn root webworm only have one generation a year. The larger sod webworm has two or three generations in the eastern United States, but only one generation in the Pacific Northwest. Depending on the weather conditions in a region, two or three generations of the silverstriped sod webworm, bluegrass sod webworm, and striped sod webworm occur each year. Populations of the tropical sod webworm are present throughout the year in southern Florida, although this species cannot overwinter north of central Florida.

Several natural enemies are beneficial in the control of sod webworm, including birds and various insect predators. The fungus Beauveria, the microsporidia Nosema, and the bacterium Bacillus thuringiensis (Bt) can also infest larvae of sod webworms. The benefit of these natural enemies will be decreased for turf requiring pesticide applications for the control of other pests. Control thresholds and recommendations for insecticide use for control of sod webworm

> **PESTPOINTER**
>
> Because sod webworms become more of a serious pest under drought conditions, proper irrigation can also help control outbreaks.

vary by region. In order to estimate a population, solutions of liquid soap can flush out larvae in a section of infested sod. Various cultural practices, including selecting a vigorous variety of turf, could assist in control of sod webworm. Because sod webworms become more of a serious pest under drought conditions, proper irrigation can also help control outbreaks.

Armyworms

The common name armyworm is derived from the feeding habits of their larvae. During population explosions of armyworms, these pests move in army-like groups from one feeding source to another, destroying all available host plants in their path.

The armyworm is native to:

- *Eastern United States*
- *Southern Canada*
- *Mexico*
- *Northwestern South America*
- *Pacific coast of the United States*
- *Western Europe*

The fall armyworm is native to Central America, tropical South America, and the West Indies. The fall armyworm migrates each year into southern regions of the United States.

Armyworms feed on a variety of grasses including:

- *Small grains*
- *Pasture grasses*
- *Corn*
- *Sugarcane*
- *Millet*
- *Sorghum*
- *Fine turfgrasses*

Turfgrass is generally not the preferred food source for armyworms, but nevertheless population explosions can occur. Growing seasons following a drought are at higher risk for armyworm infestations, and the highest populations are generally present early in the spring. Turf located near susceptible field crops is especially vulnerable.

The fall armyworm has a wide variety of potential hosts, but prefers grasses. The fall armyworm is commonly considered a pest of bermudagrass in the southeastern United States, but it also feeds on fescues, ryegrass, bentgrass, bluegrass, small grains, and grass crops (Figure 7.15). Fall armyworm infestations are generally more problematic with increased populations during late summer or fall. Dry and hot summers with consistent southerly winds can increase the likelihood of fall armyworm invasions further north.

Adults of armyworms and fall armyworms have forewings with a brown to gray color and hind wings with a whitish to brownish background. The wingspan of these species is approximately two inches. Armyworm moths have a white dot at the center of the forewing. Fall armyworm moths have a mottled appearance, and females have a white spot near the tip of the forewing.

Insect Problems and Insecticides Used in Turf   **307**

FIGURE 7.15   Fall armyworm damage in a turfgrass area.

Mature armyworm and fall armyworm larvae are approximately 1½ to 2 inches long. Fully developed armyworm larvae have a yellow or gray-colored background and dark longitudinal stripes. Young armyworm larvae are pale green. Fall armyworm body color ranges from green to almost black. Fall armyworm larvae may be similar in appearance to armyworms, but fall armyworms have a light-colored inverted Y-shape on a dark head capsule. Fall armyworm larvae also have four black dots on the dorsal side of each abdominal segment (Figure 7.16). Both armyworms and fall armyworms pupate in the soil, and the pupa changes from reddish brown to almost black prior to adult emergence. Both armyworm and fall armyworm pupae are approximately ½ inch long.

Adult armyworm moths deposit bands of a few to several hundred eggs in between the sheath and a blade of grass at night. Two to five generations per year can occur depending on climatic conditions. The duration of egg development ranges from three to 24 days.

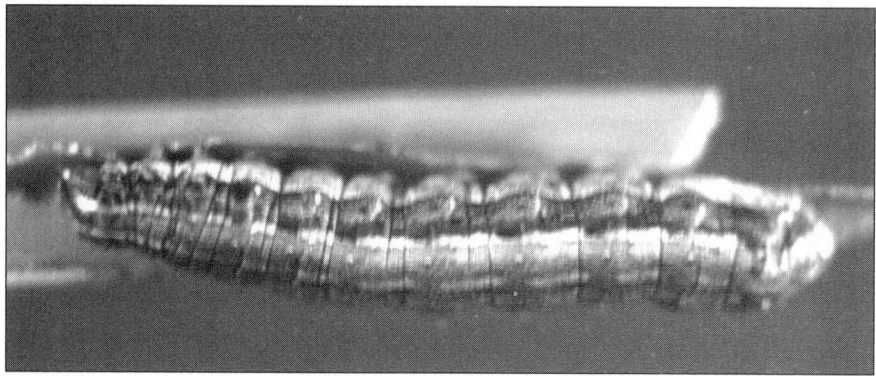

FIGURE 7.16 A fall armyworm larva, with characteristic dark color and dots on the top of each segment.

Six to nine larval instars occur, with fewer larval instars occurring at warmer temperatures. Developmental time ranges from 20 to 48 days for larvae and from seven to 40 days for pupae. In the southern portion of its range, the armyworm may overwinter as a larva or a pupa. For northern armyworm populations, the armyworm either overwinters as a larva or is present in migrant populations.

Adult female fall armyworms frequently lay clusters of 50 to several hundred eggs on light-colored objects in or nearby turf. Larval feeding activity appears to be more abundant early in the morning and late in the afternoon. The entire life cycle from egg deposition to adult emergence for fall armyworm ranges from 26 to 45 days. A continuous breeding cycle occurs in south Florida and south Texas. The fall armyworm is dependent on migration in order to infest more northern climates.

In contrast to non-stoloniferous grasses such as ryegrasses and fescues, bermudagrass and other stoloniferous turfgrasses are capable of recovering from an infestation. Dishwashing detergent can be used to estimate armyworm and fall armyworm populations. Some infestations may not be severe enough to warrant treatment. If treatment is required, mowing maximizes the surface area covered for a pesticide application and may destroy some larvae present in the turf.

PESTPOINTER

If treatment for armyworm infestation is required, mowing maximizes the surface area covered for a pesticide application and may destroy some larvae present in the turf. Control measures should be applied either early in the morning or late in the evening, whenever larval feeding activity is higher.

Control measures should be applied either early in the morning or late in the evening, whenever larval feeding activity is higher. Consult your local extension service for pesticide recommendations for your region.

Cutworms

Cutworms derive their common name from the feeding habit of their larvae. Larval cutworms cut blades at ground level. Large populations of cutworms can destroy a lawn. Circles of dead or depressed grass spots are the initial signs of an infestation. Black, bronzed, and variegated cutworms are the three species of interest to turfgrass managers. The black cutworm and variegated cutworm are widely distributed throughout North and South America, Europe, and Asia. The black cutworm is also found in Africa. The bronzed cutworm is commonly found in North America in areas east of the Rocky Mountains.

Black cutworms usually feed on grasses, and can be particularly problematic for the bentgrass turf used on golf course greens. Black cutworms are also considered to be pests of corn and vegetables. The variegated cutworm feeds on a wide variety of hosts including grasses, vegetables, and ornamental plants. Variegated cutworms may be a pest in any lawn mixture and have specifically been noted as a pest of bentgrass (Figure 7.17). Bronzed cutworms are primarily a pest of bluegrass, but can also be a pest of clover, small grains, and corn.

Adults are dark gray to black for black cutworms, and reddish-brown for variegated cutworms. Bronzed cutworm adults have a dark brown band across the forewing and their color ranges from rose, purplish-gray, to brown. The wingspan of adult cutworms ranges from 1¼ to 2 inches, and the wings fold flat over the abdomen at rest. Mature larval cutworms range in length from 1¼ to 2 inches and in width from ¼ to ½ inches. Body color of cutworm larvae ranges from gray to brown to almost black (Figure 7.18).

Black and variegated cutworms have two to seven generations per year, depending on climatic conditions. Only one generation per year occurs for the bronzed cutworm. Cutworms mate at night, and adult females deposit their eggs on a suitable host. Each adult female cutworm is capable of laying approximately 300 to 2000 eggs. Black and variegated cutworms may overwinter as either larvae or pupae in the northern extent of their range. Adult bronzed cutworms are present during late fall and the egg is the overwintering stage.

FIGURE 7.17 A cutworm larva crawling on creeping bentgrass turf.

Insect Problems and Insecticides Used in Turf **311**

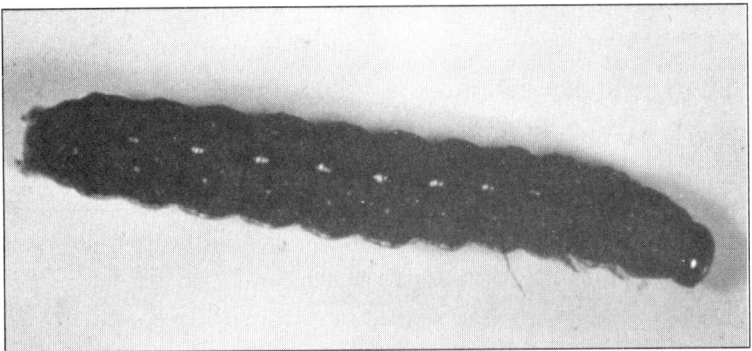

FIGURE 7.18 Cutworm larvae can range from light in color to very dark, like the one shown here.

Lawn areas near susceptible field crops are especially vulnerable to infestation. Flushing turf with a soap solution can be used to estimate population densities. Contact your local extension service for control recommendations specific to your region, but effective control can generally be obtained by using either contact or stomach insecticides. Biological control agents may also effectively control cutworm infestations. During the early stages of feeding by young larvae, the bacterium Bacillus thuringiensis (Bt) may effectively control cutworms. Cutworms have also been managed on golf greens using nematodes in the genus Steinernema.

PESTPOINTER

Biological control agents may also effectively control cutworm infestations. During the early stages of feeding by young larvae, the bacterium Bacillus thuringiensis (Bt) may effectively control cutworms. Cutworms have also been managed on golf greens using nematodes in the genus Steinernema.

Frit Flies

The frit fly is widely distributed in the Northern Hemisphere, including areas in Europe, Russia, and North America. The majority of populations occurring in North America are located in the Midwestern United States. In the United States, the frit fly is a pest of golf greens, home lawns, and reed canary grass, a common hay crop grown in areas like Virginia. In addition to damaging golf greens, the adult frit flies can be considered a nuisance pest due to their attraction to white clothing or golf balls.

Turf hosts of frit flies include bluegrass, bentgrass, fescue, and ryegrass. Shortly mowed grasses that are frequently irrigated, such as on a golf course, are especially attractive to frit flies. Larvae feed on crowns and stems of the host plant, and feeding damage can result in death of a central leaf. Initial feeding by larvae is evident by a yellowing of the center of the plant. Damage is usually first evident on collars and aprons of golf course greens. Occasionally, leaf mining by frit fly larvae may also occur.

Frit fly adults are very small, gray to black flies with a triangle pattern located on the head (Figure 7.19). Larvae are light yellow maggots with two prominent black mouth hooks (Figure 7.20). Three larval instars occur, and third instar larvae are up to 1/4 inch long. Pupae are smaller than the larvae and are usually located at the base of grasses. Adult females usually lay their eggs in between the leaf sheath and

PESTPOINTER

The adult life stage of fruit flies has been the primary focus of management efforts. Pesticide applications should be aimed at preventing egg laying by the adult female.

Insect Problems and Insecticides Used in Turf **313**

FIGURE 7.19 Frit fly adults are very small in size and gray or black in color.

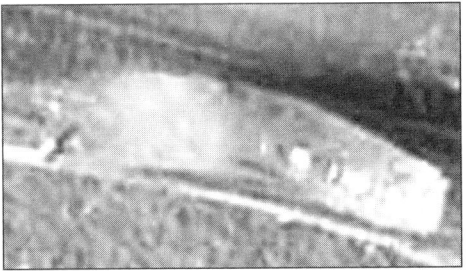

FIGURE 7.20 Frit fly larvae (maggots) have a yellowish color.

stem of a suitable host. The number of generations per year ranges from two to four or five, depending on latitude. Frit flies overwinter as larvae at the base of host leaves. The adult life stage of frit flies has been the primary focus of management efforts. Pesticide applications should be aimed at preventing egg laying by the adult female. Chemical control methods have traditionally focused on the use of contact insecticides. Region specific degree-day models have assisted in predicting adult activity.

Crane Flies

The European crane fly, native to Europe as might be expected, is a pest of lawns and golf courses in the Pacific Northwest. European crane flies prefer grasses of lawns and pastures as their host, but various crops and ornamental flowers are also susceptible. Damage to turf is caused by the larval stage. Larvae usually feed below ground on roots and crowns of host plants. Young grass seedlings are also mechanically destroyed by the movement of larvae from below to above ground at night. When larvae are feeding above ground, they will consume all of the grass plant. Upon initial infestation, some areas of the lawn may appear unhealthy and yellowing. Large infestations may completely destroy a lawn.

Adult European crane flies have two wings and a slender body. Crane flies have extremely long legs that are at least two to three times as long as the total body length (Figure 7.21). The body length of the European crane fly ranges from $\frac{1}{4}$ to $\frac{1}{2}$ inch. Four larval instars occur, and larvae are light gray to greenish-brown. Irregular

---

PESTPOINTER

Damage caused by crane flies to an unhealthy lawn may be more severe. Various birds, other predators, microorganisms, and parasites may naturally keep European crane fly populations at an acceptable level.

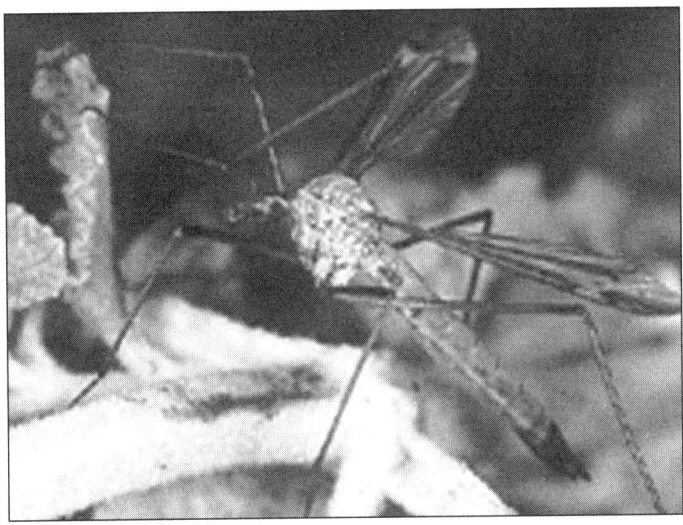

FIGURE 7.21 Crane fly adults have disproportionately long legs.

black specks occur on the larvae, and the gut canal is visible through the cuticle. The pupa forms inside the shell of the last larval instar. Unlike many homometabolous insects, crane fly pupae are capable of movement, due to spines on the last five abdominal segments.

Only one crane fly generation occurs each year. The European crane fly overwinters as a third instar larvae and larval activity is generally apparent in the Pacific Northwest from February to the middle of May. Larvae feed voraciously until they are mature. Mature larvae remain active until pupation, which usually occurs in late August. The pupal, adult, egg, first instar, and second instar stages occur from late August through September. Crane fly larvae can be sampled by digging in the soil to a depth of three inches. Consult your local extension service for specific pesticide recommendations. Generally, treatment of a healthy lawn may be necessary if more than 25 larvae occur per square foot. Damage caused by crane flies to an unhealthy lawn may be more severe. Various birds, other predators, microorganisms, and parasites may naturally keep European crane fly populations at an acceptable level.

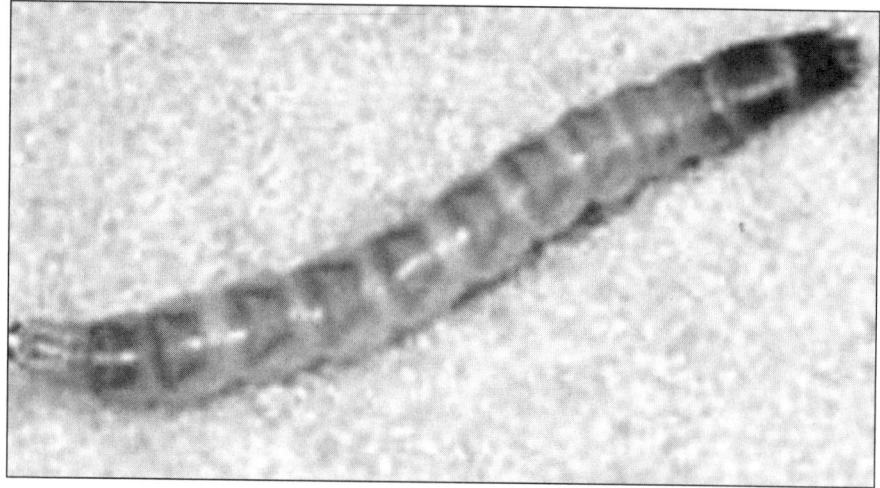

FIGURE 7.22 The wireworm larva is the most injurious to turf.

Wireworms

Wireworms are the larvae of a beetle species. Feeding by wireworm larvae on grass roots results in withered grass and dead areas in a lawn. During the spring and summer, adult females lay eggs on the roots of grasses. Developmental time for larvae ranges from two to six years. Larvae are a brownish color ranging in length from $1/2$ to $1 1/2$ inches (Figure 7.22). Larvae have characteristically robust outer shells. Mature larvae pupate in the summer and then overwinter as adults. Adults range in length from $1/2$ inch to just over 1 inch. Due to their ability to return their bodies to a right side up position with a quick motion and a clicking sound, adults are called click beetles.

Billbugs

Weevils in the genus Sphenophorus are collectively known as the billbugs. Many billbug species exist, including the bluegrass, hunting, Denver (or Rocky Mountain), and Phoenix (or Phoenician) billbug.

Even though the bluegrass billbug can be found throughout North America, it tends to be more of a pest in the northern United States. The hunting billbug is also distributed throughout the United States, but is an important pest mainly in the Southeast. Populations of the Denver billbug are only found in the Rocky Mountain region and the northern Great Plains. The distribution of the Phoenician billbug appears to be restricted to California and Arizona.

Billbug larvae damage turf by burrowing and feeding in the center of grass stems. After consuming the majority of the crown, larger larvae feed on roots. Depending on population levels, infestations may result in sporadic or extensive areas of dead turf. Early infestations may resemble dollar spot disease, but stems of turf damaged by billbugs are filled with digested plant material. Billbug damage may also superficially resemble chinch bug or white grub damage (Figure 7.23). The bluegrass billbug prefers feeding on Kentucky bluegrass, but perennial ryegrass and fescue are also suitable hosts. Zoysiagrass and bermudagrass are most susceptible to hunting billbug and Phoenician billbug infestations. Other potential hosts for hunting billbugs include:

- *St. Augustinegrass*
- *Centipedegrass*
- *Nutsedge*
- *Crabgrass*
- *Signal grass*
- *Barnyard grass*
- *Wheat*
- *Corn*
- *Sugarcane*
- *Pensacola bahiagrass*
- *Leatherleaf fern*

318 Turfgrass Chemicals and Pesticides: A Practitioner's Guide

FIGURE 7.23 Billbug damage can resemble other types of insect damage to turf.

Cool-season turfgrasses are the most common hosts for the Denver billbug.

Adult billbugs have the characteristic hardened forewings common to beetles. As member of the weevil family, adult billbugs also have elongated snouts and elbowed antennae (Figure 7.24). Billbugs are a gray, black, or brown color. Billbug size may range from 0.3 to 1/2 inches for different species. Two billbugs have distinctive marking on the prothorax in the form of raised letter-like shapes. The hunting billbug has a Y-shape, and the Phoenician billbug has an M-shape. The bluegrass billbug has straight rows of double pits on the forewings. In contrast, the Denver billbug has heart-shaped rows of double pits. Also, the Denver billbug is a shiny black color. Billbug larvae are white with brown head capsules. Larvae are robust, legless, and approximately 1/4 to 1/2 inches long at maturity. Pupae are located in the soil and are typically 1 to 2 inches long. Mature pupae are reddish-brown.

Insect Problems and Insecticides Used in Turf **319**

PESTPOINTER

Identifying the species of billbug and the level of infestation is crucial to determining control measures. Control methods for billbugs may include selection of resistant turfgrass cultivars.

Most billbugs appear to have one generation each year, but more than one generation per year may occur in some instances. The hunting, Denver, and Phoenician billbugs generally overwinter as larvae, but the bluegrass billbug overwinters as an adult. The hunting, Denver, and Phoenician billbugs complete their larval development in the spring. Billbug adults feed on grasses, mate, and then females lay their eggs in the grass stems. Females usually lay eggs for at least several weeks.

FIGURE 7.24 Adult billbugs have a characteristic snout that gives them their name.

Identifying the species of billbug and the level of infestation is crucial to determining control measures. Control methods for billbugs may include selection of resistant turfgrass cultivars. Bluegrasses, ryegrasses, bermudagrasses, and zoysias have some resistance to billbug infestations. Fertilizing and irrigating lawns may reduce billbug damage. Steinernema and Heterorhabditis nematodes have successfully controlled billbug larvae and adults. Contact your local extension service for specific pesticide recommendations for billbug control.

Fire Ants

Another key insect species worth mentioning is the fire ant. These insects don't feed on turf but can create unsightly mounds and can be very dangerous to staff and/or patrons at your facility. The red imported fire ant was introduced in the United States around 1930 and has now spread to infest more than 300 million acres in the southern United States. In addition to the potential allergic reactions of individuals stung by fire ants, this invader disrupts native ant species and habitats. The red imported fire ant is also considered a pest of agricultural crops by feeding and damaging germinating seeds and buds of developing fruits. Even though the red imported fire ant is a pest, it is also a predator which can reduce populations of other pest insects.

The red imported fire ant does not directly damage the grass, but the presence of colonies can be problematic for individuals receiving stings. The large mounds are also aesthetically displeasing for the home lawn or golf course setting. Mounds may also create problems in routine field operations such as mowing and galleries constructed by colonies can also affect moisture retention in soil (Figure 7.25).

The most prevalent diagnosis of fire ants is the presence of mounds in lawns and soil. The mounds are usually dome-shaped, and the soil on the mound will appear to be hard or "crusty" with no visible external opening present. Fire ants have tunnels that radiate (8 to 20 inches below surface) from the mound in which foraging ants may travel. Foraging by workers typically occurs in early morning, late afternoon, or at night.

FIGURE 7.25 Fire ant mounds can be very troublesome and can quickly grow in numbers.

Fire ant colonies consist of brood, workers, winged individuals, and a queen(s). The individual fire ant worker, the sterile work force caste of the colony, is rather small (up to $1/4$ inch long) with a red to reddish-black color (Figure 7.26). Workers forage for food, construct and maintain the colony, and care for the brood. The immature stages for ants can be seen if a mound is "disrupted" or busted apart. The stages are the eggs (small white-silvery elongate ovals), larva (tiny cream colored, grub-like stage), and pupae (appear as white motionless ants with legs, etc.). The queen(s) are generally larger than the other workers. A queen can live up to seven years and can produce approximately 1500 eggs/day throughout most of her life. Mature colonies may reach population levels of hundreds of thousands of individuals. The reproductive, winged individuals (females and males) emerge from the mounds once they are fully developed adults. Winged individuals can be produced throughout the year, but are more prevalent in late spring and early summer. These individuals emerge and swarm on a warm day following rainfall. After mating, either individual females or multiple females form new colonies.

There are many contact insecticides and baits available for fire ant control. Control over large areas with numerous colonies is frequently unsuccessful. Individual insecticide treatments to mounds or broadcast treatments of insecticides or baits are generally effective. Treatments with contact insecticides may yield variable results. If the mound is

322 Turfgrass Chemicals and Pesticides: A Practitioner's Guide

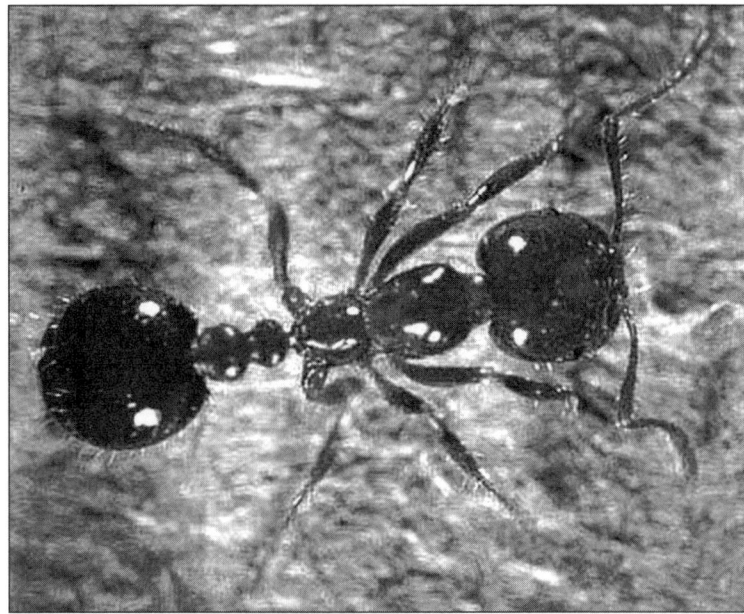

FIGURE 7.26 Fire ant workers have very narrow bodies and are generally red and black in color.

disturbed or if the chemical does not penetrate deep enough into the mound, some of the ants may relocate and construct a new mound. Broadcast treatments can be successful in suppressing foraging activ-

PESTPOINTER

Control over large areas with numerous colonies is frequently unsuccessful. Individual insecticide treatments to mounds or broadcast treatments of insecticides or baits are generally effective.

ities and establishment of new colonies, but a single application may not be sufficient. Recent developments in biological control programs using phorid flies are currently being used in many southeastern states. These phorid flies will stalk and harass fire ant workers, and deposit an egg at the base of the ant's head. Once the egg hatches, the larva bores into the head region and completes its development. The parasitized ant is rendered ineffective and overall colony work is disrupted with the presence of the flies. For further information about control methods, contact your local extension service.

Mites

Mites are a broad arthropod group with species represented in most all environmental habitats and most (if not all) crops. Mites that are damaging to ornamental plants and crops, such as the two-spotted spider mite, are well known as pests. There are many mites found in association with grasses. Most are not pests but occasionally a few species can be problematic. The clover mite is an occasional turfgrass pest. They are very small, reddish in color, and generally appear as tiny red moving spots. Clover mites feed on grasses and clover and are especially abundant in well-fertilized lawns. Insecticidal soaps used as a direct spray can be used to treat these infestations. Perhaps the best-known pest mite for turf is the bermudagrass mite (Eriophyes cynodoniensis Sayed). This mite is primarily a pest in the southern states where bermudagrass is normally grown. Coarse cultivars of bermudagrass seem to be preferred and finer textured varieties are less preferred. Bermudagrass mites are very small and cannot be seen by the naked eye, but reduction in length between host nodes and nodes with a "tufted" appearance are an indication that mites are present. Like many mites, very little is known about the life history of this pest. Management practices may include the use of fine textured varieties that are not preferred by the mite and, when possible and practical, maintenance of a low cut turf. Contact your local extension service for pesticide recommendations for your region.

> **PESTPOINTER**
>
> Management practices for mites may include the use of fine textured varieties that are not preferred by the insect and, when possible and practical, maintenance of a low cut turf.

INSECTICIDES

Insecticides are often what people think of when they hear or read the general term pesticide. Visions of cockroaches, exterminators, and vehicles with large insects on top of them are all too common in the public's perception of insecticides and the need for insect control. Within the green industry and agriculture in general, insecticides represent a much more focused and specialized effort to preserve yields and/or crop aesthetics. Among pests, weeds may cause the most crop losses worldwide but insects cause the most physical damage. Problems with insects in crops date back to biblical times, where there were stories of swarms of locusts that could wipe out an entire crop. As noted in Chapter 1, some of the earliest reported pesticides were those targeting a variety of insect pests, although these early materials were more oriented towards preservation of human health. Insecticides began being used for plant protection back in the 1800s and then exploded in popularity in the 20th century. The advent of better and better materials, coupled with easier means of applying them, made insect control more feasible than ever before.

The road to success for insecticides has not been void of a few bumps here and there. One of the most notorious pesticides of the modern era, DDT, was introduced to the world in the mid-20th century and was a mainstay for insect control for the next 30 years. However, unforeseen side effects of this insecticide drew heavy criticism from scientists like Rachel Carson and eventually both the public and politicians. The aftermath of the DDT fiasco included not only its banned use in the United States but also new waves of pesticide regulations and organizations like the EPA to help enforce them. Today, DDT is still used by many agriculturists in other countries but it has long since been replaced here by newer technologies. Today's insecticides, like any other pesticides, come under heavy scrutiny during the product development phase and are much more benign than their predecessors. The good news is that there are many quality materials to choose from and for a wide variety of insect pests.

What materials are currently out there and what are their niche uses? This is the question that practitioners really want answered when insect problems arise and require immediate action. Turfgrass, being an aesthetic crop, often has very low thresholds for the type of destructive feeding that certain insects will bring. Insecticides must conform to regulatory guidelines but must also act quickly and effectively if they are to be useful to the turfgrass practitioner. Table 7.2 lists some of the common insecticide families that include turfgrass products.

The general mode of action for most insecticides is to act as a neurotoxin in affected insects, usually resulting in loss of bodily functions before death. The differences in the chemical families listed above can, however, be significant. Issues such as how the target insect gets exposed, persistence of the insecticide, effects on other animal organisms, and toxicity to humans are all key issues that have allowed new chemistries and products to develop over the years. Let's now

PESTPOINTER

The general mode of action for most insecticides is to act as a neurotoxin in affected insects, usually resulting in loss of bodily functions before death.

briefly discuss some key characteristics of each of the families listed in Table 7.2.

Biological insecticides are naturally occurring microorganisms that can induce many of the same effects in target insects as other insecticides. Most of these agents are bacteria, the rapid development of which makes for suitable production of pesticides. Specific products containing these organisms can be obtained and applied but perhaps the greatest niche for biological insecticides has come from their unique use in genetically-engineered field crop plants. Many field crops today like cotton have been infused with organisms like Bacillus thuringiensis (Bt), enabling these crop plants to "naturally" withstand insect predators. This approach to crop production can save considerable costs associated with applying other insecticides during the growing season.

Carbamate insecticides include some of the most widely used materials worldwide, the key example being carbaryl, commonly sold as Sevin®. These materials have a similar neurological impact on target pests as other insecticides but their best attribute may be their low toxicity to animal organisms, including humans. This feature has enabled carbamates like Sevin® to permeate the over-the-counter market and thus dramatically increasing their frequency of use. Carbamates

TABLE 7.2 Common insecticide families.

Insecticide Family	Example Insecticides
Biological	Bacillus thuringiensis (Bt), Beauvaria bassiana, Heterorhabditis bacteriophora, Myrothecium verrucaria, Spinosad
Carbamate	Bendiocarb, carbaryl, fenoxycarb
Chlorinated hydrocarbon	Dicofol, lindane
Organophosphate	Acephate, chlorpyrifos, diazinon, ethion, ethoprop, fenamiphos, fonophos, isofenphos, malathion, trichlorfon
Pyrethroid	Bifenthrin, cyfluthrin, cypermethrin, lambda-cyhalothrin, permethrin, pyrethrin
Miscellaneous	Fipronil, halofenozide, hydramethylnon, imidacloprid

> **PESTPOINTER**
>
> The greatest niche for biological insecticides has come from their unique use in genetically-engineered field crop plants.

are also known for their effectiveness on a wide variety of insect species, further increasing their marketability.

Chlorinated hydrocarbons are one of the oldest classes of insecticides, with products that were first produced more than 50 years ago. Many of these products are now banned from use, the most noteworthy of these being DDT. There are few products that remain today within this chemical family. This highly lethal mode of action makes these insecticides fairly dangerous to use and is, in many ways, responsible for the limited number of available products for today's practitioner. However, low manufacturing costs make chlorinated hydrocarbons still popular to this day, especially in other countries where pesticide restrictions aren't as stiff as they are here.

Organophosphates have also enjoyed a fairly lengthy history. As the name might imply, these materials all contain phosphorus. As a group, they have a dubious historical distinction of being close analogs of many of the nerve agents, past and present, used in chemical weapons. This characteristic makes organophosphate insecticides some of the more dangerous to humans and other animal organisms. Fortunately, formulation chemists have created products that minimize this potential toxicity by reducing volatility and thus exposure to applicators and patrons. Organophosphates might have been banned altogether were it not for the fact that they are not persistent in the environment. This short shelf-life increases safety to both humans and other animals that may come into contact with these materials.

Pyrethroid insecticides are analogs of the botanical insecticide pyrethrum and have also been around for better than 50 years. However, they have perhaps benefited the most from technological developments. The trend for many insecticide families has been fewer and fewer products as time unfolds. This could be due to either better alternatives or, in many cases, to regulatory restrictions. Pyrethroids have enjoyed the opposite trend. There are now more available than ever. This is largely due to the fact that the first pyrethroid insecticides were unstable in sunlight and couldn't last long enough to be effective in field conditions. Today's products are comparatively much more stable when exposed to sunlight and this has greatly increased the usefulness of these highly effective materials.

The miscellaneous category of insecticide families represents many obscure chemistries. The most common materials in this group, fipronil, halofenozide, hydramethylnon, and imidacloprid are each chemically unique but all possess neurotoxic effects to insect pests that are similar to other insecticides. Fipronil and halofenozide have developed a niche by being effective against pests that have developed resistance to more common insecticide families like the pyrethroids. Fipronil is also particularly effective against difficult mole crickets. Hydramethylnon is not commonly used for plant predatory insect pests but is effective for control of a variety of nuisance pests, most importantly fire ants. Imidacloprid is modeled after one of the earliest known insecticides, nicotine. Since we haven't yet replaced citronella candles with lit cigars for outdoor gatherings, this material may be the closest current rendition we have of this older concept.

CONTROL STRATEGIES FOR HERBIVOROUS INSECTS

Current control measures for insects and related arthropod pests focus on the use of biological control agents, arthropod pathogens, pesticides, and cultural control methods. Use of pesticides is probably the most common type of control sought by homeowners and lawn care professionals. Both growers and homeowners alike should consult with local extension services in their states when considering pesticides for control of pests. Biological control relies on either natural enemies (including pathogens) of a pest or biological pesticides such as Bacillus thuringiensis. Examples of cultural controls include thinning of thatch layers, destruction of infested materials and proper watering/fertilization. The use of alternative control strategies and proper timing of pesticide applications reduces residual effects on the environment. A proper integrated pest management system involves the following steps:

- *Inspection of hosts and surveys for potential pests*
- *Proper identification of the pest and associated damage*
- *Understanding the behavior and biology of the pest*
- *Use of proper control methods when justified*

While there are numerous alternatives to insecticides for insect control, these pesticides are still a solid foundation to build upon when dealing with insect problems. This discussion, which will include the use of insecticides to help develop a solid insect management strategy, will be based upon the four steps outlined above.

A key first step to good insect control is to know your pests. Past experience can give you a good clue as to what pests are most likely to appear and where but proper diagnosis is always critical for both insecticide selection and application timing. Scouting for insect pests and early signs of damage can give you a valuable clue as to their activity and allow for efficient control timing. Actual sightings of the insects can work for some species but subterranean feeders are often unseen so look for other signs like declining turf quality or high numbers of foraging bird species to clue you in to an insect's presence.

Understanding the pests you are trying to control is also a key part of good insect pest management. Issues such as which life cycle stage actively feeds, where the feeding occurs, and when the

> **PESTPOINTER**
>
> Diagnosis is always critical for both insecticide selection and application timing. Scouting for insect pests and early signs of damage can give you a valuable clue as to their activity and allow for efficient control timing.

feeding is most aggressive will improve your chances of successfully controlling these pests with minimal turf damage. Many extension resources are available for this sort of information, in addition to that which is covered in this text. Activity of insects will vary by region so local resources and peer input are also good parts of your information network. Insecticide labels will also offer key insights as to when to time applications so reference this information and expect this knowledge from distributors who sell you these products.

Once you've identified a pest and a need to control it, an adequate insecticide must be selected. There are many insecticide products available for control of a wide assortment of turfgrass insect pests. Table 7.3 outlines some of these key insecticides and the corresponding trade names under which they are marketed and sold.

> **PESTPOINTER**
>
> Insecticide labels will also offer key insights as to when to time applications so reference this information and expect this knowledge from distributors who sell you these products.

TABLE 7.3 Common turfgrass insecticides and some of the most common trade names that accompany them.

Insecticide Common Name	Example Trade Names	Pests Targeted
Acephate	Orthene, Pinpoint, Velocity	Most major turfgrass insect pests
Bacillus thuringiensis	Biobit, Dipel, Caterpillar Clobber, Worm-ender, MVP II, Condor	Armyworms, cutworms, sod webworms
Beauvaria bassiana	Naturalis-T	Most major turfgrass insect pests and non-insect arthropods
Bendiocarb	Turcam	Most major turfgrass insect pests and non-insect arthropods
Bifenthrin	Talstar	Most major turfgrass insect pests and non-insect arthropods
Carbaryl	Grubtox, Insecticide V, Fluid Insecticide, RegalFate, Sevin	Most major turfgrass insect pests and non-insect arthropods
Chlorpyrifos	Dozens of trade names (Dursban the most notable)	Most major turfgrass insect pests and non-insect arthropods
Cyfluthrin	Tempo	Most major turfgrass insect pests
Cypermethrin	Cynoff	Most nuisance insect and non-insect arthropod pests
Diazinon	Diazinon, Evict	Most major turfgrass insect pests and non-insect arthropods
Dicofol	Kelthane	Mites
Ethion	Ethion	Chinch bugs, sod webworms
Ethoprop	Mocap	Billbugs, mites, mole crickets, sod webworms, white grubs
Fenoxycarb	Award	Ants, fire ants
Fipronil	Chipco Choice	Mole crickets
Fonofos	Crusade	Billbugs, chinch bugs, mole crickets, sod webworms, white grubs

Continued on next page

TABLE 7.3 *(continued)* Common turfgrass insecticides and some of the most common trade names that accompany them.

Insecticide Common Name	Example Trade Names	Pests Targeted
Halofenozide	Mach 2	Armyworms, billbugs, cutworms, sod webworms, white grubs
Heterorhabditis bacteriophora	Cruiser	Armyworms, billbugs, chinch bugs, cutworms, sod webworms, white grubs
Hydramethylnon	Amdro, Siege	Ants, fire ants
Imidacloprid	Merit (several formulations)	Billbugs, mole crickets, white grubs
Isofenphos	Insecticide IV, Oftanol	Armyworms, chinch bugs, cutworms, crane flies, fire ants, mole crickets, sod webworms, white grubs
Lambda-cyhalothrin	Battle, Scimitar	Most major turfgrass insect pests and non-insect arthropods
Lindane	Lindane	Armyworms, billbugs, cutworms, leafhoppers, mole crickets, sod webworms, spittlebugs, white grubs, most nuisance insects and non-insect arthropods
Malathion	Malathion	Billbugs, leafhoppers, most nuisance insects and non-insect arthropods
Permethrin	Astro, Dragnet, Flee, Perm-X	Most major turfgrass insect pests and non-insect arthropods
Pyrethrin	Exciter	Armyworms, billbugs, chinch bugs, cutworms, mole crickets, sod webworms, many nuisance insects and non-insect arthropods
Spinosad	Conserve	Armyworms, cutworms, sod webworms
Trichlorfon	Dylox, Grub Beater, Proxol	Cutworms, mole crickets, sod webworms, white grubs

PESTPOINTER

Insecticide selection should also include the spectrum of pests a product will control. Many facilities have multiple insect pests to contend with so an insecticide that targets multiple species may result in fewer applications that need to be made.

Insecticide products can act in one of two ways: by contact with the pest or by ingestion. Which of these means of activity is more desirable depends a lot upon the pest and where on the plant it feeds. Leaf and crown feeding pests are better candidates for contact insecticides, as long as the pests are actively feeding. Root feeders are better candidates for ingestion-based insecticides, which absorb into treated plants and move into tissues the pest will try to eat. Be sure to follow label specifications for proper timing and placement of applications. Insecticide selection should also include the spectrum of pests a product will control. Many facilities have multiple insect pests to contend with so an insecticide that targets multiple species may result in fewer applications that need to be made.

Because insects can reproduce very quickly, they have the ability to adapt and become resistant to some insecticides. This challenge becomes more likely when the same types of insecticides are used over a period of several growing seasons. Many different insecticide families and products effectively target common pests so insecticide rotation is a necessary strategy to avoid resistance. The concept of resistance was introduced in our discussion of herbicides but the problem is more likely with insects and diseases because of their shorter life cycles. As is the case for other pesticides, proper insecticide rotation means alternating families, not just products within a given family. An insect that becomes resistant to an organophosphate insecticide will likely be resistant to most or all organophosphates. Switching to a pyrethroid, carbamate, or chlorinated hydrocarbon will result in a slightly different mode of action that the resistant insect should succumb to. Resistance is a critical part of insect pest management and can help avoid unnecessary problems with insect damage and ineffective pesticide use.

Once insecticides have been used, it is critical to evaluate their effectiveness for future planning. Dead insect specimens and recovering turfgrass may both be ways to tell an insecticide application has been a success. Because insecticide applications are usually targeted towards a particular stage of an insect's life cycle, other life cycle stages may survive the effects of an insecticide. This can be due to lack of feeding by some life cycle stages or absence from the targeted areas. Most insects do follow predictable development patterns but those that have multiple generations each year may be more at risk for escapes and future damage. Planning your applications around insect life cycles is thus very critical and an important detail to acquire from either your insecticide label or from other information resources. Proper success monitoring can help you make future decisions for a particular insect pest or group of pests. By following this key sequence of steps outlined above, you will be well on your way to developing an insect management program that fits the needs of your facility.

SUMMARY QUESTIONS

- *What insect pests are most common at your facility? Do they require treatment?*

- *How would you rate your diagnostic skills for insects? If your rating is not as good as you might like, how can you improve these skills?*

- *What control measures have you used for insect pests? Have you tried or considered cultural methods instead of insecticides?*

Chapter 8
Insect Control in Trees and Landscapes

Insects are one of the oldest living organisms on the planet. It seems that "bugs" have many ways to survive hard times and they are capable to adaptation to most any environmental changes (Figure 8.1). It is this capability to adapt that makes insects a constant challenge to control in landscaping and ornamental trees. The problem with insect control in general right now is that many of the insecticides we have used and depended on for years are now in the process of being banned all together. This means that new chemicals will need to be developed along with new strategies for controlling insects so that we keep them under control. If new methods for controlling insects

PESTPOINTER

The problem with insect control in general right now is that many of the insecticides we have used and depended on for years are now in the process of being banned all together.

336 Turfgrass Chemicals and Pesticides: A Practitioner's Guide

FIGURE 8.1 Mosquitoes have been proven along with other insects to have existed since the time of dinosaurs.

are not developed, the future for ornamental tree and shrub species could change a great deal.

There are a great number of different insects that can feed on or do damage to ornamentals. Like disease pathogens, many insects are host specific to the types and varieties of trees that they invade or feed on. I should point out that in the ecosystem that exists for ornamental plants, insects can also act as carriers for diseases and viruses that spread in plants. Insects are able to reproduce at a high rate and have a great ability to build up resistance to chemicals in a short amount of time. Maybe it is the combination of the characteristics of insects that have allowed them to survive since the beginnings of time.

Insects can be categorized into many different families and further distinguished by genus and species. However, the more practical categorization for insects in landscaping plants is based on the

> # PESTPOINTER
>
> The more practical categorization for insects in landscaping plants is based on the type of damage that they do to plants or their destructive nature.

type of damage that they do to plants or their destructive nature. There are boring insects, chewing insects, and sucking insects, all of which have their own methods for destruction and the havoc they wreak (Figure 8.2). Insects can also be rated on a mobile or non-mobile characteristic which can help to determine how severe the immediate risk is to plants in the area. Insects that are limited to a larva-like stage or even a crawling stage can only move so far at a very slow rate. Many insects molt into adult stages that allow the insects to have wings and fly to new areas at the point of maturity. These are some of the more important characteristics to the level of risk to plants.

The classification of the insects is also important when it comes to the actual chemical controls that are being used. Certain types of insects respond well to the various chemical controls while other insects may not respond to the chemicals nearly as well or at all. A lot of this has to do with the mode action of the chemical versus the characteristic of the insect that is allowing it to resist treatment. A bug with a hard casing may have defenses against predators, as well as chemicals, compared to that of a soft-cased insect that could be easily killed by either. There are so many variables when it comes to the actual characteristics of insects, that it would be best to research an entomology book to find out what those differences are if you are experiencing a particularly difficult insect infestation. Insects are capable of building resistance regardless of their physical attributes but some characteristics do make certain insects more durable, so to speak.

FIGURE 8.2 Grasshoppers are a leaf feeding insect that can be destructive in nursery crop production. One reason for the rapid destruction caused by these insects is that they have wings and are very mobile.

There are several alternatives for insect control but in most situations where treatment is needed, chemicals are the most feasible scenario. Furthermore, our best resource in learning what will control certain insects would be the pesticide label. The pesticide label will list most of the insects that can be controlled by that particular product. However, the insect species is so vast that there are actually new types of insects discovered each day, so not all labels reflect each and every insect that may be affected by that particular product. Since new problems develop and insects can migrate from one side of the country to the other, it is possible to encounter a new species of insect on any given day. Some insects pose little threat to the plants in their environment while others have no known chemical control and are very destructive in nature. It is possible for an insect to be very destructive and also not be host specific when it comes to the damage that they cause. This means that an insect can migrate much faster because they are able to feed on multiple varieties of trees and shrubs.

PESTPOINTER

Insects are capable of building resistance regardless of their physical attributes, but some characteristics do make certain insects more durable.

In the last few years, the EPA, in conjunction with state agricultural institutions, has sought to ban some of the most effective products in use today for insecticide control. Dursban is probably the most commonly known insecticide that was taken off of the shelves for use in lawns and on ornamental trees and shrubs. Dursban was considered to be a very safe product for several years until it was discovered that the chemical was causing symptoms similar to lead poisoning in people. These symptoms were noticeably worse in small children. This was a scary discovery considering that Dursban was commonly used in the home for the control of cockroaches. I know when I was a golf course superintendent I relied heavily on Dursban for the treatment of almost every troublesome insect we had in greens. It was also one of the few products I found little resistance to when used in lawn care or ornamental shrubs spraying. It was considered the backup for other chemicals that were not working so well. Dursban, along with many other chemicals, is in a continually expanding group of insecticides that eventually will not be usable by any of the public. Many of the more popular and well-known chemicals that are currently being used fall into that category and may also soon be put on the list of non-usable products.

Diazinon has also been banned for most uses and Sevin®, Malathion, and several other household names potentially could also be added to that list. Each year, several chemical patents expire and these chemicals are being are being pulled from our shelves and

> **PESTPOINTER**
>
> As chemicals are removed from the list of usable products, the possibility of resistance becomes more of a problem. It is a good idea to rotate the products you are using each year for chemical controls.

we are told that we cannot use them anymore. One of the problems with chemicals being taken off the market is that insects are very quick to adapt and respond to chemicals. When chemicals are overused or not alternated, resistance is always a possibility. As chemicals are removed from the list of usable products, the possibility of resistance becomes more of a problem. It is a good idea to rotate the products you are using each year for chemical controls. Many insecticides work across the board on a wide range of different insects. However, if the same products are used from year to year, resistance is much more likely with any insect.

As with plant diseases, one of the most effective ways to prevent insect damage is to use varieties of plants or cultivars that tend to resist insects altogether. Many plants are able to build their own defenses against insects, making them a much more desirable choice for your landscaping needs. For instance, many people avoid junipers and various types of evergreens because they have a problem with insecticide damage and they feel that other types of plants should be used that do not require constant spraying. Cultivars that resist insect damage are as good a control as the best chemicals in the world. Now, using cultivars does not mean that you will never have an insect infestation for no apparent reason. As I mentioned, insects are resilient creatures that can adapt and survive in any and all environments. It is possible for an insect to spawn a whole generation of offspring that are completely adapted to a new environmental stimulus. This stimulus could be the lack of a food source, presence of a new predator, or use of a certain chemical. The commonality is that insects are able to adapt in a short period of time to anything that might have been a control mechanism before.

Control of insects can be created through cultural mechanisms or through various cultural practices. By changing an insect's "comfort zone," you can create a situation where bugs will move to other locations. Movement of insects is a dangerous strategy that can cause future problems in the environment. Now it is difficult to do such things but modification of environment can be done by changing light, wind, sound, moisture, or any other variables that make up the insects' environment. Insects respond to stimuli that many other plants and animals do not respond to. Insects also are known to use many variables for communication. To disrupt a bee hive, you simply take away the queen. The same is true for termite hives.

Insects have a chain of command and a social order that, when changed, can ruin the entire hive or colony. This is the case for many of the different insect categories. Ants, for instance, have a social order that keeps the colony going strong. Many insect groups have a social order where individuals in the group have their own responsibilities in the population. If that order is disrupted or the queen is killed, it is possible to destroy the entire population. This is one strategy or method that can be used to combat colonized insect problems.

PESTPOINTER

Modification of environment can be done by changing light, wind, sound, moisture, or any other variable that makes up the insects' environment.

PESTPOINTER

Many insect groups have a social order where individuals in the group have their own responsibilities in the population. If that order is disrupted or the queen is killed, it is possible to destroy the entire population. This is one strategy or method that can be used to combat colonized insect problems.

Another interesting thing that drives insects is the response to various pheromones that are used for communication and even mating. These pheromones can be used against the insects in traps and baits that allow the insects to be captured. This has been a proven defense system that has been valuable against many insects that were typically hard to control. In some cases, insecticides were useless or could not be used for fear of damage to less harmful insects that existed in the same ecosystem. One of the reoccurring problems with insect control is the possible damage to the other animals in the food chain.

The use of traps has been very successful in fighting gypsy moths and other hard-to-control insects that did not respond well to chem-

PESTPOINTER

Pheromones can be used against the insects in traps and baits that allow the insects to be captured. This has been a proven defense system that has been valuable against many insects that were typically hard to control.

ical controls. Since the method of trapping has been successful, it does leave options for entomologists who are struggling with the growing list of insecticides that are being taken off of the market each year. The one good thing from new insecticide development is that it is less likely to have resistance from a given species since it does take insects a period of time to adapt to a chemical control. The bad part of chemical development is finding something that will actually work.

TROUBLESOME INSECTS

One of the most troublesome insects that I fight in ornamental shrubs are bagworms (*Thyridopteryx ephemeraeformis*). Bagworms are typically associated with damage to evergreens but it is a big misconception that they do not feed on deciduous species. Actually, although they seem to prefer evergreens (Figure 8.3), bagworms are very fond of some deciduous species. I often get calls about Japanese maples that are totally devoured by bagworms.

Bagworms hatch in late May or in early June and the larva immediately start feeding on the plants in that area. These larvae will often reoccur in the same locations from year to year because the larvae are able to winter over in the mothers' bags. If any untreated carcasses are left behind, it is possible anywhere from 500 to 10,000 new eggs can be left to hatch the following year. Bagworms build a protective sack or bag that surrounds the worm which only emerges to feed on foliage or needles of the plants that they are located on. Diazinon was a good chemical control but is one that is on the list of chemicals that will no longer be available in retail trade after this year. There are many other chemical controls that can be used for bagworm control such as Sevin, Malathion, or a bacterial insecticide known as Bt (*Bacillus thuringiensis*). Older bagworms are much more difficult to control with insecticides and I have seen resistance to many of the chemicals on the market. No matter what chemical you use, make sure you rotate products from year to year if possible.

FIGURE 8.3 Bagworms feed primarily on evergreens but they do like some types of deciduous species as well.

The next very common insect you will encounter in the landscape is an insect called scale (Figure 8.4). Now most homeowners do not even realize that scale is an insect so it is not uncommon for scale to get way out of hand before it is even noticed. Scale has a variable appearance but for the most part there will be a waxy or coarse shell that will cover the bark. Sometimes scale will appear to be a blue coarse coating and other times it would have a reddish or white shiny appearance to it. The colors are variable

Insect Control in Trees and Landscapes 345

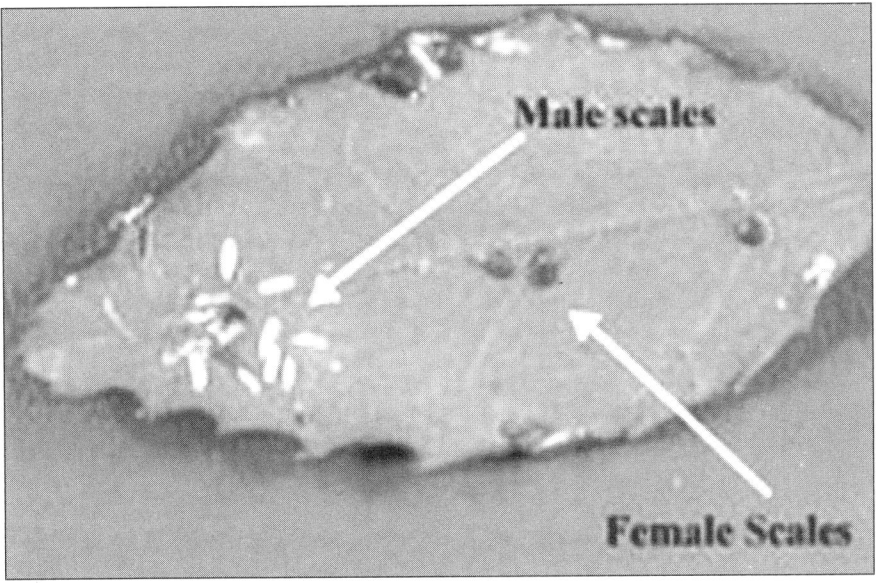

FIGURE 8.4 There are many different types of scale that feed on plants. Some are host specific while others invade many species.

depending on the type of scale and what type of plant the scale is on. Scale can be treated with a general purpose insecticide but one of the best ways to control scale is by using dormant oil in the early spring before leaf emergence on the tree or shrub. Scale can occur on many types of trees and shrubs, both deciduous and evergreen.

PESTPOINTER

Scale can be treated with a general purpose insecticide but one of the best ways to control scale is by using dormant oil in the early spring before leaf emergence on the tree or shrub.

346 *Turfgrass Chemicals and Pesticides: A Practitioner's Guide*

Spider mites are a problematic insect-like pest that occurs in some evergreens (Figure 8.5). Spider mites do not infect all types of evergreens and their presence is often mistaken for many other problems that are not even pest related, such as drought or water stress. One very common host of these pests is the dwarf Alberta spruce that is commonly used in foundation plantings and flower gardens. The easiest way to detect the presence of these pests is by placing a plain sheet of white paper under the browning areas on your

FIGURE 8.5 Although this plant appears to have a problem with a drought condition, it is actually suffering from spider mites. Home owners make the mistake of watering which can make the mites spread and survive easier.

evergreens. Then tap the plant repeatedly, knocking some material loose from the plant. You can see these small pests, even though some are smaller than a pin head, moving around on the sheet of paper. You might have to repeat this test several times to find the spider mites but typically one good shaking of the plant will cause some of the pests to fall out of an infected plant. Treatment can consist of a general insecticide such as Sevin® or Malathion.

An insect that I always hate to deal with in many parts of the Midwest are bores. There are many types of bores but they are very difficult to detect and even harder to control. There are several different types of bores and many are host-specific. One of the most common types of bores is the bronze birch bore which is responsible for wiping out most of the white birch (*Betula papyrifera*) in the Midwest (Figure 8.6). Bores seem to infect the birch around the 20th year of age. In younger trees, the bark separation is not believed to be significant enough to allow bores to enter the plant and feed. It is very difficult to tell if a birch or any other tree is suffering from bores until areas in the top of the tree may suddenly die for no apparent reason. Many types of boring insects are the larvae of beetles or moths that tunnel through bark until they strangle or girdle the conductive tissues of the tree, causing eventual death. What happens is that the tree will eventually die from either starvation or lack of water. A good control would have been an insecticide containing chlorpyrifos (Dursban). Imidacloprid is a newer possible control for some bores and it has great systemic properties.

PESTPOINTER

Many types of boring insects are the larvae of beetle or moths that tunnel through bark until they strangle or girdle the conductive tissues of the tree, causing eventual death.

FIGURE 8.6 This rare white birch has managed to reach maturity but is showing signs of damage in the top of the tree that could be signs of the bronze birch bore.

A wide range of leaf-feeding insects can occur in landscaping plants. Aphids, mites, thrips, and leafhoppers are all moderately destructive insects that can cause damaging effects to the leaves of many tree species. I am lumping these insects together because there are many different types of these insects and some are host-specific while others are not. The results from the presence of these insects will vary by species but almost all can be controlled with a general

> # PESTPOINTER
>
> Most leaf feeding insects must attack in large numbers before they even cause noticeable damage; however, if the feeding of a foliar insect is heavy enough, it can cause defoliation of the specimen or death if it occurs in repeated years.

purpose insecticide such as Malathion. Most leaf feeding insects must attack in large numbers before they even cause noticeable damage; however, if the feeding of a foliar insect is heavy enough, it can cause defoliation of the specimen or death if it occurs in repeated years. Since most of these insects only attack leaf structures, their immediate presence is only cosmetic. If there are isolated numbers of the insects or only parts of a tree infected, treatment is not a necessity. Aesthetic and economic thresholds will indicate the necessary treatment point.

Not all insects pose a threat to landscaping plants. For example, one insect symptom that poses a purely cosmetic problem in the landscape would be gall, caused by bladder gall mites. Now, I have fielded many phone calls about such leaf galls that are caused by mites or other insects. The bladder gall mites really cause no damage to the tree with the exception that the damage they cause just looks bad. Once the galls are formed on the trees there really is nothing that can be done about them anyway. The galls cause the leaves to look bad, due to the coverage of rough-looking bumps almost like a rash. These mites are too small to be seen with the naked eye so treatment before the galls occur is also unlikely unless the problem has occurred from year to year. I mentioned several other leaf-feeding insects that would not necessarily be deadly to the tree or shrub they are feeding upon. While in many cases insect damage is purely a cosmetic issue, this does not mean that you should not treat for damage (Figure 8.7).

350 Turfgrass Chemicals and Pesticides: A Practitioner's Guide

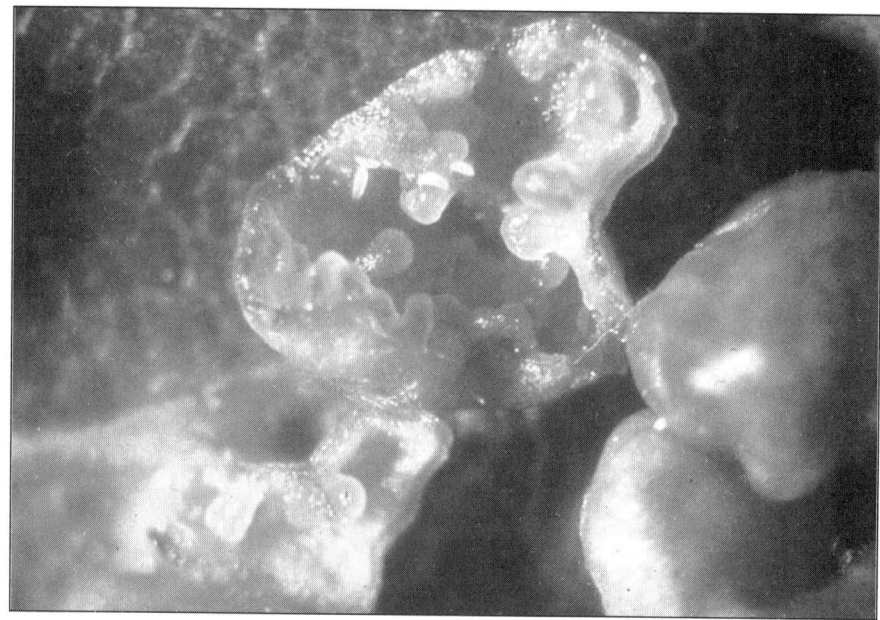

FIGURE 8.7 These bladder gall mites may look bad but they really have no negative effect to the plant.

Many responses of the tree are confused for something else when actually they are responses of the plant to an insect problem. A gall is simply a growth response of the tree which could be caused by an insect or a fungus. Do not assume that the presence of a knot, gall, or canker is caused by a fungal problem. It may still be necessary to have a sample of the injured material tested to determine what type of disease or insect is at the source of the problem in the specimen. Wasps can even cause odd growths on trees, so do not take anything for granted when diagnosing a problem in any ornamental plant.

One insect comes to mind that always gets a lot of attention from homeowners and landscapers. This would be the presence of webworms which again may not be that harmful to the specimen. It is true that webworms can defoliate a tree in a short period of time and there is a higher risk in younger specimens. Unless complete defoliation occurs in repeated years, there is seldom a risk for death of

PESTPOINTER

Many responses of the tree are confused for something else when actually they are responses of the plant to an insect problem.

the tree. Other than the obvious appearance of the insect and the fact that parts of the tree can be defoliated, it is not necessary to treat for webworms. They are unsightly but often, by the time they are noticed, it is too late to save the leaves that have been devoured.

Treatment for insects is typically performed with the use of foliar sprays which require a high pressure sprayer to apply the product. Since insecticides do pose a personal and environmental risk, this is now not necessarily the preferred method for taking care of ornamental insect problems. There is a chemical with the trade name of Merit™ (Imidacloprid) that is now being used with an insecticide injector. This chemical was originally used in turf for the control of grubs but it has since become a very useful insecticide for many other insect problems. This systemic insecticide is now one of the most commonly used and popular insecticides currently on the market.

PESTPOINTER

Other than the obvious appearance of the insect and the fact that parts of the tree can be defoliated, it is not necessary to treat for webworms. They are unsightly but often, by the time they are noticed, it is too late to save the leaves that have been devoured.

Since it is also not currently on any lists for being banned, it has become a standby for success in treating many insect problems. Not only is the systemic mode of action quite effective but the chemical also provides a wider time frame for control than many of the older products. There are some insects that have not responded well to older chemicals that can be treated with Merit™. This is a product that I highly recommend and it can be applied as a root drenching or using a tree injection method.

FIGURE 8.8 This person may be killing many beneficial insects by spraying a general insecticide that will kill many insect pests in the shrubs.

PESTPOINTER

There are some insects that have not responded well to older chemicals that can be treated with Merit™. This is a product that I highly recommend and it can be applied as a root drenching or using a tree injection method.

EVASIVE SPECIES

There are many insects that I would classify as evasive species because they are either tremendously difficult to control or they are able to reproduce at such a high rate that control measures are not as successful. A good example of an evasive species would be the gypsy moth. The gypsy moth is an insect with a history of being very difficult to control (Figure 8.9). There are now more modern treatments such as traps and the use of *Bacillus thuringiensis*, a bacterial insecticide. The gypsy moth is an insect that feeds voraciously on leaves of native hardwoods and is common to the New England states. It may not be necessary to try and control this species but hardwoods can only survive two to five years of defoliation before this insect can cause death. It is difficult to control this insect and it has many hosts. It is believed that this insect was brought into the United States back in the 1800s and it is quite surprising that the progression of this species has not been more rapid. If you live in a county where gypsy moth populations are uncommon, you should contact your extension office if you encounter this species.

One of the problems with treating large open areas for gypsy moths was the possibility of injury to other flora and fauna or non-target insects. When timber areas were sprayed with a dusting of insecticide, there was potential for damage to other species that could be quite irreversible. If another insect was wiped out that was the

354 Turfgrass Chemicals and Pesticides: A Practitioner's Guide

FIGURE 8.9 These gypsy moths cause lots of damage to hardwood tree but seldom cause death unless they infect the same areas from year to year.

primary food source for a rare bird or mammal then that animal could be at risk for extinction due to the insecticide usage (Figure 8.8). These considerations are all part of the delicate balancing act we deal with in regard to insect control in our environment in general. This is one reason that gypsy moth control is handled in less aggressive modes and even through trapping.

A new insect that is at the top of the evasive list is an insect known as the Asian long-horned beetle (Figure 8.10). This Asian beetle is an insect that is not native to the United States and which, like others,

PESTPOINTER

It is believed that the gypsy moth was brought into the United States back in the 1800s and it is quite surprising that the progression of this species has not been more rapid. If you live in a county where gypsy moth populations are uncommon, you should contact your extension office if you encounter this species.

Insect Control in Trees and Landscapes 355

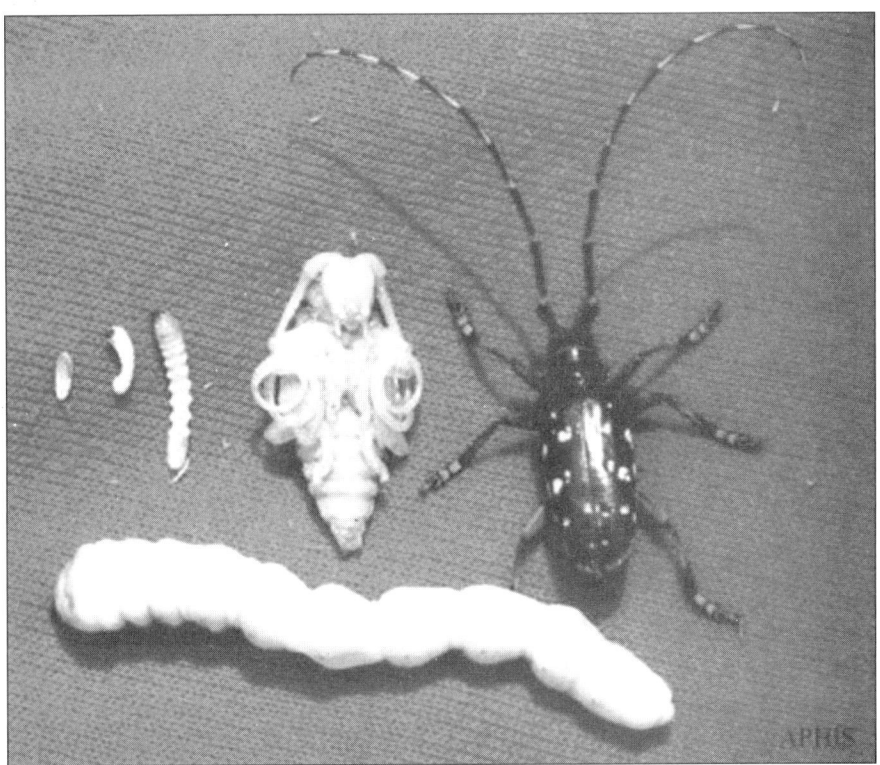

FIGURE 8.10 The Asian long horned beetle can be very destructive but is fairly immobile since mature adults do not have wings.

makes its presence here that much more questionable. Anytime a nonnative species is introduced into an environment, the result of the presence of that species is unknown. The Asian beetle is believed to have been transported into the United States in the lumber for fireworks crates. This is the reason that customs is in place to monitor such transportation of flora or fauna into and out of the United States. The Asian beetle at first had no known chemical control when it was first encountered. There has been limited success using Merit™ for treatment of Asian beetle and also the use of natural predators but a more drastic approach has been taken in most locations.

It was discovered that this insect is a fairly immobile treat especially in the larval stage. An insect's progress may be less of threat if

PESTPOINTER

The Asian beetle is believed to have been transported into the United States in the lumber for fireworks crates. This is the reason that customs is in place to monitor such transportation of flora or fauna into and out of United States.

the insect cannot fly which makes its progression a little less scary when there is no treatment of control available. The control that was used for Asian beetle was a slash-and-burn method of taking out infected specimen trees along with all other species in the immediate surroundings of the infection point. All material from these specimens is then burned to kill all of the potential living insects. The Asian beetle is not the first species to show up that has not responded to traditional chemicals, but people who are fighting this insect are finding it a difficult one to control. The progression of this insect has been slow and most occurrences have been in the New England states along with small segments of the Chicago area. Progression of this insect should be watched closely and, if you think you have encountered one of these insects, you should contact your local extension service or state agency associated with pesticide application.

PESTPOINTER

Progression of the Asian beetle should be watched closely and, if you think you have encountered one of these insects, you should contact your local extension service or state agency associated with pesticide application.

On a related note there is also a eucalyptus long-horned beetle that has been a problem in California. The problems with this insect have been similar and it is believed that this species originated in Australia. I want to point out that this insect exists because treatments and controls have been very unsuccessful with this species as well. There are a lot of similarities even though the origins of the species are much different.

INSECT CONTROL ISSUES

The primary focus of insect control in landscaping refers to the use of chemical controls to rid plants of unwanted insect pests. Although insect control is not always performed by using chemicals, this practice has become commonplace among landscaping professionals. One consideration is to what degree should insects be controlled at and when they are serious enough that chemical controls are necessary? This type of pest management falls into the guidelines of IPM or Integrated Pest Management where strategies are used and thresholds are given for when insect control should be used. The two primary thresholds would be economic or aesthetic damage that is occurring to the landscaping plants.

For instance, let's look at the average home and the location of the plants around that home. It is pretty likely that you will find a row of shrubs and shade trees around a large percentage of the homes that exist almost anywhere in the United States. Since shrubs are breeding grounds and food sources for insects, it is a pretty safe bet that insects will be present in your lawn and landscaping. The consideration here isn't "How can we rid our landscape of all of the bugs?" but rather when is a chemical control needed and what are the results of treating the plants in and around the home.

Since insecticides do pose a threat to humans and pets, caution should always be taken when applying these products. Insecticides often target the nervous systems of humans and pets when accidental exposure occurs. The point I am trying to make is that not all insects are bad and sometimes the risks of using insecticides are not worth

> **PESTPOINTER**
>
> According to the guidelines of IPM, thresholds are given for when insect control should be used. The two primary thresholds would be economic damage or esthetic damage that is occurring to the landscaping plants.

riding a potentially non-harmful insect on your property. For instance, if you should happen to find one bagworm on a plant located on your property, evaluate that situation.

1. Is this the only bagworm I have seen on the property?
2. Have we ever had problems with bagworms in the past?
3. Will treatment possibly cause exposure to children or pets?
4. Will I be harming any beneficial insects?
5. What will the financial costs of treatment be?
6. Is this insect causing severe aesthetic or financial damage?

These are questions for which you should have solid answers before making a decision to spray your entire property with an insecticide. The labels of most insecticides are quite broad and you could be getting rid of several desirable insects or predatory insects with your treatment decision. If you have one bagworm on a plant and no other signs of activity anywhere else on your property, it is perfectly acceptable to pick the bagworm off by hand and burn it. This is a simple alternative that may solve your problem.

There are many insects that exist in the landscape that could be considered beneficial from many standpoints. Praying mantis and ladybugs are two of the more common insects that actually eat on a lot of harmful insects that exist in the landscape. This is one reason why over-the-top blasting of all the insects on the property may not be

PESTPOINTER

The labels of most insecticides are quite broad and you could be getting rid of several desirable insects or predatory insects with your treatment decision.

such a good solution. In certain circumstances, however, it is too risky to leave one potentially destructive insect. These cases may involve evasive species or areas that are found to be at high risk for damage.

There are many areas where insect control may be desired in the landscaping. Golf courses are one high maintenance area where insects are controlled in both turfgrass and in ornamental plants. In some instances, pests such as mosquitoes and ticks are controlled just because they are a potential nuisance to golfers. There are other places such as college campuses, hospitals, schools, parks, and many other locations that are considered to have pristine landscapes where insect damage is not an acceptable situation. I could even include the various production areas that contain ornamental plants to some degree. Orchards, nurseries, and even greenhouses are places where insect damage is not an acceptable situation.

PESTPOINTER

Praying mantis and ladybugs are two of the more common insects that actually eat on a lot of harmful insects that exist in the landscape. This is one reason why over-the-top blasting of all the insects on the property may not be such a good solution.

PESTPOINTER

Keep in mind that, whatever your circumstance is, control should be based on the immediate threat that exists from that particular insect.

Keep in mind that whatever your circumstance is, all insects should be controlled based on the immediate threat that exists from that particular insect. Also not all insects are bad. Insects have very positive roles in the landscape such as pollination of trees and flowers or the predatory presence they may have in eating an unwanted insect. If you are experiencing an insect problem, contact your local extension office if you have discovered something you are not familiar with. If you choose to use chemical controls, make sure that you are using something that is going to target the pest you are specifically after and avoid soaking everything down that may not be infected. As chemicals continue to change and evolve, we as pest control people will need to be more cautious in our decisions of when and where we use insecticides.

SUMMARY QUESTIONS

- Why is it important not to blast a plant or tree with a large amount of insecticide?
- What precautions do you take to ensure that beneficial insects are not wiped out by chemical applications?
- Have you observed any plant responses to pests that were not harmful to the plant?
- How do non-native insect species get into the US?

Chapter 9

Other Chemicals and Pesticides Used in Turf and Landscapes

At this point in this text, we have covered the three primary types of pests that are encountered in turfgrass and landscape systems: weeds, diseases, and insects. We have covered a lot of detailed but practical information and you are to be commended for sticking with it and bringing your level of knowledge to the next level. However, we still have a lot of information to cover and, as you know, pesticides are not the only chemicals out there that you experience on a day to day basis. Many chemicals and control strategies exist for what might be termed second tier pests. These may not be as consistent or as damaging to turf or landscapes but do require attention in many cases because of potential for damage or because they are simply nuisances to your clients.

Other turfgrass and landscape chemicals fall into what could be termed a luxury class of materials. Examples are the diverse types of plant growth regulators, all of which modify the developmental properties of target plants but none of which are truly essential to the everyday operations at your facility. As will be discussed later, plant growth regulators can either stimulate or slow down growth and may have some management benefits when used properly. A final category of turfgrass and landscape chemicals is adjuvants. Adjuvants can be common additives to many pesticide or chemical products but also may be commercial products in their own right.

This chapter will complete our discussion of the types of chemicals you face in the workplace and thus prepare you for later chapters, which will involve how to apply pesticides safely and properly with a regulatory and environmentally conscious approach. You will find that the contents of this chapter are rather diverse but that they effectively tie up any loose ends that have not already been discussed. Once completed, you will have been exposed to just about every type of chemical that is available for application at your facility. With that in mind, let's charge forward!

ADJUVANTS USED IN CHEMICAL AND PESTICIDE APPLICATIONS

The success of a chemical or pesticide application often hinges upon what additives are included when an application is made. Additives can include fertilizer and a diverse group of compounds called adjuvants. If you look closely, you can see the fragment of the word "add" in this term, and that most effectively sums up the role of these compounds. Any compound added to a spray solution for the purpose of enhancing the efficacy of the application is an adjuvant. With this definition in mind, fertilizers are not adjuvants. They are usually added either to conveniently feed treated plants or to mask any potential injury a chemical may cause. They do not enhance the efficacy of the application. What compounds do fall under the definition of an adjuvant?

Adjuvants can play a number of roles in improving the efficacy of a chemical or pesticide application and they can also differ in terms of how complicated they are chemically. Typically, they are only used for liquid spray applications and they are only used when treating the foliage of target plants. Table 9.1 lists some of the different types of adjuvants and briefly describes what they do. As you read through this list, think of what circumstances you might use different adju-

PESTPOINTER

Adjuvants can play a number of roles in improving the efficacy of a chemical or pesticide application and they can also differ in terms of how complicated they are chemically.

vants for and how you may have used them in the past. The adjuvants listed in Table 9.1 can be broken down into three primary categories: activator adjuvants, spray modifiers, and utility modifiers.

Activator adjuvants are those that we most associate with the spray additive concept. Activator adjuvants, commonly called surfactants, increase the ability of the spray solution to spread evenly onto a treated

TABLE 9.1 Types of adjuvants, with a brief description of how they work.

Type of Adjuvant	Function
Anti-foam	Reduces foam produced when some chemicals are mixed with water
Compatibility agents	Helps facilitate tank-mixes of multiple chemicals
Drift retardants	Increases solution particle size to minimize drift potential
Dye	Colors spray solution so treated areas can be identified
Spray buffers	Modifies the acidity of the spray solution
Spreader	Helps evenly coat spray solution on the surface of treated plants
Sticker	Helps spray solution adhere to the surface of treated plants
Surfactant	Reduces surface tension of spray solution to enhance uptake by plants
Suspension aids	Helps chemicals to suspend more evenly in water

surface, thus enhancing the potential for the solution to be taken up by the plant. These adjuvants are most commonly used with post-emergence herbicide applications that rely upon leaf uptake for their success. The term surfactant is a composite of the three-word term surface active agent and this accurately describes what these materials do. Their usefulness has a lot to do with the chemical properties of water. We've all seen how water beads up on a recently waxed car. The wax does the opposite of what a surfactant will do. Wax repels water so its presence causes water particles to collect together, avoiding the wax and forming the beads that we see. A surfactant allows water tension to relax and form an even coat on a treated area. This is the principle behind most common soaps or detergents. By reducing water tension, the natural cleaning ability of water is enhanced and our bodies or dishes can become cleaner. The easiest and cheapest way to use a surfactant for a chemical application is to add a portion of dish detergent to the solution. You'll find that, instead of water droplets on treated plants, you'll see a thin moist film of water. How does this affect chemical efficacy? We can presume that there is an even concentration of a chemical or pesticide in a spray solution. If that solution is dispersed onto a plant leaf in a thin film, rather than in scattered droplets, the amount of the chemical or pesticide that is in contact with the leaf will increase dramatically and the efficacy of the application will increase.

Surfactants can come in a variety of forms. The simplest, as described above, is a detergent of some kind. These types of sur-

PESTPOINTER

The easiest and cheapest way to use a surfactant for a chemical application is to add a portion of dish detergent to the solution.

factants are available commercially for the agricultural and green industries and, as a group, are known as nonionic surfactants. The name comes from their ability to break down the surface tension caused by water's chemical properties. They also better enable organic chemicals to disperse evenly in a water solution. Nonionic surfactants are some of the most common that are used by both agricultural and turfgrass practitioners. This nonionic surfactant technology has been taken a step further in the turfgrass industry in a class of materials known as wetting agents or soil surfactants. Wetting agents are used to improve the ability of water to penetrate into soil and also more thoroughly wet the soil (Figure 9.1). Wetting agent technology uses many of the same chemical principles described above and has become very common in turfgrass, especially on sand-based putting greens.

FIGURE 9.1 The reduced water tension caused by soil surfactants or wetting agents can result in reduced dew formation on treated turf.

Other types of surfactants may more directly enhance the ability of a chemical, usually a herbicide, to be absorbed by a plant leaf. All surfactants will improve the distribution of a spray solution on the leaf surface, but not all are equally good at promoting plant uptake. Surfactants that are better at improving plant uptake include seed oil and organosilicone surfactants. The most common seed oil surfactant is a crop oil concentrate (COC). These materials can dramatically improve the ability of a herbicide to penetrate through the waxy layer that leaves have. While effective, COCs can sometimes cause injury to desirable plants and should be used with caution. In a way, they sometimes enhance herbicide activity too well so they should only be used in turfgrass species that can readily outgrow any injury that may occur. The positive side of this issue is that COCs also enhance the performance of herbicides on weeds so a balance of risks and rewards needs to be assessed. Other seed oils have been investigated for their potential uses but they are more common in field crop systems.

Organosilicone surfactants offer a balance between traditional non-ionic surfactants and the potentially injurious seed oils. They are founded upon silicone technology, which is best known in this context by the types of super lubricants that are used on automobiles. Organosilicone surfactants result in both enhanced coverage of spray solution on target leaves and enhanced penetration into the leaf interior. They aren't prone to cause injury like some seed oils and can result in complete herbicide uptake in fractions of the time it would

PESTPOINTER

The most common seed oil surfactant is a crop oil concentrate (COC). These materials can dramatically improve the ability of a herbicide to penetrate through the waxy layer that leaves have.

PESTPOINTER

While the effectiveness of organosilicone surfactants is without question, their superior performance and relative newness in the world of adjuvants make them more expensive than other products.

take if they weren't added to the spray solution. While the effectiveness of organosilicone surfactants is without question, their superior performance and relative newness in the world of adjuvants make them more expensive than other products. The cost issue has made organosilicones less popular than other surfactants but their usefulness makes them worthy of consideration when choosing a product.

Other adjuvants that are used with spray applications may not have the direct positive effects on product efficacy that are seen with surfactants. However, spray modifier and utility modifier adjuvants can be critical in determining the success of an application. Spray modifiers from Table 9.1 include:

- *Compatibility agents*
- *Spray buffers*
- *Spreaders*
- *Stickers*
- *Suspension aids*

This group of adjuvants is more user-friendly in that the practitioner may reap the benefits without necessarily being aware of it. This is because most spray modifier adjuvants are added to chemicals and pesticides at the manufacturing level, rather than at the product mixing stage like others discussed in this section.

Compatibility agents and suspension aids are spray modifier adjuvants that are commonly added by manufacturers during the product formulation stage. In Chapter 2, we discussed how some materials inherently do not mix well with water and thus must be formulated properly so they can be mixed and applied effectively. Part of this formulation process involves the use of compatibility agents and suspension aids. Manufacturers identify the need for these types of adjuvants and build them into the formulated product that you purchase. This helps eliminate the kinds of problems that might arise if practitioners were solely responsible for adding all the right ingredients to make a product dispersible in a water-based spray solution. While these sorts of necessary adjuvants are usually added beforehand, compatibility of multiple products cannot be assumed.

If you are planning to apply more than one product in a tank mix, first check the product labels for each material to see if there are any compatibility problems that are mentioned. This is because some products can react with each other negatively. In addition, always run a simple compatibility test to ensure the tank mix will result in an even spray solution. Compatibility tests are easy to do and can help you avoid problems before you've placed products in 200 to 300 gallon spray tanks. Fill a small jar or container about $2/3$ full of water and then add small quantities of each of the products you intend to use in a tank mix. Stir the mixture up and watch for any unexpected settling or formation of clumpy solid material in the solution.

PESTPOINTER

If you are planning to apply more than one product in a tank mix, first check the product labels for each material to see if there are any compatibility problems that are mentioned. This is because some products can react with each other negatively.

If it works in a small container, make a note of it so you know the tank mix will work for future applications and then proceed. Of course, if the mix does not appear to be compatible, you'll want to apply the materials separately.

Other spray modifier adjuvants commonly added by manufacturers during the formulation process are spreaders and stickers. Sometimes both are used for a particular product. These adjuvants are not common for herbicides but are very common for either fungicide or insecticide products. With herbicides, there is a premium on achieving uptake or absorption by the plant. The plant may also absorb some fungicide and insecticide products but many simply remain on the plant surface. Spreaders and stickers allow these pesticides to spread more evenly over the surface of the plant and also adhere to the plant. Remember that we want the target pests to consume these products so, the more evenly they coat the plants and the longer they remain there, the likelihood increases that the pest will consume the pesticide.

The spray modifier adjuvant most commonly added directly by the practitioner is a spray buffer, which alters the pH of the spray solution. This type of modification is not always necessary but, if you live in an area where hard water is prevalent, spray buffers can be extremely important. Hard water has a characteristically high pH, and can thus

PESTPOINTER

Addition of a spray buffer like ammonium sulfate strips hard water molecules like calcium off of the pesticide, leaving it in a form more accessible by the plant. If you live in an area where hard water is common, be sure to check the labels for the products you use to see if a spray buffer is recommended.

have a neutralizing effect on many of the acidic pesticides that we might choose to use. Addition of a spray buffer to lower spray solution pH can improve the availability of the chemical or pesticide to the target plants and render applications more successful. How does this work? Many commercial chemicals and pesticides have acidic properties. In hard water, these acidic molecules can bind to large molecules like calcium or magnesium, making them more bulky and thus more difficult for the plant to absorb. Addition of a spray buffer like ammonium sulfate strips hard water molecules like calcium off of the pesticide, leaving it in a form more accessible by the plant. If you live in an area where hard water is common, be sure to check the labels for the products you use to see if a spray buffer is recommended.

Utility modifiers are the bridesmaids of the adjuvant world. They have a visible presence and are important but often do not receive the credit they deserve. They improve the success of liquid applications but don't directly influence the characteristics of the spray solution or its fate once it reaches target plants. Examples of utility modifiers are drift control agents, anti-foaming agents, and indicator dyes. These adjuvants don't alter the performance of the pesticide in any way but they do benefit applicators. Drift control agents are not necessary for all pesticides or for all circumstances but can be very helpful if your pesticide or chemical can adversely affect non-

PESTPOINTER

Utility modifiers are the bridesmaids of the adjuvant world. They have a visible presence and are important but often do not receive the credit they deserve. They improve the success of liquid applications but don't directly influence the characteristics of the spray solution or its fate once it reaches target plants.

target plants in the area you are spraying. As will be discussed in the next chapter, nozzle selection and other sprayer specifications may be just as useful to you in minimizing drift. Anti-foaming agents are highly beneficial to applicators because of the tendency for certain chemicals to foam up enthusiastically when they are added to water. Waiting for foam to subside can take more time than is desired and can result in unwanted settling of the product in the spray tank. Anti-foaming agents can help eliminate this hassle and shrink the time between product mixing and product spraying.

Indicator dyes can be extremely useful for the applicator by showing where product has already been applied and improving the evenness of an application. Especially in uneven terrain, this issue can be very challenging and dyes may help overcome potential pitfalls. Dyes can also serve as evidence to your clients that an application has been made and provide a visual rationale for staying off of recently treated areas. The story comes to mind of an individual who had the habit of placing his golf ball in his mouth between holes (don't ask me why) and became sick over time through inadvertent exposure to pesticides. Dyes can help eliminate the potential for this sort of situation, one that could result in legal repercussions. Indicator dyes are usually blue, providing just enough contrast against green turf to be visible. They are also safe to turf and inexpensive, so they should be seriously considered as components of the successful applications you make.

PLANT GROWTH REGULATORS

Plant growth regulators are materials that alter the developmental properties of the plants they target. This sets them apart from nutrients that simply fuel new growth or can slow growth when withheld. Plant growth regulators (PGRs) can either stimulate or retard growth but do so most commonly through manipulation of cell division or the hormones that all plants contain. PGRs are applied with the intent of modifying the growth patterns of turfgrasses or landscape plants.

PESTPOINTER

Plant growth regulators (PGRs) can either stimulate or retard growth but do so most commonly through manipulation of cell division or the hormones that all plants contain. PGRs are applied with the intent of modifying the growth patterns of turfgrasses or landscape plants.

As noted in Chapter 4, some herbicides are classified as hormone disrupters. These herbicides are by definition plant growth regulators but differ from those that will be discussed in this section because hormone disrupters are applied with the intent to kill unwanted plants. If you also recall, some herbicides act by disrupting the cell division process. PGRs that act in this fashion are characteristically less severe to the plants they target, resulting in slowed growth without the toxic repercussions that herbicides may have. Commercial PGRs can be separated into two primary classes: growth enhancers or biostimulants, and plant growth retardants.

Biostimulants are materials that are marketed for their ability to promote growth in target plants. Usually, enhanced root development is an identified advantage of using biostimulants, as opposed to simply using conventional fertilizers. Biostimulants usually contain a number of ingredients. Nutrients are themselves common ingredients in most biostimulant products but the nutrient composition varies from conventional fertilizers. Typically, biostimulants contain more micronutrients like:

- Iron
- Zinc
- Boron

- *Molybdenum*
- *Chlorine*
- *Copper*
- *Manganese*

Micronutrients serve less to feed the plant and more to keep normal metabolism at a healthy level. All organisms contain enzymes that regulate metabolism. Plants are no different and micronutrients are often responsible for the normal function of these enzymes. Some micronutrients, like iron, can actually produce a darker green visual response in turf but visual impacts of micronutrients are the exceptions, not the norm.

Because biostimulants are often derived from natural materials like seaweed, they also naturally contain important plant pigments called carotenoids. If you think you see the word carrot in carotenoid, it's no coincidence. Carotenoid pigments are characteristically yellow or orange. To understand how carotenoids can help plants, let's consider a human medical example. It seems that every year there is some new health craze that everybody should be following. For a while, it was fiber, then oat bran, and then betacarotene. More recently, we hear about how we should be focusing upon getting our share of antioxidants. Interestingly, these "new" crazes are really not that novel but they offer catch phrases that people will ask for. Betacarotene is an example of a carotenoid pigment and is also an antioxidant. As a human vitamin or in plants, these types of materials dissipate excess energy. These energy excesses, if left unchecked, can damage cells and tissues. Carotenoids absorb this energy and render it harmless.

Plant hormones are also common ingredients in many biostimulants. The development of these products from plant extracts makes the inclusion of hormones very easy to achieve. Aside from the color effects seen with iron-containing products, hormones can result in the most noticeable effects on turfgrass growth following a biostimulant application. Most hormones will stimulate growth and their impact

> **PESTPOINTER**
>
> Most hormones will stimulate growth and their impact can thus be mistaken for that of a conventional fertilizer. However, some hormones are also more intimately involved with root development or stress responses so their positive effects can be less apparent.

can thus be mistaken for that of a conventional fertilizer. However, some hormones are also more intimately involved with root development or stress responses so their positive effects can be less apparent. Table 9.2 lists the primary hormones found in plants and their functions.

As you can see, there is a lot of overlap and it is important to recognize that hormone-altering growth regulators can affect the balance of the entire plant system, thereby affecting plant development

TABLE 9.2 Primary plant hormones and their functions.

Hormone Class	Primary Function
Abscisic acid	Closes pores on plant leaves to reduce water loss, regulates seed germination
Auxin	Controls branching in broadleaf plants, stimulates plant cell enlargement
Cytokinins	Stimulate cell division and enlargement, promote leaf senescence (fall drop)
Ethylene	Stimulates fruit ripening, promotes root growth, induces activity of other hormones
Gibberellins	Stimulate cell division and elongation, promote seed germination
Polyamines	Stimulate plant growth, slow the natural breakdown of green plant pigments

> **PESTPOINTER**
>
> Essentially, plant defense activators help plants confine injured areas, reduce the spread of injury, and accelerate healing. The most common type of plant defense activator is salicylic acid, most commonly used as an ingredient in common aspirin.

in a unique way. A final class of biostimulant ingredients is a plant defense activator. Essentially, plant defense activators help plants confine injured areas, reduce the spread of injury, and accelerate healing. The most common type of plant defense activator is salicylic acid, most commonly used as an ingredient in common aspirin. Used in plants, salicylic acid can help reduce the effects of mechanical or predatory injury and allow the plant to recover more quickly. Inclusion of these compounds in commercial products is a newer concept, so expect to see more of this in the future.

Because turfgrass biostimulants have multiple ingredients, I often teach people to remember them as being like multi-vitamin products. The micronutrients can improve color and act as vitamins for turf. Carotenoid pigments are the antioxidants we see stressed so much in the human diet. Plant defense activators can help localize and minimize injury to the plant. Boosted levels of hormones can result in both greater leaf and root growth in many turfgrass species. This all sounds pretty good, but are biostimulants right for every situation? The answer to that question is, of course, no.

Among turf specialty products, biostimulants are relatively new and researchers don't know as much yet about them as other turf products. I ran a quick Internet search recently and it was amazing how many of these products are now available. Among these products are ones called Launch®, Roots®, Rutopia™, GroWin®, and CPR™. These brand names don't represent the entire market of

> **PESTPOINTER**
>
> Many commercially available biostimulants have known attributes but some are more mysterious. Researchers have struggled with how to properly investigate products for which detailed ingredient lists are sometimes unavailable. Identifying turfgrass responses to a product can be easy but explaining why a certain response occurs can be difficult if key ingredients or their amounts in a product are unknown. The best approach to take is to regard biostimulants as the luxury applications that they are.

biostimulants but they do reflect the target market for these products. Clearly, better root growth and improved plant health are two common claims for biostimulants, but what should we, as practitioners, expect? The bottom line is that the practitioner should recognize the potential benefits of biostimulant use but also recognize that not all materials are created equal. Many commercially available biostimulants have known attributes but some are more mysterious. Researchers have struggled with how to properly investigate products for which detailed ingredient lists are sometimes unavailable. Identifying turfgrass responses to a product can be easy but explaining why a certain response occurs can be difficult if key ingredients or their amounts in a product are unknown. The best approach to take is to regard biostimulants as the luxury applications that they are. We can survive without them but, if circumstances and budgets allow for it, you may want to consider use of these materials to see what they are capable of at your facility.

While biostimulants can increase plant growth by increasing levels of certain hormones, what happens when these hormones are suppressed? The response we see is the type of growth reduction

> **PESTPOINTER**
>
> Type I growth retardants act by inhibiting cell division, similar to some herbicides.

characteristic of most commercial growth retardants. Growth suppression with chemicals has been available for some time but only in the past 10 to 15 years has it been an acceptable option for high quality turf. Older growth retardants like maleic hydrazide were prone to cause unacceptable turf injury. In some areas like roadsides, this sort of response can be tolerated and more phytotoxic materials are still used for this type of turf. However, injury is not an acceptable option for fine turf and newer products had to be developed before chemical growth suppression was a tool available for golf courses or other commercial facilities. There are a number of issues that warrant discussion with growth retardants but let's start with how they are classified. The simplest and best way is to separate them out by mode of action. Type I growth retardants act by inhibiting cell division, similar to some herbicides. Type II growth retardants act by suppressing hormone levels in target turf plants. A third category, herbicide growth retardants, includes an assortment of different products common for weed control. The difference is that the herbicides

> **PESTPOINTER**
>
> Type II growth retardants act by suppressing hormone levels in target turf plants.

> ## PESTPOINTER
>
> A third category, herbicide growth retardants, includes an assortment of different products common for weed control. The difference is that the herbicides in this context are used at low doses, so that turf is stunted but not killed.

in this context are used at low doses, so that turf is stunted but not killed. Each of these types of growth retardants will now be discussed in more detail.

The most well known growth retardants are those that fit into the Type I or Type II categories. A quick summary of Type I and Type II growth retardants is given in Table 9.3. These products are very commonly used in turfgrass but some also have a history in the commercial horticulture industry, especially in fruit crop production. Their use in turf has been more recent than for many other chemical and pesticide products but their impact has been significant. Modern Type I and Type II growth retardants have achieved market status in high quality turf, something that their predecessors could not attain. This is because current Type I and Type II growth retardants have few, if any, phytotoxic effects on turfgrass. In fact, some of these growth retardants may actually improve visual turf quality.

TABLE 9.3 Common Type I and Type II growth retardants used in the turfgrass industry.

Common Name	Trade Name	Type I/ Type II
Mefluidide	Embark, Embark Lite	Type I
Ethephon	Proxy	Type II
Flurprimidol	Cutless	Type II
Paclobutrazol	TGR, Trimmit, Turf Enhancer	Type II
Trinexapac-ethyl	Primo, Primo Maxx	Type II

PESTPOINTER

There is not a magic formula for applying sublethal doses of herbicides. For liability reasons, many product labels do not outline such a strategy, so practitioners need to and should experiment with product rates in a test or nursery area before making a larger scale application.

Herbicidal growth retardants are usually not specific products designed solely for that function. Rather, they represent standard herbicide products for which an extra niche use has been discovered. The first herbicidal growth retardants may have been discovered by accident. Imagine a practitioner inadvertently applying a low herbicide dose to a non-target turf area and finding the turf to be stunted, not killed. It can be amazing how the discovery process works. Of course, once a discovery is made, it is then broadened to include other materials that should produce a similar response. Today, there are several herbicides that are used as growth retardants and they are listed in Table 9.4. There is not a magic formula for applying sublethal doses of herbicides. For liability reasons, many product labels

TABLE 9.4 Common herbicides that are used as growth retardants, when applied at low doses.

Herbicide Common Name	Herbicide Trade Name
Chlorsulfuron	Telar, TFC, Corsair
Diquat	Reward
Ethofumesate	Progress
Glyphosate	Glyfos, Roundup Pro, Touchdown Pro, others
Imazapic	Plateau
Metsulfuron-methyl	Escort, Manor
Sethoxydim	Vantage

do not outline such a strategy, so practitioners need to and should experiment with product rates in a test or nursery area before making a larger scale application.

Selection of a plant growth retardant is usually not a difficult task since most products have specific niches and there are relatively few products to choose from. Being specific about what you expect the product to do will help you both identify a product and also plan how to use it. Remember that there is more research that has been done with growth retardants than with biostimulants. However, there is still a lot we have to learn so use all the available resources you have when making key decisions about products and product uses. University researchers can be great resources but only your peers have used growth retardants for extended periods of time. Consulting with your colleagues about potential long-term effects of growth retardants can thus be an excellent additional source of information.

MATERIALS AND STRATEGIES FOR RODENT AND BURROWING PEST CONTROL

Rodents and turfgrass clearly don't mix. Although today's golf course superintendents are a far cry from the burned-out greenskeeper played by Bill Murray in the movie *Caddyshack*, the gopher plight in that movie was a fictional but memorable example of a rodent problem (Figure 9.2). Sure, most cases we may have to deal with aren't that severe, but rodents can be a serious nuisance to both turfgrass and landscapes. Rodents and other burrowing animals can disrupt the smoothness of a surface, can uproot plants, and even create dangerous situations with holes or depressions that can result in twisted ankles or worse. Common animals that fit into this category are badgers, gophers, prairie dogs, and moles. Other rodents like muskrats and beavers can live in turf settings but are more aquatic in nature and are generally not disruptive.

Rodent problems are usually isolated to a particular area of a lawn or other turf and landscape facility. However, the areas where they do become problems can be particularly troubling because target-

Other Chemicals and Pesticides Used in Turf and Landscapes 381

FIGURE 9.2 The famous and destructive gopher from the movie *Caddyshack*.

ing the rodent pest can be a real challenge. We usually don't see the pests, and the type of damage they do can vary. Moles, prairie dogs, and badgers are all pests that are physically disruptive but generally don't actually feed upon the turf or landscape plants in the damaged area. Ground squirrels or traditional gophers can actually feed upon roots or other subterranean plant parts. Moles (sometimes called pocket gophers) are perhaps the most widespread rodent pests in turf settings (Figure 9.3). Their aggressive and shallow burrowing tunnels can result in surface undulations and the classic mole hills that result from burrowing are clear nuisances to a turf setting. Control measures can range from poisonous baits to conventional mole traps. The key is to place the control device in or near areas of recent mole activity to improve the chance of success. Patience may be required but reasonable results are attainable.

Similar strategies may be pursued when gopher or badger damage is evident. Problems with these pests are not as widespread as

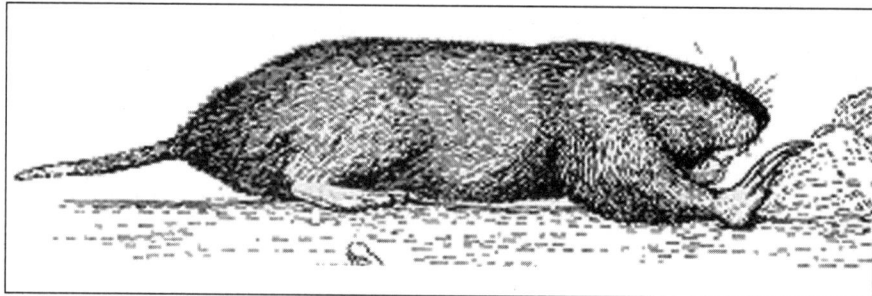

FIGURE 9.3 A common mole or pocket gopher.

with moles but the pests are larger so the control measure may need to be modified. Badgers can cause considerable damage and should be controlled as soon as possible (Figure 9.4). However, they can be vicious so passive control measures like poisonous baits or involvement of professional animal control personnel are more appropriate in the rare cases where problems exist. Gophers pose less of a hazard to applicators but their deeper burrowing characteristics may make poisoned baits the best way to get rid of them.

Prairie dogs, as the name implies, are native to the central U.S. and their large colonies can actually be something of a tourist attraction (Figure 9.5). The underground tunnel network to support a large prairie dog colony can be very extensive and proliferations of these

PESTPOINTER

Control measures can be as simple as moving ground and destroying the underground colony or as unique as large suction devices that suck the prairie dogs out of the ground into a collection tank for removal.

Other Chemicals and Pesticides Used in Turf and Landscapes 383

FIGURE 9.4 Badger damage to a golf course tee area.

FIGURE 9.5 The cute behavior of prairie dogs can create dilemmas about controlling them.

rodents can be risky in other ways, due to the propensity for these creatures to carry diseases like the plague. Control of prairie dogs is often an inadvertent consequence of recent urban sprawl that removes portions of their habitat. This habitat displacement has created some controversy over the future of these animals and there are vocal supporters of these creatures. Bear this in mind if prairie dogs have been or are a problem at or near your facility. There will be those who oppose the decision to control these pests so tread carefully. Control measures can be as simple as moving ground and destroying the underground colony or as unique as large suction devices that suck the prairie dogs out of the ground into a collection tank for removal.

AQUATIC PLANT MANAGEMENT

Aquatic plant management is not a necessity in all turfgrass or landscape environments but, as readers of this text who have experience with commercial facilities or golf courses realize, bodies of water are something that practitioners need to manage. Plant life and water are naturally attracted to one another but only certain plant species are adequately equipped to thrive in true aquatic settings. Aquatic plants can vary considerably from traditional weeds and so too can control options. Unlike ground applications where plants effectively

---

PESTPOINTER

Unlike ground applications where plants effectively intercept and filter the chemicals we use, aquatic surfaces have no such defense and chemicals can be more risky to beneficial organisms like fish.

intercept and filter the chemicals we use, aquatic surfaces have no such defense and chemicals can be more risky to beneficial organisms like fish. Aquatic weeds can include a variety of true plants but also plant-like organisms such as algae. In this section, we will quickly cover some of the key weeds you may have to contend with in aquatic settings and some different control strategies to consider. Aquatic environments support plant life and also other nuisance pests like mosquitoes, which will be covered later in this chapter.

Aquatic plant life falls into one of two primary categories: higher plants or primitive plant-like organisms. Algae are the most common examples of this second category. Algae are also one of the most widespread problems faced in aquatic settings. They can develop over a broad range of environmental conditions and can do so quite rapidly. Algae come in a variety of sizes and forms but we usually think of them as the slimy films that form on pond surfaces (Figure 9.6). Algal colonies can form most rapidly when bodies of water are very still. Therefore, ponds where air movement across the surface is

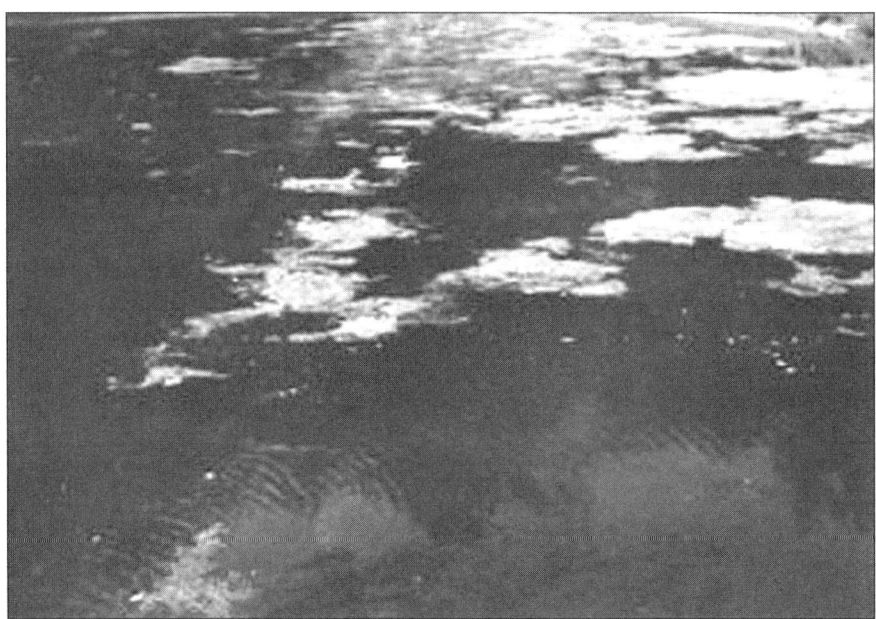

FIGURE 9.6 An algal bloom on a body of water.

PESTPOINTER

Algal colonies can form most rapidly when bodies of water are very still. Therefore, ponds where air movement across the surface is restricted are more prone to algae infestations.

restricted are more prone to algae infestations. Algae are very small by themselves but can form into large aggregate colonies when conditions for development are good (Figure 9.7). What other conditions promote algae and what are the consequences of algae formation?

Turfgrass management practices can influence the formation and proliferation of algae. Ponds tend to be positioned to receive flowing water and what it carries from turfgrass surfaces. Therefore, nutrients and chemicals applied to turf can end up in ponds and influence the pond ecosystem. Nutrients like nitrogen and phosphorus can harm small bodies of water by stimulating the development of algae or other plant life. This promotion of aquatic plant life can be at the expense of other aquatic organisms. Aggressive algal colonies can block sunlight and consume valuable oxygen in the water. Over time, high oxygen consumption by plants or algae can harm fish by gradually suffocating them.

PESTPOINTER

Judicious fertilizer use, proper incorporation of nutrients into turf, and the use of vegetative buffers can all minimize nutrient impact on aquatic areas.

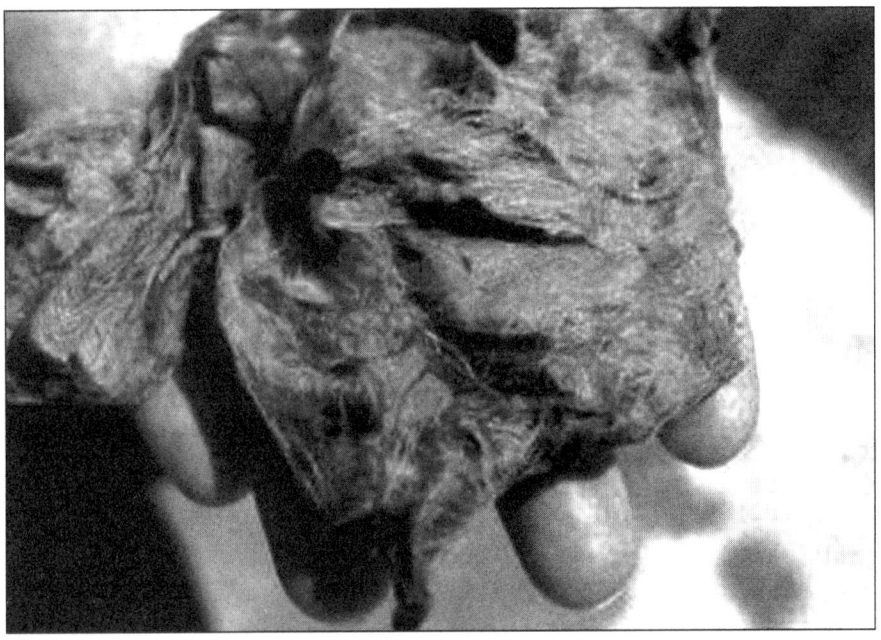

FIGURE 9.7 A large clump of algae removed from a body of water.

Judicious fertilizer use, proper incorporation of nutrients into turf, and the use of vegetative buffers can all minimize nutrient impact on aquatic areas. These types of practices are all known as Best Management Practices (BMPs) and are the subject of the last chapter of this text.

How to best control algae is subject to some debate. In addition to the turfgrass BMPs mentioned above, there are numerous other options available. Because algae are not true plants, commercial pesticides are typically not used for control of algae. Figure 9.8 shows how control of algae can improve a pond's appearance. Various natural and chemical alternatives can be used to discourage algae formation:

- *One cosmetically appealing option is a fountain or some other means of stimulating water movement. This helps oxygenate the water and disrupts the ability of algae to aggregate and form*

FIGURE 9.8 Partial algae control in a medium-sized pond.

colonies. The downside to this option is the cost associated with installing such a structure but any method that helps continuously agitate the water will help you achieve this goal, regardless of the cosmetic appeal.

- *Another common option for algae control is the use of biological agents like grass carp. These fish can aggressively consume algae but are not natives to the United States and there is some concern that these aggressive fish can disrupt ecosystems that normally support multiple fish species.*

- *Other natural materials like bales of barley straw are believed to exude algae-suppressing materials as they decompose in water but results can be inconsistent.*

- *Blue dyes have become fairly common for control of algae as well. The blue color can help filter incoming sunlight and effectively block the types of light that algae need to thrive. Copper sulfate or other copper-containing compounds can provide excellent algae control with little potential risk to other aquatic life when used properly.*

- Lastly, there are numerous chemical options like bleaches or algaecides that can effectively suppress algae but their potential impact on other aquatic life should be considered before using these options.

Higher plants that can occur in aquatic settings may not develop as quickly as some algae but they can certainly be more robust when they do develop. Some aquatic plants like cattails and rushes can add scenic value to bodies of fresh water while others can be more invasive, showing the propensity to take over large spaces of open water. Generally speaking, more aquatic plant species exist the further south one moves in the country. Northern climates may have little more than algae to contend with while a plethora of nuisance aquatic species can prevail in warmer climates. Figures 9.9 to 9.14 illustrate some of the common aquatic weeds that may be found at your type of facility. For

FIGURE 9.9 Cattails are some of the most common aquatic plants found across the country.

390 Turfgrass Chemicals and Pesticides: A Practitioner's Guide

FIGURE 9.10 Water hyacinth, a common flowering aquatic plant.

FIGURE 9.11 Horsetails are found in similar areas to cattails but have a noticeably different "tail" appearance.

Other Chemicals and Pesticides Used in Turf and Landscapes 391

FIGURE 9.12 Water lilies, sometimes referred to as lily pads, can exclude valuable sunlight from bodies of water.

FIGURE 9.13 Alligatorweed is an aggressive aquatic weed common in Southern areas.

392 Turfgrass Chemicals and Pesticides: A Practitioner's Guide

FIGURE 9.14 Purple loosestrife is a highly invasive aquatic weed, particularly in Northern climates.

TABLE 9.5 Herbicides commonly used for control of aquatic weeds other than algae.

Common Name	Trade Name(s)
Diquat	Reward, Weedtrine-D
Endothall	Aquathol, Hydrothol
Fluridone	Sonar
Glyphosate	Rodeo
2,4-D	various trade names

complete lists of what aquatic weeds you should be most concerned with, always check with your local extension resources. Control of higher plant aquatic weeds can be conventionally achieved with herbicides but we need to be careful as to which ones are chosen. Herbicides with any known toxicity to fish species should certainly be avoided, as should those that have higher persistence. A list of common herbicides registered for aquatic use is shown in Table 9.5. As you might expect with a more sensitive environment, few herbicides are available for weed control in aquatic settings.

As you can see, some of these materials are ones we've seen mentioned earlier in this text and some are more unique to the aquatic environment. Proper control will include correct chemical selection but also proper knowledge of the volume of the body of water to be treated. Unlike for a ground surface, treating water means knowing area and depth so that the applied rate is most accurate. Know the facts and treat a specialty application like this with the care it deserves and your results will reflect the effort.

CONTROL OF BIRDS AND OTHER HERBIVORES

Birds and other herbivores (plant eaters) can add significant beauty to turfgrass or landscape environments (Figure 9.15). They represent the attributes of a true natural setting and can offer significant appeal to patrons or clients. When these same animals

394 *Turfgrass Chemicals and Pesticides: A Practitioner's Guide*

FIGURE 9.15 Many aquatic birds are graceful complements to a natural setting.

become disruptive, challenges can arise and these creatures can assume pest characteristics. Many animals that fall under this category might also be deemed nuisance pests but their ability to directly affect the plant systems we manage makes it worthwhile to mention them separately. Common examples of herbivorous or plant damaging animal pests are birds, deer, and rabbits. Birds are usually thought of in a more positive light and most bird species are completely harmless. However, many migratory birds can collect at turfgrass facilities in large numbers and cause damage, usually by leaving droppings that can burn turf (Figure 9.16). High nitrogen content in bird droppings is the cause of the damage and can result in these species being viewed negatively. Larger

Other Chemicals and Pesticides Used in Turf and Landscapes 395

FIGURE 9.16   Large populations of migratory birds like geese can damage turf with their droppings.

species like ducks and geese are the most common culprits and can be regular but unwanted occupants at your facility when their migration patterns reach your area. Facilities without bodies of water are less prone to experience these bird pests, since they naturally gravitate to bodies of water they can land and swim in. Simply

---

PESTPOINTER

Facilities without bodies of water are less prone to experience these bird pests, since they naturally gravitate to bodies of water they can land and swim in.

running these birds off is the easiest means of diminishing the problem but it can be temporary at best. Many larger facilities and golf courses have started using herding dogs like border collies to more consistently discourage these types of birds from staying where they are not wanted. Irrigation of areas where goose and/or duck droppings have been deposited can help reduce the injury potential in sensitive turfgrass.

True herbivorous pests may or may not affect turfgrass but usually have a greater impact on landscape plants. Rabbits can be a significant threat to low-growing flowers and ornamental plants and can decimate planted areas. They usually are most active at night, so it can be difficult to catch them in the act of eating plants. The habitat for rabbits is often in secluded areas like dense brush or within larger shrubs in a managed landscape. This sometimes gives them very little distance to travel to reach the plants they are looking to eat. They are prone to multiply very quickly so controlling them effectively can be important to address as soon as a problem is identified. Deer are naturally more nervous around populated areas but continued exposure to humans in their habitat can make them bolder (Figure 9.17). Like rabbits, they can target low-growing flowers and ornamentals in landscapes. However, they also can pose serious threats to small trees, either by eating leaves or physically damaging the stems with antler rubs. Using plastic or PVC tubes to protect young tree stems is recommended if deer are in your area. Deer don't feed on turf but can damage turf areas with their hooves if many deer travel over a common area.

PESTPOINTER

Using plastic or PVC tubes to protect young tree stems is recommended if deer are in your area.

Other Chemicals and Pesticides Used in Turf and Landscapes 397

FIGURE 9.17 Deer are beautiful creatures but can be disruptive in some turfgrass settings.

Control strategies for these types of pests can vary. Use of predator urine around ornamental beds can help discourage herbivore pests, as can the subtle placement of traps for smaller pests like rabbits. Deer can't be trapped but running them off consistently may help persuade them to move to another area away from your facility. Deer hunting

PESTPOINTER

Use of predator urine around ornamental beds can help discourage herbivore pests, as can the subtle placement of traps for smaller pests like rabbits.

> **PESTPOINTER**
>
> Reducing the habitat for herbivorous pests can also be helpful. Trimming low-growing shrubs can eliminate rabbit habitat and force them to live elsewhere.

is a popular sport but is disallowed in most suburban and urban areas so think twice before simply opting to kill these pests if you live and work in such an area. Reducing the habitat for herbivorous pests can also be helpful. Trimming low-growing shrubs can eliminate rabbit habitat and force them to live elsewhere.

CONTROL OF NUISANCE PESTS

To call a pest a nuisance might seem to be redundant. After all, aren't all pests a nuisance to our efforts to maintain quality turf and landscapes? What sets apart a true nuisance pest from traditional pests is the effect a nuisance pest has on its surroundings. Most traditional pests have potentially damaging effects on the turf or landscapes they inhabit. For example, weeds, insects, diseases, aquatic plant life, and even rodents can all damage turf or landscapes to a significant degree. The potential for this damage is the driving force behind the control measure we choose. By contrast, nuisance pests represent a threat or annoyance to the patrons or clients we serve. Sure, our clients may see all pests as nuisances but are they really threatened by our traditional pests? What are nuisance pests? Examples include:

- *Mosquitoes*
- *Ticks*
- *Chiggers*

- Bees
- Ants
- Spiders
- Snakes
- An occasional alligator if you live in the right part of the country

Naturally, our plant material we manage may not be at risk from these pests but clients may be more worried about nuisance pests than any of the pests we spend so much money on.

The fact that our clients perceive nuisance pests as threats means that particular emphasis must be placed on them. We may not need to control the pests but they do need to be addressed in some shape or form. Handling a nuisance pest problem may be as simple as placing strategically located signs to warn patrons of a possible nuisance pest threat. If you manage or work at a facility where snakes or alligators are prevalent, you know what I'm talking about. Nothing would make a patron more willing to take their business elsewhere or even file a lawsuit than if they were bitten or even mauled by a wild animal on your premises. Necessary precautions must be taken to warn patrons of any such risks. Sure, we can't monitor patrons or clients all the time but posted warnings are usually sufficient to cover your legal liability. Control of dangerous nuisance pests can be warranted in the rare cases where patron injury or worse does

PESTPOINTER

Handling a nuisance pest problem may be as simple as placing strategically located signs to warn patrons of a possible nuisance pest threat.

occur. Bravery is an admirable trait but, in these cases, my practical suggestion would be to allow animal control professionals to handle such a problem.

Other nuisance pests may not pose as large a threat to your patrons or to your employees and/or coworkers but they still may warrant significant attention. Allergies to bee or fire ant stings and some of the diseases carried by mosquitoes or ticks can pose serious health threats (Figure 9.18). At the very least, the bites or stings from these nuisance pests can be painful or irritating. Control strategies for these types of pests can vary. Bees and fire ants are insects that can be controlled with specific products and the control measure can usually be isolated to the hills or hives where the pests are concentrated. Staff members at your facility should be able to handle these necessary measures but adequate caution is always needed to prevent injury.

Mosquitoes seem to be an annual scourge in most parts of the country but targeted control measures may not be warranted unless the problem is dense enough to pose a significant deterrent to your patrons. In severe cases, applications of insecticides like malathion can help keep mosquito populations in check. Biting flies can be serious nuisances but require the correct habitat to really become prob-

FIGURE 9.18 Ticks come in a variety of sizes and some can pose human health risks.

PESTPOINTER

In severe cases, applications of insecticides like malathion can help keep mosquito populations in check.

lems. Ticks and chiggers are common in many areas, especially at warmer times of the year. They are most often encountered in tall grassy or wooded areas so encouraging patrons and employees/coworkers to avoid such areas at your facility can be helpful. Little can be done to control these pests but removing low-growing tree growth and maintaining grass areas at a respectable height can reduce the likelihood of human contact with these pests. Nuisance pests will always seem like that itchy spot in the middle of your back that you just can't seem to reach. Manage them prudently. Offer warnings where necessary, try to modify your less maintained areas so they don't serve as good pest habitats, and control pests in extreme cases. Remember that nuisance pests appeal to the nervousness and fears people carry with them. You may not always need to control these pests but keep customer service in mind, too. Keep an extra can of bug spray in your work area and your unprepared patrons will thank you.

SUMMARY QUESTIONS

- *Do you use adjuvants with any of your chemical applications? If so, what types and what do you expect them to do?*
- *Do you use plant growth regulators of any kind at your facility? If not, have you ever considered their use and what would you expect of them?*

- What alternative pests outlined in this chapter pose the greatest challenges to maintenance at your facility? Do they comprise a regular part of your budget?
- Think back over the last nine chapters and ask yourself if there are any pests or similar challenges that haven't been covered in at least some detail. You should now be armed with the basics to develop a plan of attack for any sort of chemical use. If you need a refresher, I encourage you to now review these topics before we proceed into details of how to safely, efficiently, and responsibly apply the materials we have covered so far.

Chapter 10

Chemical and Pesticide Application Equipment and Calibration

Now that we have covered the different types of chemicals and pesticides that are applied to turfgrass and landscapes, it is time to address methods and theories that are involved in applying these materials. These topics, which will comprise the remainder of this text, include the following:

- *Proper use of equipment for chemical and pesticide applications*
- *Proper mixing and application techniques for chemicals and pesticides*
- *Safety and regulatory issues for chemical and pesticide applications*
- *Strategies for environmentally responsible and efficient chemical and pesticide use*

The first of these listed topics concerns equipment used for turfgrass chemical and pesticide applications. Many types and sizes of equipment, ranging from aerosol cans to airplanes, are used to apply pesticides to lawns, golf courses, and other turf areas. Despite the many possible variations and combinations of equipment, however, most pesticides are applied to turf with manually operated sprayers, power-operated spray booms or spray guns, and granular applicators. Each of these sprayers or applicators has its distinct uses and features. This equipment and how sprayers and spreaders are calibrated are the primary topics of this chapter.

EQUIPMENT USED FOR APPLYING CHEMICALS AND PESTICIDES

Chemical and pesticide applications can be made using two primary types of equipment. Sprayers are used for liquid materials and for water soluble solid materials, while spreaders are used for fertilizers and granular chemicals or pesticides. Each of these two primary application methods will be discussed, using numerous examples of each technique. We will then focus on proper means of calibrating these tools so that applications are precise and uniform. There are several types of sprayers that are used commonly in turfgrass and landscape settings: manual and power sprayers. Each of these will now be discussed in more detail.

Manual Sprayers

Manual sprayers, such as compressed air and knapsack sprayers, are designed for spot treatment and for areas unsuitable for larger units. They are relatively inexpensive, simple to operate, maneuverable, and easy to clean and store. Compressed air or carbon dioxide is used in most manual sprayers to apply pressure to the supply tank and force the spray liquid through a nozzle (Figure 10.1). Several types of small manual sprayers are available that can deliver one to three gallons per minute at pressures up to 300 pounds per square inch (psi). Adjustable spray guns are commonly used with these units, but spray booms are available on some models.

Pressure for most compressed-air sprayers is provided by a manually operated air pump that fits into the top of the tank and supplies compressed air to force the liquid out of the tank and through

Chemical and Pesticide Application Equipment and Calibration 405

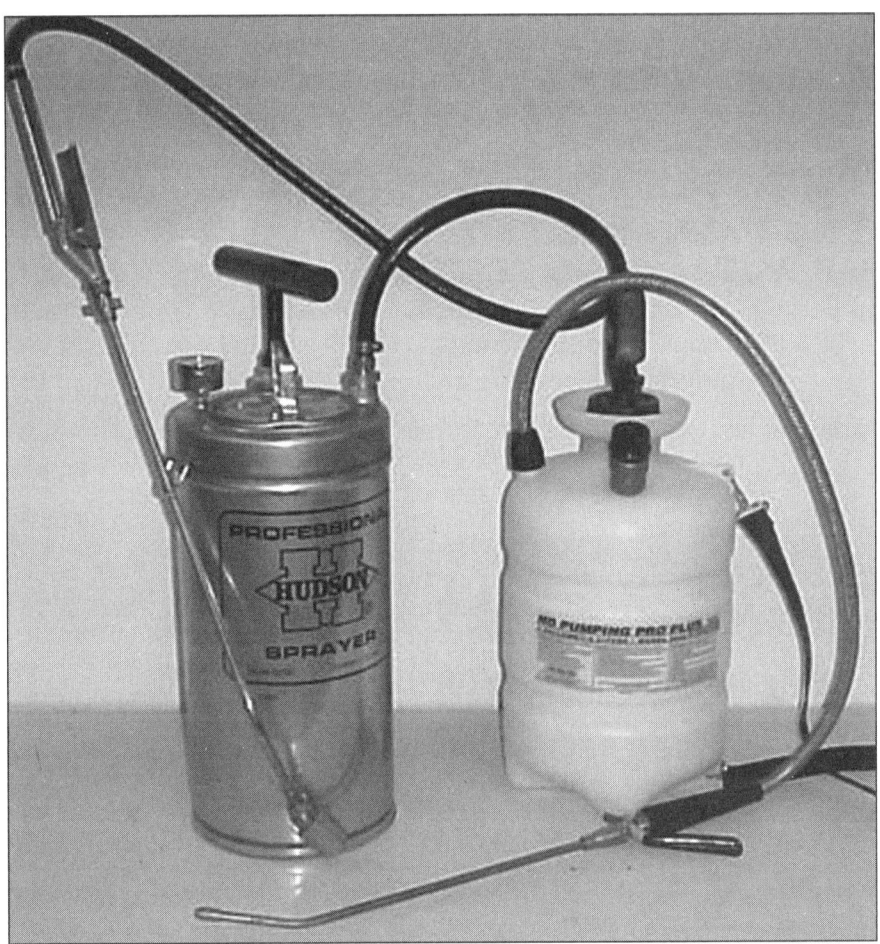

FIGURE 10.1 Compressed-air sprayer.

a hose. A valve at the end of the hose controls the flow of liquid. Agitation is provided by shaking the tank. The capacity of the tank ranges from one to five gallons. Normal spraying pressure is between 20 and 60 psi, and is maintained by occasional pumping. Because the pressure varies so easily, manual sprayers have a tendency to apply in a non-uniform manner. A pressure control valve should be used to help maintain a constant pressure. Pressure control valves are designed to manage the operating pressure and will stop the

406 *Turfgrass Chemicals and Pesticides: A Practitioner's Guide*

> # PESTPOINTER
>
> Because the pressure varies so easily, manual sprayers have a tendency to apply in a non-uniform manner. A pressure control valve should be used to help maintain a constant pressure.

spray process once a preset pressure is reached (Figure 10.2). Typical pressure presets are in the 15, 30, and 45 psi range. If a pressure control valve is not available, equip the sprayer with a pressure gauge to monitor the pressure. In some compressed-air sprayer units, a precharged cylinder of air or carbon dioxide is used to provide pressure. These units include a pressure-regulating valve to maintain uniform spray pressure, and may be mounted on wheels for easy portability. Pesticides may be applied through a spray gun or short boom.

FIGURE 10.2 Pressure control valve.

Chemical and Pesticide Application Equipment and Calibration   407

FIGURE 10.3   Knapsack sprayer.

As the name indicates, a knapsack sprayer is carried on the operator's back (Figure 10.3). The hose and nozzle are similar to those used on compressed-air sprayers. Pressure is maintained by a piston or diaphragm pump that is either operated by hand or by a small two-cycle engine. An air chamber helps smooth out pump pulsation. Tank capacity ranges from two to six gallons, and pressures up to about 150 psi can be developed. Spray material in the tank is agitated by a mechanical agitator or by bypassing part of the pumped solution back into the tank. A pressure gauge, pressure control valve, or spray control valve should be used to help maintain uniform application and minimizing spray drift.

Power Sprayers

Most sprayers for applying pesticides to turf areas use a power source to develop the pressure required to meter and distribute the spray solution. Spray pressures range from nearly zero to over 500 psi, and application rates vary from less than one quart per 1,000 square feet to over 100 gallons per acre. All power sprayers used by turf applicators have several basic components: a pump, a tank, an agitation system, a flow-control assembly, and a distribution system including the spray nozzle. Some systems will also contain electronic components to help improve the accuracy of application (Figure 10.4).

Power Sprayer Pumps

The pump is the heart of the sprayer. The common types of pumps available for applying pesticides include:

- *Piston*
- *Roller*
- *Diaphragm*
- *Centrifugal*

FIGURE 10.4 Self-propelled boom sprayer equipped with electronic controller.

These pump types can be divided into two general categories: positive displacement and non-positive displacement. Positive displacement pumps maintain a flow output directly proportional to the pump speed. Pumps in this category would include roller, diaphragm, and piston, and would require a pressure relief valve and a by-pass line for proper performance. Non-positive displacement pumps do not have a proportional output flow to pump speed and thus would not require a relief valve and by-pass line. The centrifugal pump is an example of a non-displacement pump style. A summary of pump characteristics can be found in Table 10.1.

For low-pressure sprayers, such as boom sprayers, the centrifugal and roller pumps are the most common. Diaphragm and piston pumps are more commonly used on high-pressure systems typically found with systems equipped with hose reels and hand guns. Regardless of the type of pump you choose, it must provide the necessary flow rate at the desired pressure. It should pump enough spray liquid to supply the gallons per minute (GPM) required by the nozzles and the tank agitator, with a reserve capacity of 10 to 20 percent to allow for some flow loss as the pump becomes worn. The reserve is also necessary to overcome friction loss due to flow resistance through the plumbing system extending from the pump to the

PESTPOINTER

Regardless of the type of pump you choose, it must provide the necessary flow rate at the desired pressure. It should pump enough spray liquid to supply the gallons per minute (GPM) required by the nozzles and the tank agitator, with a reserve capacity of 10 to 20 percent to allow for some flow loss as the pump becomes worn.

TABLE 10.1 Common sprayer pump types and characteristics.

Characteristics	Roller	Centrifugal	Diaphragm	Piston
Cost	Low	High	Medium	High
Displacement	Positive, self-priming, requires relief valve	Non-positive, needs primed, relief valve not required	Positive, self-priming, requires relief valve	Positive, self-priming, requires relief valve
Drive Mechanism	PTO, gas engine drives, electric motors	PTO, hydraulic, gas engines, electric motors	PTO, hydraulic, gas engines	PTO, gas engines, electric motors
Adaptability	Compact and versatile	Good for abrasive materials, handles suspensions and slurries well, needs higher RPMs	Compact for amount of flow and pressure developed	Wide range of applications, dependable
Durability	Parts to wear, replace	Very durable, not much wear	No corrosion of inner parts	Parts to wear, replace
Serviceability	Easy to work on, repair	Simple maintenance extends life	Low maintenance	Potential for high maintenance
Pressure Range	Up to 300 psi	Up to 180 psi	Up to 725 psi	Up to 400 psi
Output Volume	2-74 gal/min (GPM), high volumes for size, proportional to pump speed	Up to 190 GPM, high volumes for size and weight	3.5-66 GPM, proportional to pump speed	Low, up to 10 GPM, proportional to pump speed, independent of pressure
RPM	540, 1,000	Requires speed up mechanism, very efficient at higher speeds, up to 6,000 RPM	540	540

nozzle. Pump flow and pressure requirement charts are available from equipment manufacturers to be used as a guide for selecting the proper size and style of pump.

Roller pumps are popular for smaller sprayers because of their low initial cost, compact size, ease of repair, and efficient operation at power-take-off (PTO) speeds. Roller pumps are also self-priming. There is a wide assortment of roller pumps with maximum outputs ranging from 5 to 30 GPM and maximum pressures ranging from 100 to 250 psi. Roller pumps are usually constructed with:

- *Cast iron or corrosion-resistant housings and rotors*
- *Nylon, Teflon, or rubber rollers*
- *Viton, rubber, or leather seals*

The type of material selected depends on the chemical being pumped.

Centrifugal pumps are the most popular pump for low-pressure sprayers. They are durable, simply constructed, and can readily handle wettable powders and abrasive materials. Because of the high output of centrifugal pumps (70 to 130 GPM), the spray solution can be agitated sufficiently even in large tanks. The initial cost of a centrifugal pump is somewhat higher than that of a roller pump, but its long life and low maintenance make it an economical choice. Figure 10.5 shows a typical centrifugal pump plumbing diagram.

PESTPOINTER

The initial cost of a centrifugal pump is somewhat higher than that of a roller pump, but its long life and low maintenance make it an economical choice.

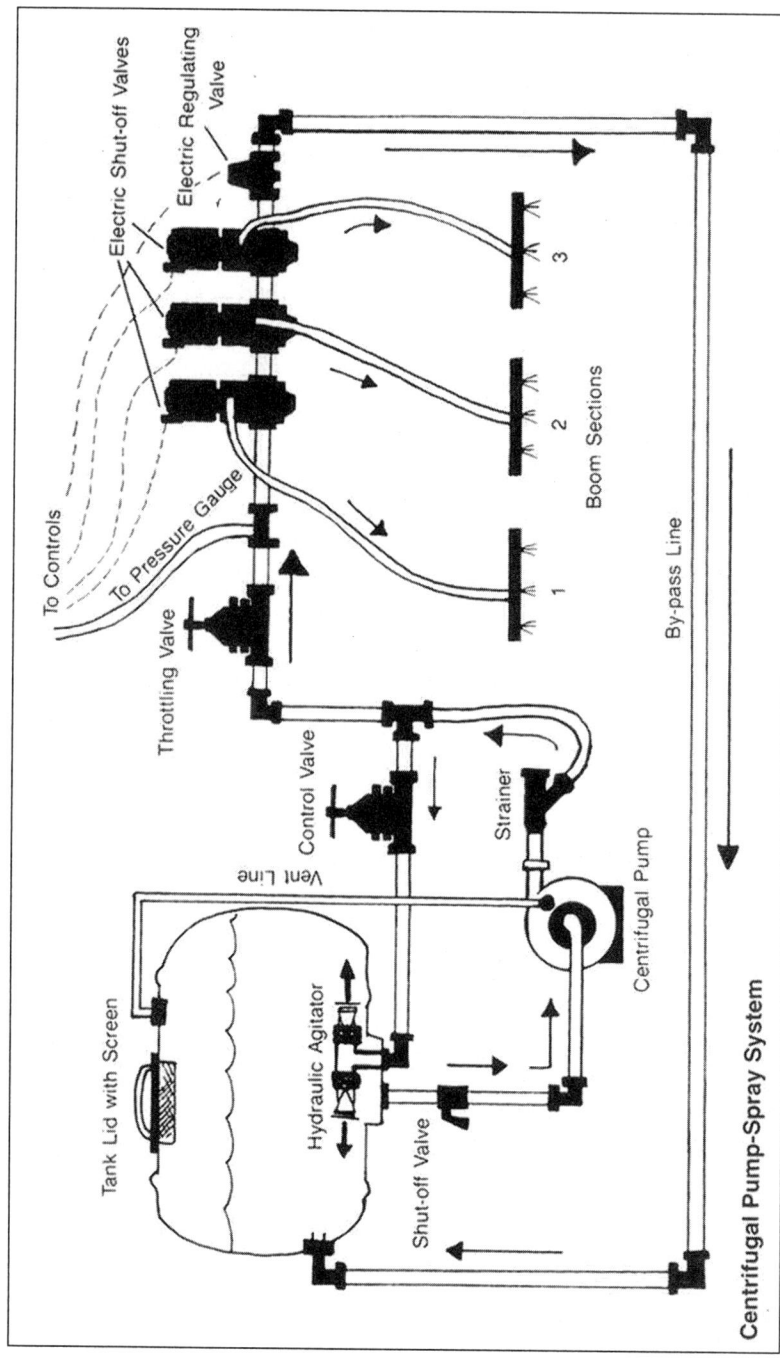

FIGURE 10.5 Centrifugal pump plumbing diagram. *(Illustration courtesy of the University of Illinois)*

Since centrifugal pumps are not self-priming, they should be mounted below the supply tank to aid in priming. In addition, a small vent tube should be installed from the top of the pump housing to the supply tank. This vent line allows the pump to prime itself by *bleeding off* trapped air upon starting and when the pump is not operating.

The inlet of a centrifugal pump should never be restricted. A partially clogged suction strainer, a collapsed suction line, or a suction line with insufficient capacity will result in a loss of pressure control and possible damage to the pump. Since centrifugal pumps can handle small pieces of foreign material without damage, a suction strainer is not always required. If a suction strainer is used, it must be capable of handling the large capacities of the pump with a minimum drop in pressure across the strainer. It must also be cleaned frequently.

Diaphragm and piston pumps are popular when higher pressures are needed for applying lawn care products. Both are easier to prime, maintain higher pressures, and can be used with more abrasive and corrosive materials because all moving parts are sealed in an oil bath, protecting them from corrosive and abrasive spray solutions. Some models provide maximum outputs ranging from 6 to 50 GPM and maximum pressures ranging from 200 to 700 psi. These pump types are more expensive to purchase and maintain compared to roller and centrifugal pumps.

Power Sprayer Tanks

Sprayer tanks for power sprayers should be large enough so that they need not be refilled frequently. They should also be clean, corrosion-resistant, easy to fill, and suitably shaped for mounting and effective agitation. The openings on the tank should be suitable for pump and hydraulic or mechanical agitation connections.

The tank capacity at various levels should be clearly marked on the outside of the tank. If the tank is not transparent, it should have

> **PESTPOINTER**
>
> If the tank is not transparent, it should have a sight gauge or other external means of determining the fluid level. Sight gauges should have shutoff valves at the bottom to permit closing in case of failure.

a sight gauge or other external means of determining the fluid level. Sight gauges should have shut-off valves at the bottom to permit closing in case of failure. The opening of the tank should be fitted with a cover that can be secured to avoid spills or splashes. It also should be large enough to facilitate cleaning. The drain should be located at the bottom so that the tank can be completely emptied.

Fiberglass tanks are widely used on all types of sprayers and applicators. Although fiberglass is strong and durable, it can break or crack under impact. One advantage of fiberglass tanks is that you can buy repair kits to repair cracks that may occur. Fiberglass tanks are moderate in cost and can be used with most chemicals. They may, however, be affected by certain kinds of solvents.

Stainless steel tanks are strong, durable, and resistant to corrosion by chemicals. Because stainless steel is the most expensive material used for pesticide and fertilizer applicator tanks, these tanks are generally used only with equipment that gets high annual use.

Polyethylene tanks are relatively inexpensive and come in many shapes and sizes. High-quality polyresins provide improved control of stress cracking and ensure compatibility with commonly used agricultural chemicals. However, if cracking does occur, there are kits available to repair the tank. Although polyethylene is tough and durable, it does break down under ultraviolet light. Even though inhibitors are added to protect against ultraviolet rays, it is best to keep these tanks out of direct sunlight when not in use.

> **PESTPOINTER**
>
> One advantage of fiberglass tanks is that you can buy repair kits to repair cracks that may occur. Fiberglass tanks are moderate in cost and can be used with most chemicals. They may, however, be affected by certain kinds of solvents.

When barrels or small metal tanks are used, the tank mounting is not critical. Polyethylene and fiberglass tanks, however, must be properly mounted on a *saddle* that supports the tank over a large area. Without a saddle, the weight of the liquid could break the tank as the sprayer bounces over obstructions or rough terrain.

POWER SPRAYER AGITATION AND FLOW CONTROL

Agitation requirements for power sprayers depend largely on the formulation of the chemical being applied. Soluble liquids and powders do not require special agitation once they are in solution, but emulsions, wettable powders, and liquid and dry flowables will usually separate if they are not agitated continually Separation causes the concentration of the pesticide spray to vary greatly as the tank empties. For this reason, thorough agitation is essential. Hydraulic jet agitation is the most common method used with low-pressure sprayers. The fluid is circulated by returning a portion of the pump output to the tank and discharging it under pressure through holes drilled in a pipe that runs the entire length of the tank or through special agitator nozzles. Jet agitation is simple and effective as long as the device is correctly installed and flow is sufficient. The agitator orifices should receive fluid from a separate line on the discharge side of the pump and not merely from the bypass line.

PESTPOINTER

Foaming can occur if the agitation flow rate remains constant as the tank empties. You can prevent foaming by using a control valve to gradually reduce the amount of agitator flow.

The amount of flow needed for agitation depends on the chemical used, as well as on the size and shape of the tank. For a simple orifice jet agitator, a flow of 6 GPM for every 100 gallons of tank capacity is usually adequate. There are several types of siphon attachments available that will help stir the tank with less flow. If these are used, the agitator flow from the pump can be reduced to two to three GPM for every 100 gallons of tank capacity. Foaming can occur if the agitation flow rate remains constant as the tank empties. You can prevent foaming by using a control valve to gradually reduce the amount of agitator flow.

Roller, diaphragm, and piston pumps usually have a flow control assembly consisting of a bypass-type pressure regulator or relief valve, a control valve, a pressure gauge, and a boom shut-off valve. Bypass pressure relief valves usually have a spring-loaded ball, disk, or diaphragm that opens with increasing pressure so that excess flow is bypassed back to the tank, thus preventing damage to the pump and other components when the boom is shut off.

When the control valve in the agitation line and the bypass relief valve in the bypass line are adjusted properly, the spraying pressure will be regulated. To adjust the system properly, follow these steps:

1. Close the control valve in the agitation line and open the boom valve.

2. Start the sprayer and run the pump at operating speed. Then adjust the relief valve until the pressure gauge reads about 10 psi above the desired spraying pressure.

3. Slowly open the control valve (agitation line) until the spraying pressure is reduced to the desired level. If the pressure cannot be lowered significantly, even with the control valve open, use larger orifice caps in the jet agitator, or use an agitator tube with larger orifices.

If there is insufficient agitation even when the spraying pressure is correct and the relief valve is closed, install a smaller orifice in the agitator. A smaller orifice allows you to increase agitation because the control valve can be opened wider at the same pressure. Because the output of a centrifugal pump can be reduced to zero without damaging the pump, a pressure relief valve and separate bypass line are not needed. The spray pressure can be controlled with simple gate or globe valves. It is preferable, however, to use special throttling valves designed to accurately control the spraying pressure. Electric-controlled throttling valves are becoming popular for remote pressure control.

The spray pressure is controlled with two throttling or control valves, one in the agitation line and the other in the spray boom line that permit agitation flow to be controlled independently of nozzle flow. To adjust for spraying, follow these steps:

1. Prime the pump with all valves open.

2. Close both valves and, with the pump running, open the boom control valve until the pressure gauge indicates the desired spray pressure.

3. Open the agitation line valve until you have sufficient agitation. If the agitation flow has lowered the pressure, readjust the boom control valve to restore the desired pressure.

Because nozzles are designed to operate within certain pressure limits, a pressure gauge must be included in every sprayer system.

The importance of a good pressure gauge cannot be overemphasized. The pressure gauge must be used for calibrating and while operating in the field. Select a gauge for the pressure range that you will be using. A range of 0 to 60 psi is adequate for most pesticides. When a 150 psi gauge is used for operating at 20 psi, accurate pressure adjustment is difficult, if not impossible.

A quick-acting boom cut-off or control valve allows the sprayer boom to be shut off while the pump and the agitation system continue to operate. Electric solenoid valves, which eliminate inconvenient hoses and plumbing, are also available.

POWER SPRAYER STRAINERS

Three types of strainers are commonly used on low pressure sprayers: tank filler strainers, line strainers, and nozzle strainers. The strainer numbers (20-mesh, 50-mesh, etc.) indicate the number of openings per inch. Strainers with high mesh numbers have smaller openings than strainers with low mesh numbers.

Coarse basket strainers set in the tank filler opening prevent twigs, leaves, and other debris from entering the tank as it is being filled. A 16- or 20-mesh tank filler strainer will retain lumps of wettable powder until they are broken up, helping to provide uniform tank mixing.

PESTPOINTER

A suction line strainer should be used between the tank and a roller pump to prevent rust, scale, or other material from damaging the pump. A 40- or 50- mesh strainer is recommended.

A suction line strainer should be used between the tank and a roller pump to prevent rust, scale, or other material from damaging the pump. A 40- or 50-mesh strainer is recommended. A suction line strainer is not usually needed to protect a centrifugal pump, except against large pieces of foreign material.

The inlet of a centrifugal pump must not be restricted. If a strainer is used, it should have an effective straining area several times larger than the area of the suction line. It should also be no smaller than 20-mesh and should be cleaned frequently A line strainer (usually 50-mesh) should be located on the pressure side of the pump to protect the spray nozzles and agitation nozzles.

Small capacity nozzles must have a strainer of the proper size to stop any particle that might plug the nozzle orifice. Nozzle strainers vary in size depending on the size of the nozzle tip used, but they are commonly 50- or 100-mesh. Nozzle catalogs list a recommended mesh size for each nozzle tip.

In general, 100-mesh strainers are recommended for most nozzles with a flow rate below 0.2 GPM, and 50-mesh strainers for nozzles with a flow rate between 0.2 and 1GPM. If a good line strainer is used, no nozzle strainer is needed for nozzles with flow rates above 1GPM. When applying wettable powders, do not use nozzles with a flow rate of less than 0.2 GPM, and use 50-mesh or larger strainers to prevent the powder from clogging the screens. Finer strainers, such as 100-mesh, can be used to protect small nozzles when applying liquid concentrates, emulsions, or soluble powders.

POWER SPRAYER HOSES AND FITTINGS

All hoses and fittings should be of a suitable quality and strength to handle the chemicals at the selected operating pressure. A good hose is flexible and durable and resistant to sunlight, oil, and chemicals. It should also be able to hold up under the rigors of normal use, such as twisting and vibration. A special reinforced hose must be used for suction lines to prevent collapsing.

> **PESTPOINTER**
>
> All hoses and fittings should be of a suitable quality and strength to handle the chemicals at the selected operating pressure. A good hose is flexible and durable and resistant to sunlight, oil, and chemicals.

Sometimes the pressure rises high above the average operating pressures. These peak pressures usually occur as the spray boom is shut off. For this reason, the sprayer hoses and fittings must always be in good condition to prevent a possible break that could cause the operator to be sprayed with the chemical.

As liquid is forced through the spray system, the pressure drops due to the friction between the liquid and the inside surface of the hoses, pipes, and fittings. The pressure drop is especially high when a large volume of liquid is forced through a small diameter hose or pipe. Pressure drop also occurs in fittings, valves, or any component through which flow is forced. It is not uncommon to have a drop in pressure of 10 to 15 psi between the outlet of the pump and the end of the boom.

To minimize pressure drop, spray lines and suction hoses must be the proper size for the system. The suction hoses should be airtight, noncollapsible, as short as possible, and as large as the opening on the intake side of the pump. A collapsed hose can restrict flow and "starve" a pump, causing decreased flow as well as damage to the pump or the pump seals. If you are having trouble maintaining spray pressure, check the suction line to be sure that it is not restricting flow.

Other lines, especially those between the pressure gauge and the nozzles should be as straight as possible with a minimum of restric-

PESTPOINTER

If you are having trouble maintaining spray pressure, check the suction line to be sure that it is not restricting flow.

tions and fittings. The proper size of these lines varies with the size and capacity of the sprayer. A high but not excessive fluid velocity should be maintained throughout the system. If the lines are too large, the velocity will be low enough for the pesticide to settle out and clog the system. If the lines are too small, an excessive drop in pressure will occur.

POWER SPRAYER BOOMS AND NOZZLES

Boom stability is important in achieving uniform spray application. The boom should be relatively rigid in all directions. It should not be allowed to swing back and forth or up and down. Dampen the breakaway hinge arrangement of the boom so that it is rigid in the fore and aft directions. The boom should be constructed to permit folding for transport. The boom height should be adjustable.

Regardless of the type of application system and cost, the selection of the correct type and size of spray nozzle is essential. The nozzle determines the amount of spray applied to an area, the uniformity of the application, the coverage of the sprayed surface, and the amount of drift. Drift can be minimized by selecting nozzles that give the largest droplet size while providing adequate coverage at the intended application rate and pressure. All nozzles develop a range of droplet sizes. Those that develop the least amount of fines are least drift prone. Although nozzles have been developed for practically every kind of spray application, only a few types are commonly used in the

> **PESTPOINTER**
>
> Drift can be minimized by selecting nozzles that give the largest droplet size while providing adequate coverage at the intended application rate and pressure.

application of turf protection products with boom sprayers. Emphasis in nozzle design during the past few years has resulted in a vast improvement in spray quality. A few of the commonly used nozzle types are described below.

Extended range flat fan nozzles are frequently used for soil and foliar applications when better coverage is required. Extended range flat fan nozzles are available in 80 and 110 degree fan angles. The spray pattern produced by this nozzle has a tapered edge distribution. The outer edges of the spray pattern have reduced volumes. This makes it necessary to overlap adjacent patterns along a boom to obtain uniform coverage (Figure 10.6). 80 degree flat fan nozzles are usually mounted on 20-inch centers at a boom height of 17 to 19 inches. 110 degree nozzles could be mounted on 20- or 30-inch centers at boom heights of 16 to18 inches, and 20 to 22 inches, respectively. To achieve maximum uniformity in the spray distribution, regardless of the spacing and height, the spray patterns should overlap 50 to 60 percent of the nozzle spacing (25 to 30 percent on each edge of the pattern). Foam markers and computer-aided guidance systems are commonly used to help operators with swath width overlap requirements on multiple passes.

For soil applications, the recommended pressure range is 10 to 30 psi. For foliar application in which smaller drops are required to increase coverage, pressures from 30 to 60 psi may be required. The incidence of drift may increase when operating pressures exceed 30 psi. Nozzle wear rate is also increased at higher pressures.

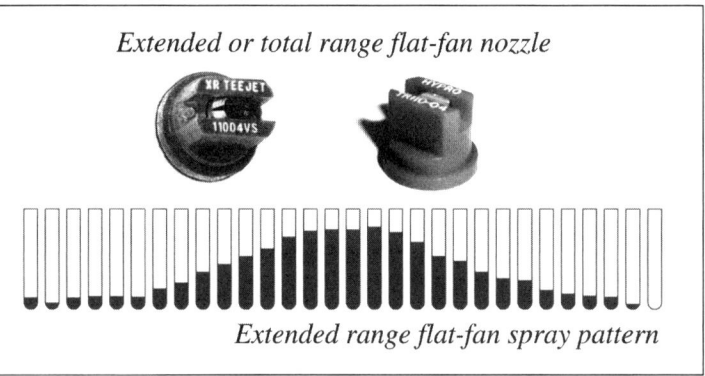

FIGURE 10.6 Extended range nozzle pattern distribution. *(Illustration courtesy of Kansas State University)*

Flooding flat fan nozzles produce a wide angle, flat fan pattern, and are used for applying herbicides, herbicide mixtures, and liquid fertilizers. The nozzle spacing should be 40 inches or less for common sprayer application. These nozzles are most effective in reducing drift when they are operated within a pressure range of 10 to 30 psi. Pressure changes affect the width of the spray pattern more with the flooding flat fan nozzle than with the extended range flat fan nozzle. In addition, the distribution pattern is usually not as uniform as that of the extended range flat fan tip (Figure 10.7). The best distribution is achieved when the nozzle is mounted at a height and angle that obtains at least double coverage or 100 percent overlap. Uniformity of application depends upon the pressure, height, spacing, and orientation of the nozzles. Pressure directly affects the droplet size, nozzle flow rate, spray angle, and pattern uniformity. At low pressures, flooding nozzles produce large spray drops; at high pressures, these nozzles actually produce smaller drops than flat fan nozzles at an equivalent flow rate.

The spray distribution of flooding nozzles varies greatly with changes in pressure. At low pressures, flooding nozzles produce a fairly uniform pattern across the swath, but at high pressures the pattern becomes heavier in the center, tapering off toward the edge. The width of the spray pattern is also affected by pressure. To obtain

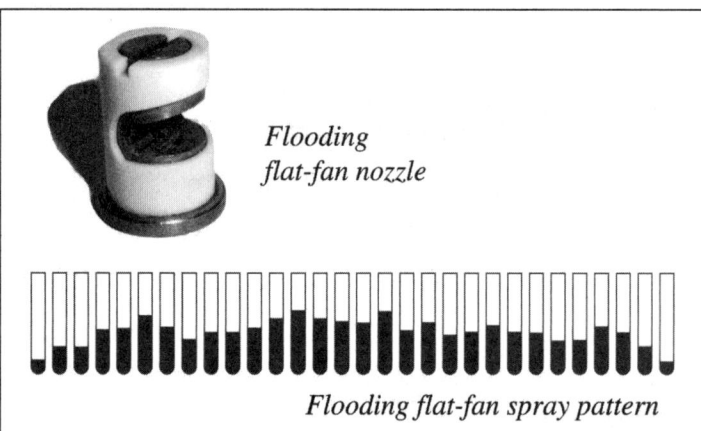

FIGURE 10.7 Flooding flat fan nozzle pattern distribution. *(Illustration courtesy of Kansas State University)*

an acceptable distribution pattern and overlap, you should operate flooding nozzles within a pressure range of 10 to 30 psi.

Nozzle height is critical in obtaining uniform application when using flooding nozzles. Flooding nozzles can be mounted vertically to spray backwards, horizontally to spray downward, or at any angle between vertical and horizontal. When the nozzle is mounted horizontally to spray downward, heavy concentrations of spray tend to occur at the edges of the spray pattern. Rotating the nozzles 30 to 45 degrees from the horizontal will usually increase pattern uniformity over the recommended pressure range of 10 to 30 psi. For the

PESTPOINTER

The best distribution is achieved when the nozzle is mounted at a height and angle that obtains at least double coverage or 100 percent overlap.

PESTPOINTER

The width of the spray pattern is also affected by pressure. To obtain an acceptable distribution pattern and overlap, you should operate flooding nozzles within a pressure range of 10 to 30 psi.

uniform distribution over a range of pressures, mount the nozzles to obtain double coverage at the lowest operating pressure.

The most recent improvements in nozzle design have incorporated a pre-orifice concept with an internal turbulation chamber. These design changes have resulted in larger, less driftable droplets and improved spray pattern uniformity. Turbulation chamber nozzles are available in flood and flat fan tip designs.

Turbo Flood® nozzles combine the precision and uniformity of extended range flat fan spray tips with the plugging resistance and wide-angle pattern of older style flooding flat fan nozzles. The design of Turbo Flood® nozzles results in larger droplets and improved distribution uniformity. Turbulence in the chamber portion of the spray tip lowers exit pressure, reducing the formation of small driftable droplets. Exit orifice design changes improve pattern uniformity over older style flooding nozzles (Figure 10.8). Turbo Flood® nozzles are designed to operate at pressures of 10 to 40 psi. Turbo Flood® nozzles require at least 50 to 60 percent of the nozzle spacing (25 to 30 percent on each edge of the pattern). The relationship between nozzle pressure, height, and spacing is critical for obtaining uniform application. Nozzles can be mounted vertically to spray backwards, horizontally to spray downward, or any angle between vertical and horizontal. For uniform distribution, proper overlap is required regardless of the nozzle-mounting angle. Turbo Flood® nozzles are highly recommended when applying tank mix combinations of fertilizer

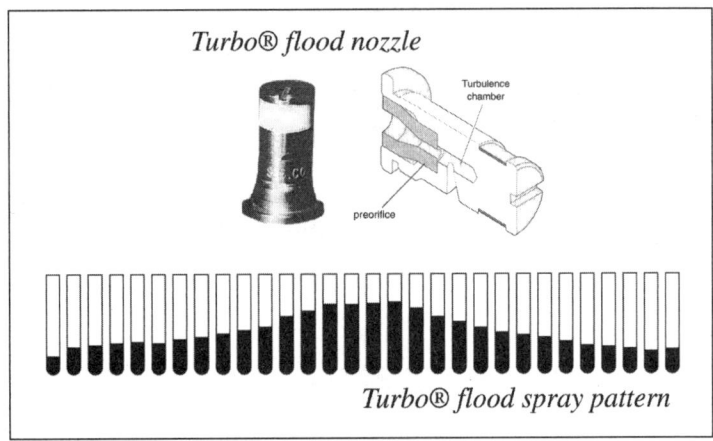

FIGURE 10.8 Turbo® flooding flat fan nozzle pattern distribution. (Illustration courtesy of Kansas State University)

and herbicides. Turbo Flood® nozzles produce larger droplet spectrums than standard flooding nozzles and would work well in drift-sensitive applications.

The Turbo® flat fan nozzle design develops a greatly improved spray pattern when compared to the extended range flat fan and other drift reduction flat fan nozzles (Figure 10.9). This nozzle was modeled after the Turbo® Flood, but for use in the application of postemergence products. Turbo® flat fan nozzles are wide-angle, pre-orifice nozzles that create larger spray droplets across a wider pressure range (15 to 90 psi) than comparable low-drift tips reducing the amount of driftable particles. The unique design of the nozzles allows them to be mounted in a flat fan nozzle body configuration. The wide spray angle will allow for a 20 or 30 inch nozzle spacing and requires an overlap of at least 50 to 60 percent of the nozzle spacing (25 to 30 percent on each edge of the pattern) to achieve uniform application across the boom. Position the tip so that the pre-set spray angle is directed away from the direction of travel. The Turbo® flat fan nozzle is recommended for use with electronic spray controllers where speed and pressure changes occur regularly.

Chemical and Pesticide Application Equipment and Calibration 427

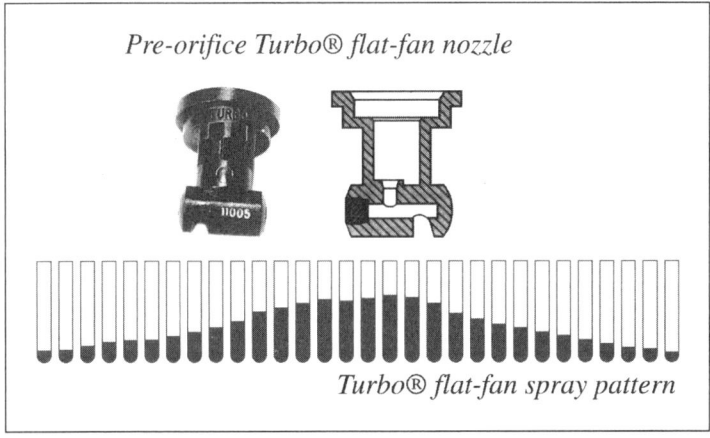

FIGURE 10.9   Turbo® flat fan nozzle pattern distribution. *(Illustration courtesy of Kansas State University)*

The Turbo® turf flood is a new nozzle designed for the turf industry. It is modeled after the Turbo® flood nozzle, which is used extensively in the application of crop protection products for agricultural field crops. The major difference is the Turbo® turf flood nozzle incorporates a larger orifice to accommodate heavier application volumes, which are common in the turf boom sprayer industry. Otherwise this nozzle exhibits the same high quality spray pattern when placed on the boom from 20 to 30 inches apart and at a height above the turf at 14 to 20 inches (Figure 10.10). Actual spacing should overlap at

PESTPOINTER

The Turbo® flat fan nozzle is recommended for use with electronic spray controllers where speed and pressure changes occur regularly.

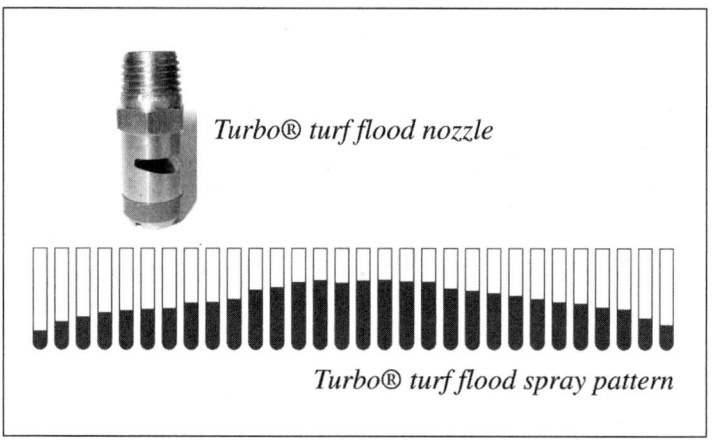

FIGURE 10.10 Turbo® turf flooding flat fan nozzle pattern distribution. *(Illustration courtesy of Kansas State University)*

least 50 to 60 percent of the nozzle spacing (25 to 30 percent on each edge of the pattern) for uniform application. As with the field crop version of this nozzle, the Turbo® turf flood has excellent drift control, resulting from the turbulation chamber creating larger spray droplets and less driftable fines. This nozzle may have use in applying certain agricultural products on soil as a replacement for the Raindrop® nozzle.

A recent trend in drift reduction nozzle design is incorporating air into the spray mixture to produce an air-fluid mix. Several different designs are currently being marketed and are commonly referred to as air-induction or venturi nozzles. Air is entrapped into the spray solution within the nozzle. To accomplish the mixing, an inlet port venturi is typically used to draw the air into the tip under a reduced pressure. The air-fluid mixture forms a larger spray droplet to help transport the droplets to the target. By increasing the size of the spray droplet, spray drift is reduced by minimizing smaller driftable fines. Current design of these tips requires a higher operating pressure to maximize performance. Most all venturi nozzles are designed to spray a wide-angle flat spray pattern (Figures 10.11 and 10.12).

PESTPOINTER

Venturi nozzles, which are currently more expensive, dramatically reduce the potential for drift. In addition to providing good protection against drift, research indicates they also provide adequate efficacy.

Venturi nozzles, which are currently more expensive, dramatically reduce the potential for drift. In addition to providing good protection against drift, research indicates they also provide adequate efficacy. The efficacy levels achieved relate closely to coverage and mode of action for the turf protection products being used. It is also important to maintain at least 40 psi as an operating pressure to maintain uniform pattern development while properly atomizing the spray solution.

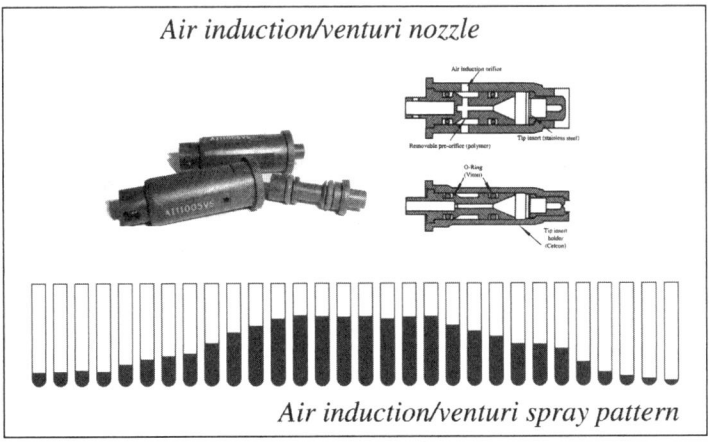

FIGURE 10.11 Air induction nozzle pattern distribution. *(Illustration courtesy of Kansas State University)*

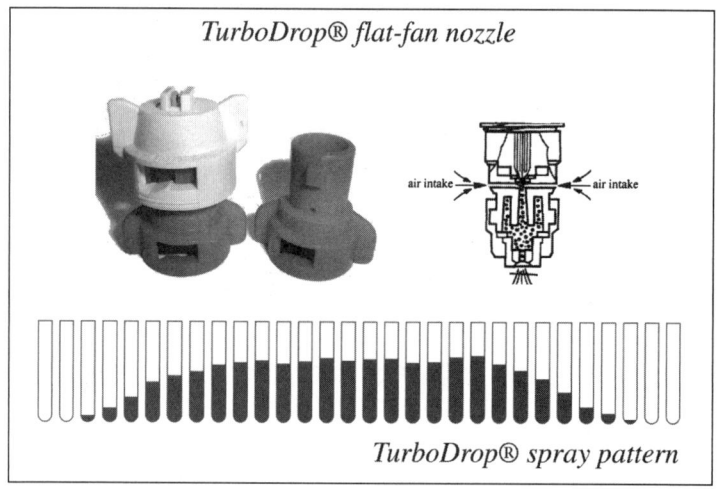

FIGURE 10.12 TurboDrop® nozzle pattern distribution. *(Illustration courtesy of Kansas State University)*

Please note any special calibration requirements for the venturi nozzles. For example, Greenleaf, designer of the TurboDrop® venturi two-piece nozzle, requires the exit orifice to be two times (2X) the size of the venturi orifice. Otherwise the exit orifice may create a negative pressure effect in the venturi area, resulting in failure of the nozzle to create the proper spray quality (actually reversing flow from the air inlets). Therefore, you will need to select and calibrate the TurboDrop® nozzle based on the venturi orifice, which is color-coded to meet manufacturing standards. A chart is available from the manufacturer for this purpose. Other venturi nozzle styles are one piece and do not have this precaution.

Spray nozzle assemblies consist of a body, cap, check valve, and nozzle tip. Various types of bodies and caps (including color-coded versions) and multiple nozzle bodies are available with threads as well as quick-attaching adapters.

Nozzle tips are interchangeable or molded in the nozzle cap and are available in a variety of materials (Figure 10.13) including:

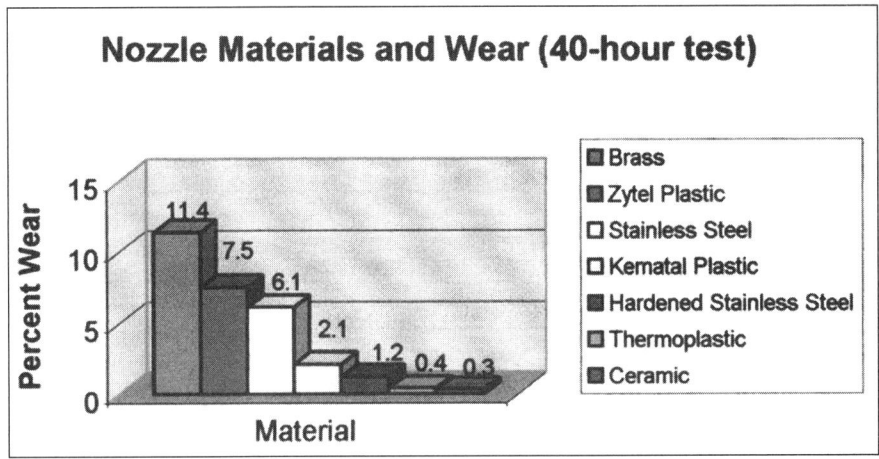

FIGURE 10.13 Materials used to make nozzle tips and their wear characteristics.

- *Hardened stainless steel and ceramic—the most wear-resistant materials, but also the most expensive. Stainless steel tips have excellent wear resistance with either corrosive or abrasive materials.*
- *Plastic—resistant to corrosion and abrasion and are proving to be very economical tips for applying crop protection products*
- *Brass—very common, but are not recommended for use. They wear rapidly when used to apply abrasive materials, such as wettable powders, and are corroded by some liquid fertilizers.*

Typically, smaller tips with elongated orifices are impacted greatest by wear. Spray tip life is dependent on pressure, how abrasive the spray solution is and other factors, such as corrosion. A good rule of thumb is to change tips when the flow becomes 10 percent greater than in new tips.

Venturi nozzles are typically designed from polymers and may incorporate stainless or hardened stainless orifice inserts for the actual tip. Many of these nozzles are designed as two-piece units with the pre-orifice removable for easier cleaning.

> **PESTPOINTER**
>
> Venturi nozzles are typically designed from polymers and may incorporate stainless or hardened stainless orifice inserts for the actual tip.

No single tip will perform well in all applications currently being used. Refer to manufacturer catalogs and web pages for selection and setup assistance. Proper selection and setup will enhance the efficacy and safety of all spray applications. Many chemical labels may specify a droplet quality classification in the near future. These are excellent resources to ensure compliance.

SPRAY GUNS

Spray guns for spraying turf areas range from those that can produce a low flow rate with a wide-cone spray pattern, a flooding or showerhead nozzle pattern, or those that can produce a high flow rate with a straight-stream spray pattern (Figure 10.14). Spray guns are not usually recommended for spraying turf areas such as lawns or golf course greens. However, much of the commercial lawn applications are currently being applied with turf guns equipped with showerhead type nozzles. If used in this manner, four factors are critical for delivering the correct rate uniformly over the application area:

- *Exact pressure must be monitored.*
- *Proper walking speed or pace must be maintained.*
- *Uniform hand/arm motion technique must be used.*

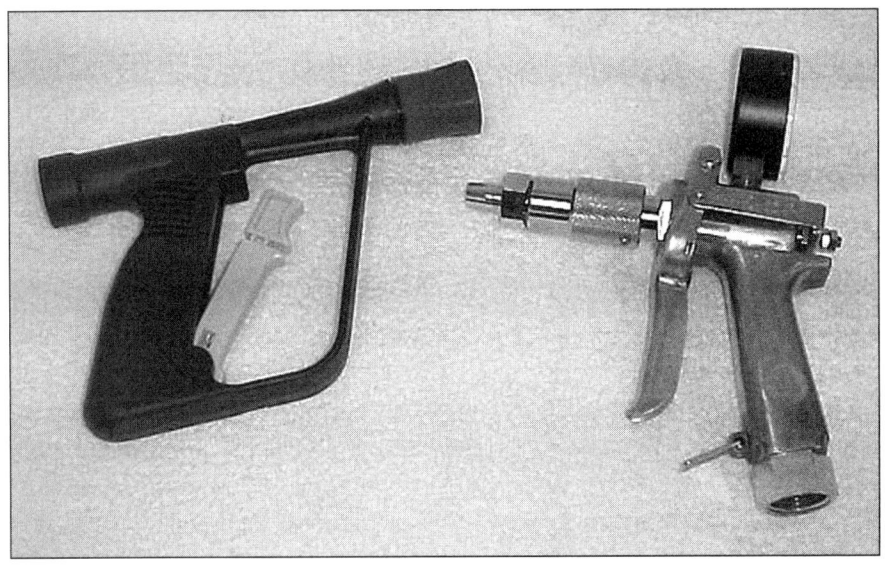

FIGURE 10.14 Turf spray gun with shower head.

- Constant nozzle height and angle in reference to the ground should be maintained.

It is quite difficult to obtain uniform coverage from a spray gun on turf areas. If it is not possible to use a conventional sprayer with a boom, use a hand or walking boom with conventional nozzles. When the spray gun must be used because of rough or irregularly shaped areas, you should be aware of the difficulty in obtaining a

PESTPOINTER

Spray guns are not usually recommended for spraying turf areas such as lawns or golf course greens.

uniform spray in the correct amount over the area. Pressure gauges at the pump and on the spray gun are recommended to indicate line pressure loss and give some indication of output.

CALIBRATING SPRAYERS

The performance of any pesticide depends upon the proper application of the correct amount of chemical. Most performance complaints involving pesticides are directly related to errors in dosage or to improper application. Failure to calibrate a sprayer can injure your turf, cause pollution, and waste money. The purpose of calibration is to insure that your sprayer is applying the correct amount of material uniformly over a given area. Accurate calibration is the only way to know how much chemical is applied. Even with the current widespread use of electronics to monitor and control the application of pesticides, a thorough sprayer calibration procedure is essential to ensure against misapplication. In addition to calibrating the sprayer at the start of the season, you should recalibrate regularly. Abrasive pesticide formulations can wear nozzle tips, resulting in an increased nozzle flow rate and the development of poor spray patterns.

PESTPOINTER

Even with the current widespread use of electronics to monitor and control the application of pesticides, a thorough sprayer calibration procedure is essential to ensure against misapplication. In addition to calibrating the sprayer at the start of the season, you should recalibrate regularly.

Calibrating Manual Sprayers and Spray Guns

Because hand sprayers are generally used to spray limited areas, the amount of spray volume should be determined on small areas such as 1,000 square feet. Most manual compressed-air sprayers do not have pressure gauges or pressure controls. The pressure in the tank will drop as the material is sprayed. This pressure drop can be partially overcome by:

1. Filling the tank only one-half full with spray material so that considerable air space remains for initial expansion.

2. Re-pressurizing the tank frequently. If the sprayer has a pressure gauge, re-pressurize when the pressure drops approximately 10 psi from the initial reading.

As mentioned earlier, equipping with a pressure control valve will also help maintain a constant flow of material. When spraying, either hold the nozzle steady at a constant height and walk back and forth, or swing the nozzle in a sweeping, overlapping motion. Maintain a uniform walking speed during application.

Step 1. Measure and mark off an area of 1,000 square feet (for example, 20 feet by 50 feet). Practice spraying the area with water. To obtain the most uniform application, spray an area twice, spraying the second time by adjusting the swath by one-half and spraying in the same direction or in the middle of the previous pass. This technique means that each application is at a one-half rate. To monitor for a uniform application, spray on a fast drying surface, such as a concrete or asphalt driveway, and observe the drying pattern. A uniform application will dry evenly across the sprayed area. If not applied evenly, the drying pattern will exhibit streaks across the sprayed area.

Step 2. Once you are able to maintain a uniform spray, add a measured amount of water to the tank, spray the area, then measure the amount of water remaining in the tank. The difference between the amount in the tank before and after spraying is the amount used. For example, 2 gallons added to the tank minus ½ gallon remaining equals 1½ gallons used per 1,000 square feet. Thus, the application rate for the above conditions is 1½ gallons per 1,000 square feet.

To calibrate a hand held spray gun, spray a known area (i.e. 1,000 square feet) and record the time. Be sure to use the proper technique to maintain adequate coverage. Then catch the spray from the hand gun into a bucket for the length of time it took to spray the known area. The application rate per 1,000 square feet is the amount collected in the bucket for that same time period.

Calibrating Power Sprayers

Proper power sprayer calibration is all the more critical because everything is being done on a larger scale. Three variables affect the amount of spray mixture applied per 1,000 square feet or per acre:

1. The nozzle flow rate
2. The ground speed of the sprayer
3. The effective sprayed width per nozzle

To calibrate and operate your sprayer properly, you must know how each of these variables affects sprayer output. The nozzle flow rate varies with the size of the tip, the nozzle pressure, and the density of the spray liquid. Installing a nozzle tip with a larger orifice, increasing the pressure, and decreasing the density of the spray liquid all increase the flow rate. To increase the nozzle output, you must multiply the pressure by the square of the desired increase in flow rate. In other words, doubling the pressure will not double the nozzle flow rate. To double the flow rate, you must increase the pressure four times. For example, to double the flow rate of a nozzle from 0.2 gallons per minute at 10 psi to 0.4 gallons per minute, the pressure must be increased to 40 psi (4 x 10).

Pressure changes should not be used to make major adjustments in the application rate. To obtain a uniform spray pattern and minimize drift, you should maintain the operating pressure within the recommended range for each nozzle. The pressure can be changed, however, to correct for minor variations in flow rate resulting from nozzle wear. Spray pressure loss through spray lines from the pump to the nozzle is common. Also, if you use check valves to prevent nozzle drip, the pressure at the nozzle is 5 to 7 psi lower than the boom pressure indicated on the pressure gauge. Always monitor application pressure and flow at the nozzle to maintain output accuracy.

Many low-pressure boom sprayers have a metering system that maintains a constant application rate while operating over a range of travel speeds. All metering systems now in use, such as ground-driven piston pumps, electronic feedback control systems, and various centrifugal pump arrangements, vary the nozzle pressure to compensate for changes in travel speed, keeping the application rate constant. Although all the systems work over a wide range of travel speeds, the spray nozzle limits the range of speeds at which precise application can be obtained. Because of the possibilities for dramatic pressure increases while using such systems, there is a serious potential for spray drift occurrence.

To regulate the flow in proportion to travel speed, the rate of increase in nozzle pressure must vary with the square of the rate

PESTPOINTER

To obtain a uniform spray pattern and minimize drift, you should maintain the operating pressure within the recommended range for each nozzle.

of increase in speed. For example, if the sprayer is traveling at 2.5 MPH at a nozzle pressure of 20 psi, increasing the speed to 5 MPH will require increasing the nozzle pressure to 80 psi to maintain the same flow volume. Remember, a fourfold change in pressure drastically reduces the droplet size, which may result in increased drift. The pattern width and distribution pattern may also be affected. For uniform application, the travel speed should be held as nearly constant as possible, even when using controlled metering systems.

To apply chemicals accurately, you must maintain the proper ground speed. Do not rely on a conventional speedometer as an accurate indicator of speed. Slippage of the drive wheels can result in speedometer errors of 20 percent or more. Electronic kits and radar guns give more accurate readings since they do not depend on the drive wheels for speed measurements. Changes in tire size also affect speedometer readings, and the accuracy of all speedometers should be checked periodically.

The effective width sprayed per nozzle also affects the spray application rate. Doubling the effective sprayed width per nozzle decreases the application rate by one-half. For example, if you are applying 40 gallons per acre (GPA) with flat fan nozzles on 20-inch spacing and change to flooding nozzles with the same flow rate on 40-inch spacing, the application rate decreases from 40 GPA to 20 GPA.

PESTPOINTER

Do not rely on a conventional speedometer as an accurate indicator of speed. Slippage of the drive wheels can result in speedometer errors of 20 percent or more. Electronic kits and radar guns give more accurate readings since they do not depend on the drive wheels for speed measurements.

To obtain uniform coverage, you must consider the spray angle, spacing, and height of the nozzle. The height must be readjusted for uniform coverage with various spray angles and nozzle spacings. Do not use nozzles with different spray angles on the same boom for broadcast spraying. Be sure the nozzle tips are clean. If necessary, clean with a soft bristle brush. A nail, wire, or pocket knife can damage the tip and ruin the uniformity of the spray pattern. While the sprayer is running, observe each spray tip for any distortions in the patterns.

Worn or partially plugged nozzles produce erratic spray patterns. Misalignment of nozzle tips is a common cause of uneven coverage. The boom must be level at all times to maintain uniform coverage. Skips and uneven coverage will result if one end of the boom is allowed to droop. A good method for determining the exact nozzle height that will produce the most uniform coverage is to spray on a warm surface, such as a road, and observe the drying rate. Streaks in the spray pattern should be obvious. Replace nozzles that are not performing correctly.

When you are convinced that the sprayer is operating properly, you are ready to calibrate. There are many methods for calibrating low-pressure boom sprayers, but they all involve the use of the variables in the equation that follows. Any technique for calibration that provides accurate and uniform application is acceptable. No single method is best for everyone.

PESTPOINTER

A good method for determining the exact nozzle height that will produce the most uniform coverage is to spray on a warm surface, such as a road, and observe the drying rate.

The calibration method described below has three advantages:

- First, it allows you to select the number of gallons to apply per acre or per 1000 square feet.
- Second, it provides a simple means for frequently adjusting the calibration to compensate for changes due to nozzle wear.
- Third, it allows the spray operator to complete most of the calibration ahead of spraying season. This method requires knowledge of nozzle types and sizes and the recommended operating pressure ranges for each type of nozzle used.
- Finally, when using the method below, the applicator will have a better understanding of how each variable will affect the application rate.

As each of the variables change, the influence on the rate (gallons per acre or per 1000 square feet) is apparent.

The gallons of spray applied per acre or 1000 square feet can be determined by using the following equation:

$$\text{Gallons per acres} = \frac{\text{GPM} \times 5940}{\text{MPH} \times \text{W}}$$

or

$$\text{Gallons per 1000 square feet} = \frac{\text{GPM} \times 136}{\text{MPH} \times \text{W}}$$

In these equations, the numbers or abbreviations and their meanings are as follows:

GPM = output per nozzle in gallons per minute
MPH = ground speed in miles per hour
W = effective width sprayed per nozzle in inches

5,940 = a constant to convert gallons per minute, miles per hour, and inches to gallons per acre

136 = a constant used to convert gallons per minute, miles per hour, and inches to gallons per 1,000 square feet.

As mentioned previously, the size of the nozzle orifice will depend on the application rate (GPA or gallons per 1000 sq. ft.), ground speed (MPH), and effective width sprayed (W) that you plan to use. Some manufacturers advertise "gallon-per-acre" nozzles, but this rating is useful only for standard conditions (usually 30 psi, 4 MPH, and 20-inch spacing). The gallons-per-acre rating is useless if any one of your conditions varies from the standard.

A more exact method for choosing the correct orifice size is to determine the gallons per minute (GPM) required for your conditions. Then select nozzles that provide this flow rate when operated within the recommended pressure range. By following the five steps described below, you can select the nozzles required for each application well ahead of the spraying season.

Step 1. From the label, select the spray application rate in gallons per acre (GPA) or gallons per 1000 square feet. Pesticide labels recommend ranges for various types of equipment. The spray application rate is the gallons of carrier (water, fertilizer, etc.) and pesticide applied per treated acre.

Step 2. Select or measure an appropriate ground speed in miles per hour (MPH) according to existing turf conditions. Do not rely on speedometers as an accurate measure of speed. Slippage and variation in tire sizes and spray loads can result in speedometer errors of 20 percent or more. If you do not know the actual ground speed, you can easily measure it. (Instructions for measuring ground speed are given below).

Step 3. Determine the effective width sprayed per nozzle (W) in inches. For broadcasting spraying, W = the nozzle spacing.

Step 4. Determine the flow rate required from each nozzle in gallons per minute (GPM) by using a nozzle catalog, tables, or the following equation. Using the equation below allows the applicator to determine flow rates for each application scenario needed for the application season. This can be done before the application season begins, thus not interfering with critical time available during the application time.

$$GPM = \frac{GPA \times MPH \times W}{5940}$$

or

$$GPM = \frac{\text{Gallons per 1000 square feet} \times MPH \times W}{136}$$

Again, the numbers or abbreviations and their meanings are as follows:

GPM = gallons per minute of output required from each nozzle
GPA = gallons per acre from Step 1
MPH = miles per hour from Step 2
W = inches sprayed per nozzle from Step 3
5,940 = a constant to convert gallons per minute, miles per hour, and inches to gallons per acre
136 = a constant used to convert gallons per minute, miles per hour, and inches to gallons per 1,000 square feet.

Step 5. Select a nozzle that will give the flow rate determined in Step 4 when the nozzle is operated within the recommended pressure range. You should obtain a catalog of available nozzle tips. These catalogs can be obtained free of charge from equipment dealers or nozzle manufacturers. If you decide to use nozzles that you already have, return to Step 2 and select a speed that allows you to operate within the recommended pressure range. Proceed to that discussion.

Example for Steps 1 through 5 on the previous page. You want to broadcast a fungicide at 60 GPA (Step I) at a speed of 3.5 MPH

Chemical and Pesticide Application Equipment and Calibration 443

(Step 2), using Turbo® Turf Flooding (TTJ) flat fan nozzles spaced 20 inches apart on the boom (Step 3). What size orifice should you select? First, you need to determine the required flow rate for each nozzle by using the proper equation in Step 4:

$$GPM = \frac{GPA \times MPH \times W}{5940} \qquad GPM = \frac{60 \times 3.5 \times 20}{5940} = 0.71$$

The orifice size that you select must have a flow rate of 0.71 GPM when operated within the recommended pressure range of 20 to 30 PSI. Table 10.2 or a manufacturer's tip guide shows the GPM at various pressures for the Spraying Systems Turbo Turf Flooding flat fan tip. For example, the TTJ-04 sprayed at 32 psi or the TTJ-05 sprayed at 20 psi will provide an output of 0.71 GPM when sprayed at a travel speed of 3.5 MPH (Step 5). Either of these nozzles would be suitable for this application.

TABLE 10.2 Turbo Turf® flooding flat fan tip selection chart.

Spraying Systems Designation	Liquid Pressure (psi)	Gal/min (GPM)	Oz./minOrifice (OPM)
TTJ02	10	0.10	13
TTJ02	20	0.14	18
TTJ02	30	0.17	22
TTJ02	40	0.20	26
TTJ04	10	0.20	26
TTJ04	20	0.28	36
TTJ04	30	0.35	45
TTJ04	40	0.40	51
TTJ05	10	0.25	32
TTJ05	20	0.35	45
TTJ05	30	0.43	55

Continued on next page

TABLE 10.2 *(continued)* Turbo Turf® flooding flat fan tip selection chart.

Spraying Systems Designation	Liquid Pressure (psi)	Gal/min (GPM)	Oz./minOrifice (OPM)
TTJ05	40	0.50	64
TTJ06	10	0.30	38
TTJ06	20	0.42	54
TTJ06	30	0.52	67
TTJ06	40	0.60	77
TTJ08	10	0.40	51
TTJ08	20	0.57	73
TTJ08	30	0.69	88
TTJ08	40	0.80	102
TTJ10	10	0.50	64
TTJ10	20	0.71	91
TTJ10	30	0.87	111
TTJ10	40	1.00	128
TTJ15	10	0.75	96
TTJ15	20	1.06	136
TTJ15	30	1.30	166
TTJ15	40	1.50	192

To measure ground speed, lay out a known distance in the area to be sprayed or in an area with similar surface conditions. Suggested distances are 100 feet for speeds up to 5 MPH, 200 feet for speeds from 5 to 10 MPH, and at least 300 feet for speeds above 10 MPH. At the engine throttle setting and gear that you plan to use during spraying with a loaded sprayer, determine the travel time between the measured stakes in each direction. Average these speeds and use the following equation or the table on this page to determine ground speed.

$$\text{Speed (mph)} = \frac{\text{distance (feet)} \times 60}{\text{time (seconds)} \times 88}$$

(1 mph = 88 feet in 60 seconds)

Example: You measure a 200 foot course, and discover that 22 seconds are required for the first pass and 24 seconds for the return pass, making for an average of 23 seconds.

$$MPH = \frac{200 \times 60}{23 \times 88} = \frac{12000}{2024} = 5.9$$

Once you have decided upon a particular speed, record the throttle setting and drive gear used.

If you plan to use a set of nozzles that you already have on the sprayer then you will need to determine the nozzle flow rate so a ground speed can be calculated. This calculated speed will then allow you to achieve the proper application rate for the pesticide product being used. Let's say that the pesticide you are applying requires a carrier rate of 20 GPA and you are using Turbo® flat fan tips spaced on 20-inch centers. Measure the flow rate (gallons per minute) from the existing nozzles at a pressure within the recommended range for that nozzle type. Use a collection container or other suitable flow measuring device (Figure 10.15). Then using that flow rate, insert the measured gallons per minute flow rate into the formula as shown below.

In this example, 0.30 gallons per minute was collected at a pressure of 25 pounds per square inch (psi).

$$MPH = \frac{GPM \times 5940}{GPA \times W \text{ (inches)}} = \frac{0.3 \times 5940}{20 \times 20} = 4.5$$

The calculated speed is 4.5 MPH. If you travel 4.5 MPH using the existing nozzles on 20-inch spacings at 25 psi, then the application rate will be the desired 20 GPA.

Selecting Nozzles for Hand-Held Booms

The size of the nozzle for a hand-held sprayer will depend upon the application rate (gallons per 1,000 square feet), walking speed

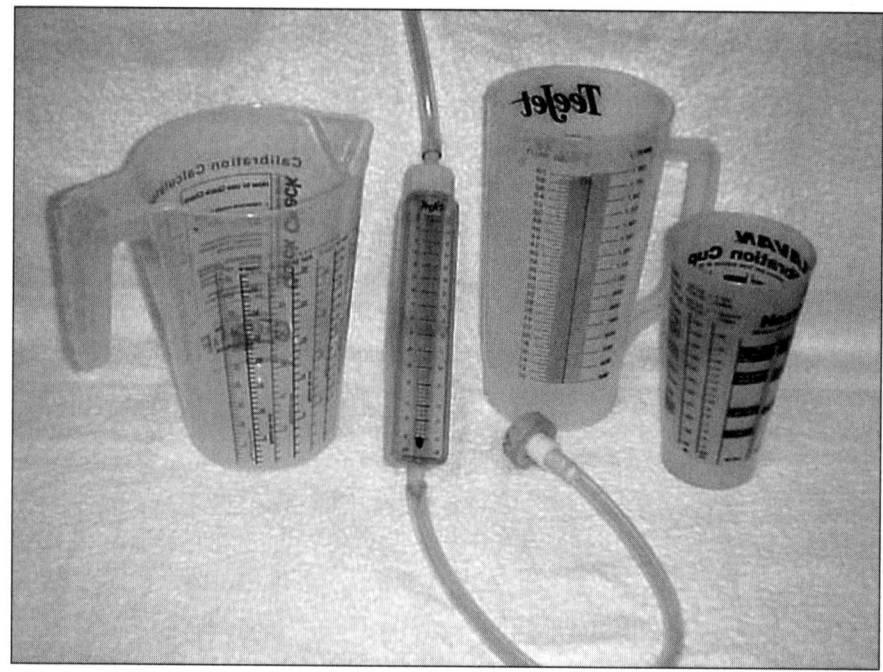

FIGURE 10.15 Flow rate indicator and calibration container.

(minutes per 1,000 square feet), and effective spray width (the number of feet between the nozzles multiplied by the number of nozzles).

An exact method for choosing the correct orifice is to determine the gallons per minute required for your conditions; then select nozzles from a nozzle manufacturer's catalog that, when operated within the recommended pressure range, provide this flow rate. By following the steps described below, you can select the nozzles required for each application well ahead of the spraying season.

Step 1. Determine the application rate in gallons per 1,000 square feet. Pesticide labels indicate recommended ranges for various types of equipment and pests. The spray application rate may be given in gallons of carrier (water, fertilizer, etc.) and pesticide applied per acre rather than in gallons per 1,000 square feet. Gallons per acre (GPA) can be converted to gallons per 1,000 square feet by the following equation:

$$\text{Gallons per 1000 square feet} = \frac{GPA \times 1000}{43{,}560}$$

where: 43,560 = number of square feet per acre

Example: A rate of 50 GPA is recommended on the herbicide label. What is the application rate in gallons per 1,000 square feet?

$$\text{Gallons per 1,000 square feet} = \frac{50 \times 1000}{43{,}560} = \frac{50{,}000}{43{,}560} = 1.15$$

Rule of thumb: One gallon per 1,000 square feet equals 43.56 gallons per acre.

Step 2. Determine the effective swath width in feet. For hand-held booms, the swath width is the distance between the nozzles multiplied by the number of nozzles on the spray boom.

Example: Your walking boom has 3 turbo flat fan nozzles spaced 20 inches apart. (20 inches equals 1.67 feet). What is the effective swath width?

Swath width = 3 nozzles x 1.67 feet = 5 feet

Step 3. Measure the time in minutes required to spray 1,000 square feet. The time can easily be measured by using your swath width from Step 2 and laying out a course measuring 1,000 square feet. To lay out a single pass course that contains 1,000 square feet, use the following equation:

$$\text{Distance of course} = \frac{1000 \text{ square feet}}{\text{swath width in feet}}$$

Example: Your boom has an effective swath width of 5 feet. What length of course is required for a 1,000 square-foot, single-pass course?

$$\text{Distance of course} = \frac{1000 \text{ square feet}}{5 \text{ feet}} = 200 \text{ feet}$$

Mark off a 200 foot course and time your walking speed along this course in both directions; then calculate your average time. For example, if one timing took 44 seconds and a second took 46 seconds, the average time would be 45 seconds (44 + 46/2). To convert seconds to minutes, divide the average time by 60 seconds per minute. In this example, the time required to spray a 200-foot course with your three-nozzle boom equals:

$$\frac{45 \text{ seconds}}{60 \text{ seconds per minute}} = 0.75 \text{ minute}$$

Step 4. Use the following equation to determine the flow rate required from each nozzle in gallons per minute (GPM):

$$\text{GPM} = \frac{\text{gallons per 1000 square feet}}{\text{minutes per 1000 square feet}}$$

Example: If the application rate is 1.15 gallons per 1,000 square feet (Step 1) and the time required to spray 1,000 square feet is 0.75 minute (Step 3), what is the required nozzle flow rate from the boom using Equation 6?

$$\text{GPM} = \frac{1.15}{0.75} = 1.5$$

Since there are 3 nozzles per boom, the required flow rate from the boom (1.5 GPM) would be divided by 3:

$$1.5/3 = 0.5$$

The required flow rate per nozzle would be 0.5 gallons per minute.

Step 5. Select a nozzle that will give the flow rate determined in Step 4 when the nozzle is operated within the recommended pressure range. The nozzle you select must have a flow rate of 0.5 gallons per minute when operated within the recommended pressure range of 15 to 40 psi. Table 10.3 provides gallons-per-minute rates at various pressures for turbo flat fan nozzles. For example, the TT10006 nozzle has a rated output of 0.5 gallons per minute at near 29 psi. This nozzle would be suitable for your boom.

After making sure that your sprayer is clean, install the selected nozzle tips, partially fill the tank with clean water, and operate the sprayer at a pressure within the recommended range. Place a suitable container under each nozzle or use a nozzle flow rate device. Check to see whether all of the containers fill at about the same time or the flow rate measured from each nozzle is equal. Replace any nozzle that has an output of five percent more or less than the average of all the nozzles, an obviously different fan angle, or a nonuniform appearance in spray pattern. To obtain uniform coverage, you must consider the spray angle, spacing, and height of the nozzle. The height must be readjusted for uniform coverage with various spray angles and nozzle spacing (see previous discussion of nozzle set up). Do not use nozzles with different spray angles on the same boom for broadcast spraying.

Worn or partially plugged nozzles and misalignment of nozzle tips will produce non uniform patterns. Skips and uneven coverage will also result if one end of the boom is allowed to raise or droop. A practical method for determining the exact nozzle height that will produce the most uniform coverage is to spray on a warm surface such as a road and observe the drying rate. Adjust the height to eliminate excess streaking.

TABLE 10.3 Turbo flat fan selection chart.

Spraying Systems Orifice Designation	Liquid Pressure (psi)	Gal/min (GPM)	Oz./min (OPM)
TT11001	15	0.061	8
TT11001	20	0.071	9
TT11001	30	0.087	11
TT11001	40	0.10	13
TT11001	50	0.11	15
TT11001	60	0.12	15
TT11001	75	0.14	18
TT11001	90	0.15	19
TT110015	15	0.09	12
TT110015	20	0.11	14
TT110015	30	0.13	17
TT110015	40	0.15	19
TT110015	50	0.17	22
TT110015	60	0.18	23
TT110015	75	0.21	26
TT110015	90	0.23	27
TT11002	15	0.12	15
TT11002	20	0.14	18
TT11002	30	0.17	22
TT11002	40	0.20	26
TT11002	50	0.22	28
TT11002	60	0.24	31
TT11002	75	0.27	33
TT11002	90	0.30	36
TT11003	15	0.18	23
TT11003	20	0.21	26
TT11003	30	0.26	33
TT11003	40	0.30	38
TT11003	50	0.34	44
TT11003	60	0.37	47
TT11003	75	0.41	51

Continued on next page

TABLE 10.3 (continued) Turbo flat fan selection chart.

Spraying Systems Orifice Designation	Liquid Pressure (psi)	Gal/min (GPM)	Oz./min (OPM)
TT11003	90	0.45	54
TT11004	15	0.24	31
TT11004	20	0.28	36
TT11004	30	0.35	45
TT11004	40	0.40	51
TT11004	50	0.45	58
TT11004	60	0.49	63
TT11004	75	0.55	68
TT11004	90	0.60	73
TT11005	15	0.31	37
TT11005	20	0.35	45
TT11005	30	0.43	55
TT11005	40	0.50	64
TT11005	50	0.56	72
TT11005	60	0.61	78
TT11005	75	0.68	84
TT11005	90	0.75	91
TT11006	15	0.37	47
TT11006	20	0.42	54
TT11006	30	0.52	67
TT11006	40	0.60	77
TT11006	50	0.67	86
TT11006	60	0.73	93
TT11006	75	0.82	105
TT11006	90	0.90	115
TT11008	15	0.49	63
TT11008	20	0.57	73
TT11008	30	0.69	88
TT11008	40	0.80	102
TT11008	50	0.89	114
TT11008	60	0.98	125
TT11008	75	1.10	141
TT11008	90	1.20	154

Completing Sprayer Calibration

Now that you have selected and installed the proper nozzle tips (Steps 1 through 5) for either manual or power sprayers, you are ready to complete the calibration process (Steps 6 through 10 below). Check the calibration every few days during the season or when changing the pesticides being applied. New and old nozzles should both be flow checked for accuracy. Once you have learned the following calibration method, you can check application rates quickly and easily.

Step 6. Determine the required flow rate for each nozzle in ounces per minute (OPM). To convert the gallons per minute (Step 4) to OPM, use the following equation:

$$OPM = GPM \times 128$$

where: 128 = the number of ounces in 1 gallon

Example: If the required nozzle flow rate = 0.56 GPM, what is the required OPM?

$$OPM = 0.56 \times 128 = 71.7$$

Step 7. Collect the output from one of the nozzles in a container marked in ounces. Adjust the pressure until the amount you collect is within plus or minus 5 percent of the desired number of ounces per minute. If it is impossible to obtain the desired output within the recommended range of operating pressures, select larger or smaller nozzle tips, and then recalibrate. It is important that spray nozzles be operated within the recommended pressure range. (The range of operating pressures indicates pressure at the nozzle tip. Pressure losses in hoses or booms, nozzle check valves, etc. may cause the pressure gauge to read much higher in order to obtain the proper flow rate).

Step 8. Determine the amount of pesticide needed for each tankful or for the acreage to be sprayed. This step is covered in greater detail in the next chapter. Add the pesticide to a tank partially filled with a carrier (water, fertilizer, etc.); then add more carrier to the desired level with continuous agitation.

Step 9. Operate the sprayer at the measured speed and at the pressure that you determined in Step 7. You will be spraying at the application rate that you selected in Step 1. After spraying an area of known size, check the liquid level in the tank to verify that the application rate is correct.

Step 10. Check the nozzle flow rate frequently. Adjust the pressure to compensate for small changes in nozzle output resulting from nozzle wear or variations in other spraying components. Replace the nozzle tips and recalibrate when the output has changed 10 percent or more from that of a new nozzle, or when the pattern has become uneven.

You may already have a set of nozzle tips in your boom, and you want to know the spray rate (GPA) when operating at a particular nozzle pressure and speed. Add water to the spray tank and make a precalibration check to be sure that all spray components are working properly. Remember the type, size, and fan angle of all nozzle tips must be the same, and the flow rate from each nozzle must be within five percent of the average flow rate from the other nozzles. Next, use the following steps:

Step 1. Operate the sprayer at the desired operating pressure. Use a container marked in ounces to collect the output from a nozzle for a measured length of time such as one minute. Check several other nozzles to determine the average number of ounces per minute of output from each nozzle.

Step 2. Convert ounces per minute (OPM) of flow to gallons per minute (GPM) by dividing the OPM by 128 (the number of ounces in one gallon).

Step 3. Determine the spraying speed. For mounted boom sprayers, the speed in MPH can easily be measured by using the sprayer's speed gauge. For hand-operated booms, lay out

1,000 square feet and record the time required to cover the area uniformly.

Step 4. Determine the sprayed width per nozzle (W) in inches. For broadcast spraying, W equals the nozzle spacing.

Step 5. For mounted boom sprayers, use the equation below to calculate the sprayer application rate in gallons per acre (GPA):

$$GPA = \frac{GPM \times 5,940}{MPH \times W}$$

Example: If the measured nozzle output is 54 ounces per minute (OPM), the measured ground speed is 6 MPH, and the nozzle spacing (W) is 20 inches, what is the sprayer application rate? First, convert OPM to GPM (Step 2) by dividing the OPM by 128 (the number of ounces in 1 gallon):

$$GPM = \frac{54}{128} = 0.42$$

Next, calculate the application rate in GPA as follows:

$$GPA = \frac{0.42 \times 5,940}{6 \times 20} = 20.8$$

For hand-operated walking booms, use the following equation to calculate the application rate:

Gallons per 1,000 square feet = Gpm × minutes per 1,000 square feet

Example: The measured nozzle output is 0.6 GPM, and two minutes are required to spray 1,000 square feet. The application rate equals 1.2 gallons per 1,000 square feet (0.6 × 2).

The application rate (Step 5) can be adjusted by changing the ground speed or nozzle pressure and recalibrating. Changes in nozzle pressure should be used only to make small changes in output, and these changes must be within the recommended pressure range.

SPRAYER MAINTENANCE

Proper maintenance and cleaning are important to keep foreign materials out of the sprayer. These materials can clog the nozzles and damage the pump and other sprayer components. Some pesticides will cause the equipment to deteriorate if they remain in the sprayer for an extended period of time. Clean your spray equipment immediately after use. Possible damage from contamination left from previous sprayings may occur. It is usually wise to have separate spray equipment for herbicides, insecticides, and fungicides. The following practices will help you maintain and clean your spray equipment properly.

Step 1. Use only water that appears clean enough to drink. Small particles often found in water from ditches, ponds, or lakes can clog nozzles and strainers. If you are in doubt, filter the water as you fill the tank.

Step 2. Check and clean strainers daily. Partially plugged strainers will create a pressure drop and reduce the nozzle flow rate. Most sprayers have three different strainers: one on the suction hose to protect the pump, another in the line between the pump and the boom, and a third, which has the smallest openings, in the nozzle body.

Step 3. Do not use a metal object for cleaning nozzles. Metal objects will destroy the orifice. When a nozzle becomes clogged, always remove it for cleaning.

Step 4. Flush a new sprayer before using. A new sprayer invariably contains metal chips and dirt from the manufacturing process. Remove the nozzles and strainers. Then flush the sprayer and boom with clean water. Thoroughly clean each nozzle before reinstalling.

Step 5. You can clean the sprayer quickly with a suspension of activated charcoal in water. Triple rinse the sprayer, and then spray the water on a target area. Use at least a third of a tank of water. Then suspend activated charcoal in water for subsequent rinsing. For each ten gallons of water, add $1/4$ pound of activated charcoal and $1/4$ to $1/3$ pound of laundry detergent. Agitate this mixture vigorously to distribute the charcoal throughout the water. Wash the equipment for two minutes by swirling the charcoal suspension around in the tank so that it reaches all the parts. Pump some of the liquid through the hose and nozzles. Then drain the tank and rinse the equipment with clean water.

For phenoxy herbicides, a one percent ammonia-water solution is effective. Follow the general instructions above, but soak the equipment for 24 hours. Refer to the product label for specific cleaning requirements. Some herbicides may require special procedures to prevent inadvertent contamination.

Step 6. When it is time to store your sprayer, add one to five gallons of lightweight oil, depending on the size of your tank, before the final flushing. As water is pumped from the sprayer, the oil will leave a protective coating inside the tank, pump, and plumbing. To prevent corrosion, remove the nozzles, strainers, and diaphragm check valves, dry them, and store in a sealed container. An alternative would be to pump RV antifreeze through the system, leaving it in the system until preparing for use the next season.

Step 7. Corrosive fertilizer solutions should not be used in certain sprayers. Liquid fertilizers are corrosive to copper, galvanized surfaces, brass, bronze, and steel. You can irreparably damage an ordinary sprayer after one use with a liquid fertilizer. Use sprayers made completely of stainless steel or aluminum for applying liquid fertilizers. Aluminum is satisfactory for some nitrogen fertilizers, but not for mixed fertilizers. For weed control products containing glyphoste, specially constructed pumps with special rollers and seals are recommended to increase the life of the pump.

SPREADERS USED FOR CHEMICAL AND PESTICIDE APPLICATIONS

Granular products are a critical part of a turf pest-control program. Proper selection, care, calibration, and use of granular applicators can minimize costs and maximize the results obtained. Improper use of granular applicators can reduce pest control, cause injury to turf, increase costs, and damage the spreaders. The two available systems for applying granular products to turf are drop (gravity) and rotary (centrifugal) spreaders. Drop spreaders are usually more precise and deliver a more uniform pattern than rotary spreaders (Figures 10.16

FIGURE 10.16 Drop or gravity spreader.

and 10.17). Because the granules drop straight down with drop spreaders, there is also less chemical drift. Some drop spreaders will not handle larger granules, however, and ground clearance in wet turf can be a problem. Moreover, because the edges of a drop-spreader pattern are sharp, any steering error will cause missed or doubled strips. Drop spreaders also usually require more effort to push than rotary spreaders.

A review of basic operating procedures for granular application devices will help ensure safe and efficient application of granular turf

FIGURE 10.17 Rotary or centrifugal spreader.

care products. First, read the operator's manual and follow the manufacturer's instructions carefully. Second, read the product label, and select the appropriate rate and pattern settings for specific conditions. However, because of variations in turf care products and application devices, proper calibration is critical for each application.

To improve application uniformity and reduce waste, use header strips to provide an area in which to turn around and realign the spreader (Figure 10.18). An operator should always move the spreader at normal operating speed on the header strip, and then activate the spreader as it enters the untreated turf area. When the operator reaches the other end, the spreader should be shut off while moving, and then stopped and turned while in the header strip. A spreader should never be open while stopped because an excessive amount of product will be applied to a small area. In addition, the end turns should not be made with the spreader open because the pattern will be very irregular while the spreader is turning.

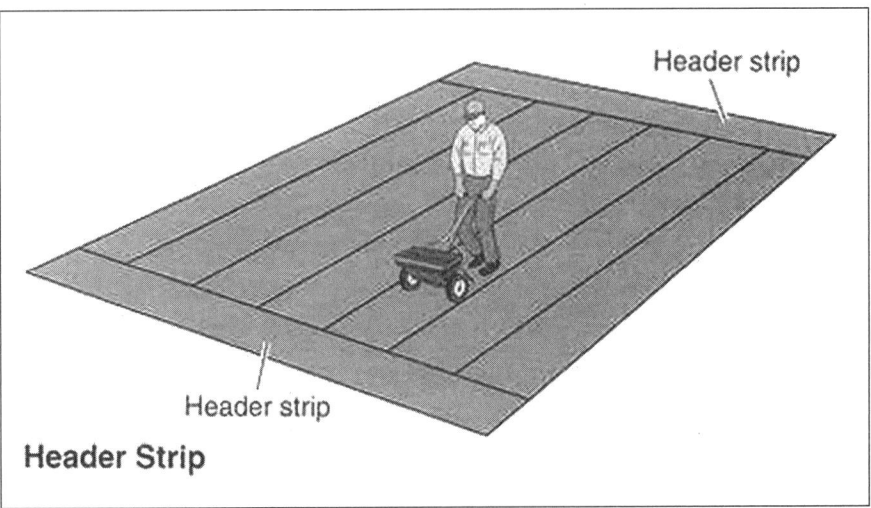

Header Strip

FIGURE 10.18 Header strips. *(Illustration courtesy of the University of Illinois)*

Occasionally, it may be impossible to obtain a completely acceptable pattern with a rotary spreader, and streaking of the turf may result. A common reaction to this problem is to reduce the setting by 1/2 and go over the area twice at right angles. Instead of "averaging out" the pattern, as is generally believed, this procedure usually changes the streaks into a diagonal checkerboard. It is better to reduce the setting and swath width by 1/2 and go back and forth in parallel swaths (Figure 10.19). Do not operate your spreader backwards. When pulled backwards, most rotary spreaders deliver an unacceptable pattern. Drop spreaders will not maintain a constant application rate if operated backwards.

Finally, set and fill the spreader on a paved surface rather than on the turf area. If a spill occurs, a driveway is much easier to sweep clean than turf. Some rotary spreaders are provided with a means of shutting off one side of the pattern. This feature is desirable when edging a driveway, sidewalk, or other non-turf area.

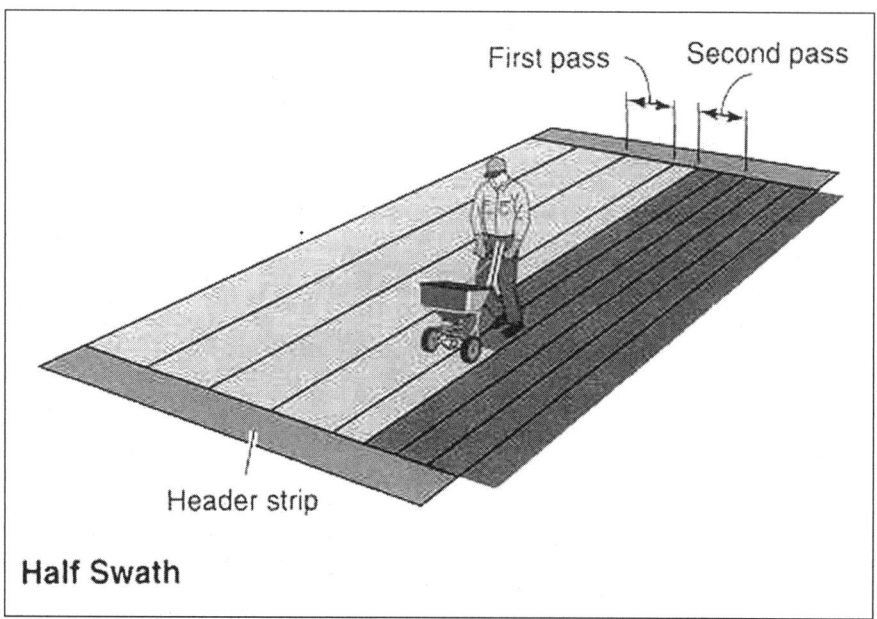

FIGURE 10.19 Proper overlap technique. *(Illustration courtesy of the University of Illinois)*

Making Applications Using Granular Spreaders

There are two important aspects to the precision application of granular products. The first is the product application rate (the amount of product applied in pounds per 1,000 square feet). Granular turf care products are designed and recommended for application at a specific rate. Over application is costly, increases the risk of plant injury, and may be illegal if label recommendations are exceeded. Under application can reduce pest control. The flow rate from granular applicators will not change in the same proportion as changes in speed. For example, doubling the speed will not double the flow rate. A constant ground speed is necessary to maintain a uniform application rate.

Even distribution of the product is as important as the application rate. For example, the pesticide label may indicate an application rate of four pounds per 1,000 square feet, and a spreader may apply that amount; however, the pesticide may not be applied uniformly over the 1,000 square feet. It is especially important to obtain uniform distribution of granules on turf because even small differences in application rate can result in obvious streaks.

The pattern applied by a rotary spreader depends upon impeller characteristics (height, angle, speed, shape, and roughness), ground speed, the drop point of product on the impeller, the physical properties of the product, and environmental conditions (temperature, humidity, wind, etc.). Methods for adjusting the pattern include blocking off part of the metering port or ports on some units and moving the metering point or impeller on other units. If pattern skewing cannot be fully corrected by following the manufacturer's recommendations, other means, such as varying the speed or tilting the impeller, can be used. When a product is so heavy or so light that skewing cannot be eliminated, it may be necessary to use a wider swath width on one side than on the other.

Calibration of Granular Spreaders

Many suppliers recommend spreader settings and swath widths for their products, but these should be used only as initial guides

> **PESTPOINTER**
>
> Calibration should be checked and corrected according to the manufacturer's directions at least once a week, and more often if the spreader has received mechanical damage.

for calibration runs prior to actual use. Every drop or rotary spreader should be calibrated for proper delivery rate with a particular product and operator because of variability in the product, the operator's walking speed, and environmental conditions. Calibration should be checked and corrected according to the manufacturer's directions at least once a week, and more often if the spreader has received mechanical damage.

The easiest method for checking the delivery rate of a spreader is to spread a weighed amount of product on a measured area (at least 1,000 square feet for a drop spreader and 5,000 square feet for a rotary spreader), and then weigh the product remaining in the spreader to determine the rate actually delivered.

To avoid contamination of the turf area during initial calibration, place the spreader on a support and try to spin the drive wheel at your walking speed but with the spreader remaining stationary. Another method is to hang a catch pan or bag under the spreader and push the spreader a measured distance at the proper speed. The container must be hung so that there is no interference with the shut-off bar or rate-control linkage.

It is necessary to check both the distribution pattern and the flow rate of a rotary spreader. The product manufacturer may recommend a particular swath width, but you should verify this width before treating a large number of turf areas. You can check the pattern by laying out a row of shallow pans on a line perpendicular to (at right angles to) the direction of travel. For commercial, push-type rotary spreaders, the pans should be one to two inches high, with an area of about one square foot, and be spaced on one foot centers (Figure 10.20). The row of pans should cover 1½ to 2 times the anticipated effective swath width. Add granules to the spreader, set it at the recommended setting for rate and pattern, and make several passes over the pans, operating in the same direction each time. The material caught in the individual pans can then be weighed and a distribution pattern plotted.

A simpler method for checking the distribution pattern is to pour the material from each pan into a small bottle. When the bottles are placed side by side in the proper order, a plot of the pattern becomes apparent. This pattern can be used to detect the correct skewing and to determine the effective swath width (twice the distance out to the point where the rate is ½ the average rate at the center). For example, if the center three or four bottles have material two inches deep, and the bottles six feet to the left and six feet to the right of the spreader centerline have material one inch deep, the effective swath width is 12 feet (Figure 10.21).

FIGURE 10.20 Pattern collection layout.

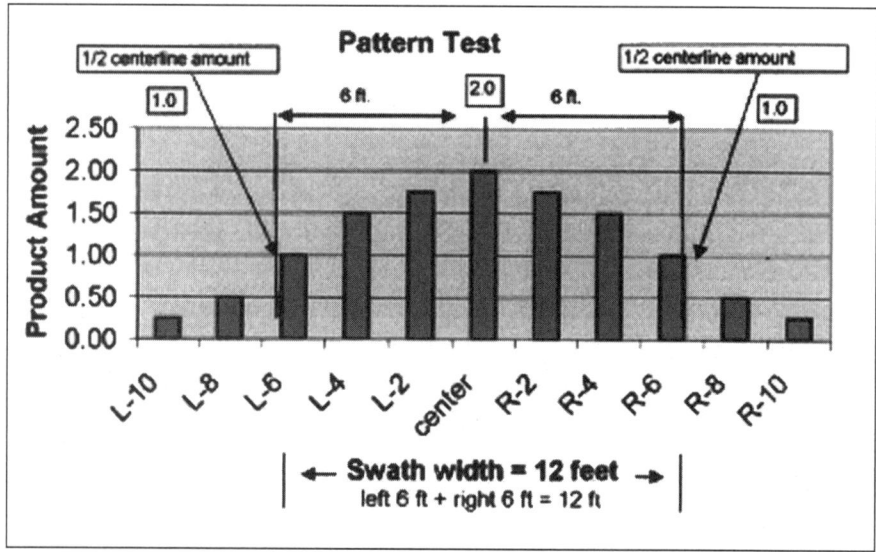

FIGURE 10.21   Spreader pattern width graphic.

SUMMARY QUESTIONS

- What types of chemical and pesticide products are more common at your facility, liquids or granules?

- What types of sprayers and spreaders do you have to make these applications?

- What procedures do you follow for sprayer and spreader calibration? Do you have "preset" equipment or is calibration performed for each application?

- Are there certain staff members at your facility who are specifically trained for sprayer and/or spreader calibration or is everyone expected to know how this is done?

Chapter 11
Pesticide Application

One of the more dangerous aspects of turfgrass care is the use of pesticides. This portion of the lawn care industry is watched very closely by government agencies and local and state agricultural offices. The state agricultural departments and the Environmental Protection Agency (EPA) have jurisdiction to monitor and prosecute individuals for any misuse of pesticides and chemicals. This is why you should learn how to apply pesticides correctly and be aware of the laws that govern their usage (Figure 11.1). Without the enforcement of pesticide use, the potential for losing the ability to use chemicals becomes more of a threat for the rest of us each day.

The technology behind the use of pesticides has evolved considerably over the last few years, and many new facts are learned about chemicals every day. The use of biotechnology is one point of change that is limiting and changing our thinking about chemical usage. In some cases the fear of chemical usage is being overshadowed by the fear of genetically engineered plants. Many people who once worried about the pesticides in their foods now worry about the genetic changes within the food they are now consuming. The use of biotechnology is not new to the plant production industry, but the rapid proliferation of these types of plants will eventually change the general use of pesticides due to the built-in resistance to pests. I only mention

> **PESTPOINTER**
>
> Without the enforcement of pesticide use, the potential for losing the ability to use chemicals becomes more of a threat for the rest of us each day.

this because it is an alternative to pesticide usage in many parts of the agriculture industry, and soon it will affect daily maintenance routines for turfgrass as well.

The most widespread examples of plants that were influenced by biotechnology are Roundup™ Ready soybeans and corn. These seeds were limited in use when first introduced into the agricultural market but soon snowballed when the response from farmers was so positive. The costs of chemical usage and applications were greatly reduced. This was the beginning of a partial acceptance of biotechnology in the plant industry. Currently the same type of technology is being used to develop new turfgrass strains with more desirable characteristics. These changes range from

> **PESTPOINTER**
>
> The use of biotechnology is not new to the plant production industry, but the rapid proliferation of these types of plants will eventually change the general use of pesticides due to the built-in resistance to pests.

FIGURE 11.1 Pesticide turf and ornamental book for Missouri (It is recommended that you purchase such a study guide to help you in the study and testing process).

disease-resistant grasses to the same Roundup™ type products as are used in the agricultural market. The idea is the to spray over the top of your bluegrass, for instance, and eradicate 100% of the foreign weeds that are present in the lawn. This may sound crazy to some of you, but I assure you that this type of breakthrough is just around the corner.

Technology continues to change, and the use of pesticides will also change with each new development. There is always the unanswered question of what these changes mean to turfgrass management for the future, but for the time being we should look at what is at hand and how we can control pest problems with the least amount of damage to our environment. The reason there are tests and licenses for the use of pesticides is only to protect the people, flora, and fauna that could be affected by the use of chemicals. Pesticides can cause problems in our environment by contaminating drinking water and by other forms of off-site movement in the form of drift. By using pesticides in a safe manner according to the label, we are helping to save our environment and yet eliminate target pest problems.

TESTING AND LICENSING

If you are planning on using pesticides in your operation, you must be aware of all aspects of their application. You will need to pass tests in order to get your license and the insurance that is necessary in case of an accident. Every state has different guidelines, but most require a test for lawn care professionals and golf course people with the designation "ornamental and turf category." This is a test that is usually given in addition to the standardized test that must be taken by anyone who sprays for pests, which is commonly known as the "core" test for basic pesticide safety. By passing these two tests you will earn your pesticide applicator's license. You must still provide proof of liability insurance and also pay a fee for your license, but the testing is the most difficult part for most people. Most states do require you to have liability insurance for pesticides, and the costs for this type of insurance are often quite high.

PESTPOINTER

If you are planning on using pesticides in your operation, you must be aware of all aspects of their application. You will need to pass tests in order to get your license and the insurance that is necessary in case of an accident.

Many states offer a more standardized test that does not have a specific category for each of the licenses offered, but category tests are more common. Most of the tests given are a standardized 50-question multiple-choice test. Many tests are offered in combination with the agriculture department and the university extension offices, which helps to promote the educational aspect of the testing. There are courses that can be taken and training given prior to taking the exams to ensure a successful outcome and provide some sort of training with the actual pesticide applications. It is still an ongoing problem for state agencies to catch the numerous people who are spraying pesticides illegally.

There may be required training prior to taking your pesticide exam, but this is not true for all states. This training is only for your benefit, and it will help inexperienced operators who are trying to get started in the lawn care business. Remember that everyone needs to start somewhere and that training will only eliminate problems for all pesticide users. The more experienced people there are who are applying pesticides correctly, the better off the entire lawn care and golf course industry will be. This will mean fewer accidents, fewer problems, and lower insurance costs for everyone involved.

Another part of the training process is the retraining or recertification process. This part of the license process will vary the most from state to state. In many states a one-day training program must be attended every two to three years. In other states a point system is

> **PESTPOINTER**
>
> Another part of the training process is the retraining or recertification process. This part of the license process will vary the most from state to state. In many states a one-day training program must be attended every two to three years.

used in which training programs and seminars must be attended throughout the license holder's career and points are awarded. Each person must maintain a certain point count of certified training per year.

SAFETY

Pesticide application should not be taken lightly, and those who pursue a career in turfgrass should learn as much about pesticide safety as possible. You should consider not only the health risks of exposing yourself to pesticides, as well as anyone who may come into contact with the chemicals or their residues. This includes the people in your own house who may be exposed to your clothing or foot traffic. I am not trying to scare people, and certainly I am a firm believer in pesticide usage. I do, however, feel that many who pass a 50-question test underestimate the importance of pesticide safety and also overlook the risks involved when using chemicals. You need to understand and respect the products that are being applied to turfgrass and also know why, how, and when chemicals need to be applied.

I always have believed that chemical usage should be used in moderation and that there are often many alternatives to chemical controls. I find that many of the misuses or overuses of chemicals come about from improper diagnosis of diseases and pest problems. When misdiagnosis of a pest problem occurs, overuse of chemicals can result from treating a problem that does not exist. For instance, if a turfgrass plot has leaf discoloration, it may be suffering from a lack of nutrition and not a disease of some sort (Figure 11.2). Many would see this as an opportunity to spray the turf without first checking to see if a disease problem was actually present. This is why pest diagnosis is as important as the application of the products used to control the disease or pest. There is something to be said for a preventive application of a chemical to avoid disease occurrence, but such treatments can still be performed based on environmental factors that lead you to believe there is going to be a disease problem.

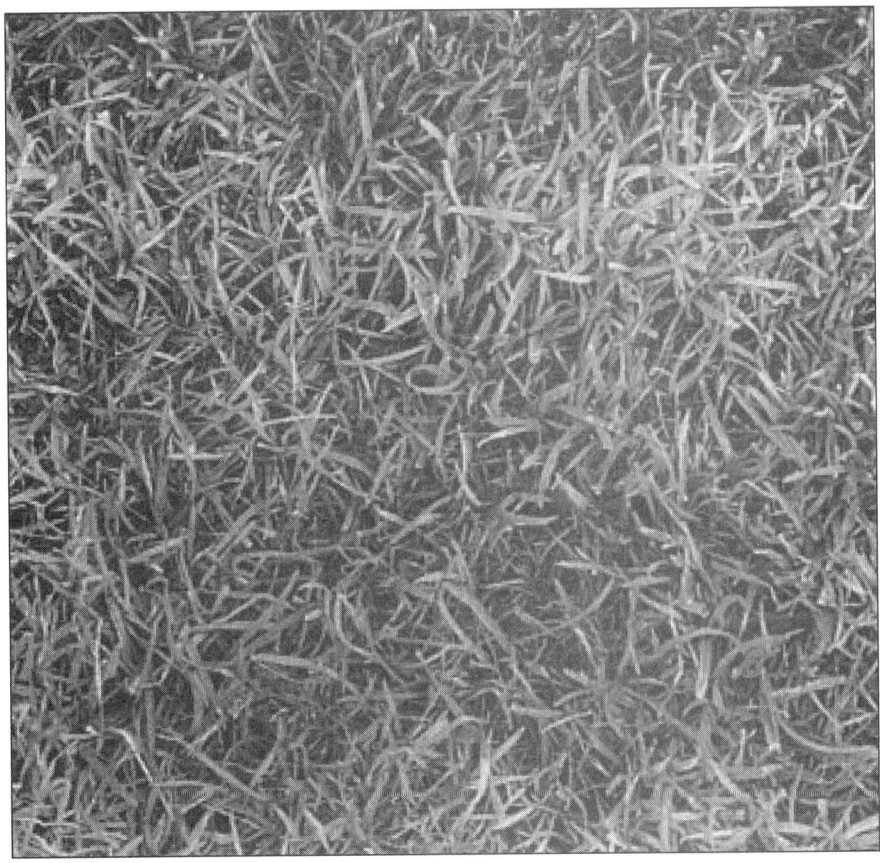

FIGURE 11.2 Nutrient-deficient turf.

The most common types of pesticide applications with turfgrass are weed treatments. These treatments are almost always done in a program that calls for several chemicals to be applied throughout the growing season. There are some who fear the use of pesticides in the lawn, but for the most part there are few risks to those who apply these pesticides safely and according to the law. When using a weed control program, you are treating for some weeds on a preventive basis but this is because they cannot be controlled in any other fashion. There sometimes cannot be the same considerations for threshold levels as there are when dealing with other pests in the lawn care industry. This is still not to say that a lawn cannot look very nice without the use of chemicals or even with some sort of organic approach. It is just more difficult to achieve that plush carpet look that has become so expected within the lawn care industry.

Pesticide safety starts with the handling of the chemicals and the strategies used for pest control. One thing I like to do is to apply granular products just because of their ease of application and the control I have when spreading the product. In most cases granular products can provide just as good or better results than what you can get by using liquids. However, there are some applications that still give more desirable results when liquids are used (Figure 11.3). It is through a combination of granular and liquid applications along

PESTPOINTER

One thing I like to do is to apply granular products just because of their ease of application and the control I have when spreading the product. In most cases granular products can provide just as good or better results than what you can get by using liquids.

Pesticide Application   473

FIGURE 11.3   Patch of clover (It is easier to cover this with a liquid product).

with some reasonable caution that a beautiful looking lawn and landscape are created. Certain broadleaf plants are harder to get rid of when using a granular product because the weeds may be in patches, and complete coverage is more difficult.

I try to provide sufficient control and not necessarily the overkill that many would believe to be necessary. The next safety issues occur with chemical preparation. There is just as large a risk to a chemical applicator when you are mixing chemicals as there is when you are actually applying them. Start thinking about safety before you ever apply any product at all. Transportation, storage, and mixing of the product can be just as big a safety hazard.

If you are handling a granular product, the safety concerns involved with transport and application are a little less severe. In most cases there is no mixing of the product at all, and the chances for drift when using dry chemicals are also very low. You should always consider off-site movement of any chemical, but with a granular product you can usually see where the product is being applied, with the exception of some dust. The granular product is much easier to transport and it is certainly easier to dispose of the empty containers or packing. There is a higher risk for small spills of product, because damage often occurs from tears or punctures in the bags. Since most chemicals are not applied in straight, active form, there is often a carrier such as corn cob, clay, or fertilizer so the product can be easily spread. Granular products do clean up much more easily than liquid products and can often be swept.

Liquid chemicals should be handled with a totally different approach. They are more likely to be absorbed into the ground, clothes, or skin. The dust is a concern with a dry product but drifting is not as likely as with liquids. A liquid chemical spill may need to have some sort of absorbent product applied to it to make it easier to clean up, whereas a dry product spill would be easy to sweep up. Liquids also need to be mixed in a tank, which means that pouring is necessary unless the chemical is in a water-soluble package (Figure 11.4). Many spills occur during mixing and often these spills are not taken care of as they should be. There are very specific mixing instructions that should be followed which can be found on every pesticide label.

Spraying or applying pesticides always requires some sort of personal protective equipment, which is also known as PPE (Figure 11.5). PPE will vary based on the requirements of the pesticide label. The pesticide label will provide you with all the information pertinent to the application of the product you intend to spread or spray. You will find mixing instructions, PPE requirements, and all the other important information about the chemical you are applying.

With granular applications the required PPE may be as simple as a long-sleeve shirt, long pants, and protective shoes. With liquid applications, the PPE requirements may include a spray suit, mask, gloves, and even goggles in some instances. There may also be a requirement for wearing an apron when mixing up certain liquid chemicals. The

Pesticide Application 475

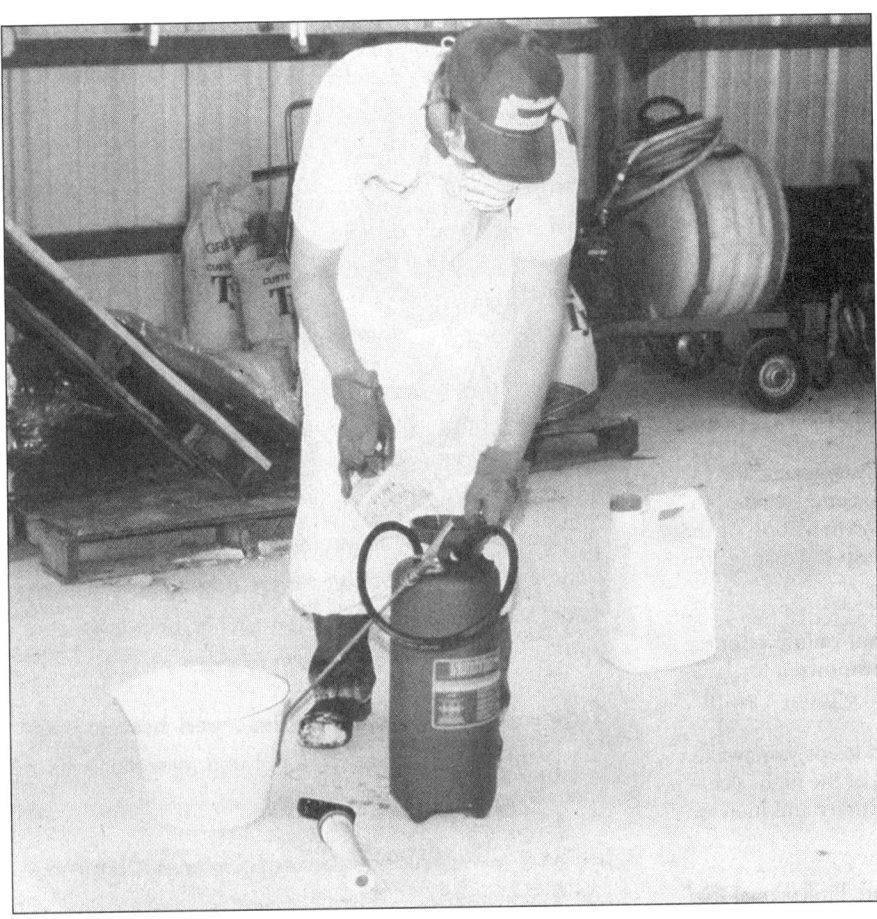

FIGURE 11.4 A worker pouring pesticides into a spray tank (Although this person is not spraying, he is actually at a higher risk for injury because he is dealing with the concentrated product).

PPE is required for the applicator's protection and should always be carefully followed. I like to use more PPE than suggested instead of less. It is always good to limit your exposure to any pesticide as much as possible. For instance, I will wear rubber gloves when applying any type of chemical and sometimes even when I am transporting chemicals or storing them to limit my exposure to product residues.

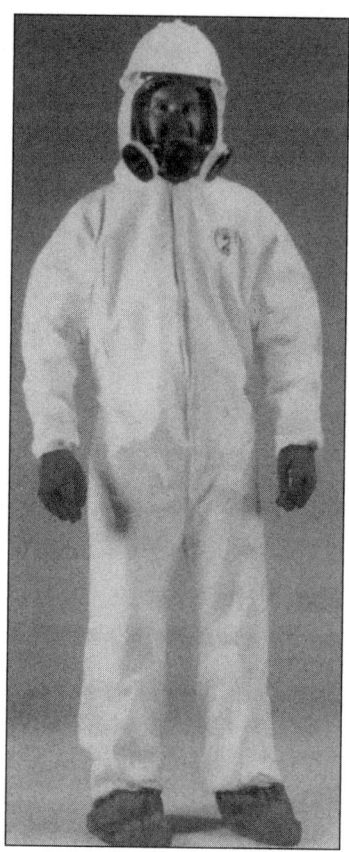

FIGURE 11.5 Personal protective clothing.

As I mentioned, most of the information that pertains to the application of a product can be found on the pesticide label. The pesticide label holds almost all the critical information not only about using the product, but also about what to do if misuse of the product were to occur. The label will also have an EPA registration number for each product. This number means that a record exists for this product so that information is readily available and on file in case of an emergency or an inquiry. The EPA number should be recorded by commercial applicators for each product they are applying every time they apply that product.

PESTPOINTER

The pesticide label holds almost all the critical information not only about using the product but also about what to do if misuse of the product were to occur.

The pesticide label also contains safety information, which will include signal words or warnings about the type of product contained in the package or jug (Figure 11.6).

There will also be a child hazard warning, which will read "Keep out of reach of children." This should be present on all pesticide labels and is there for your protection. Children are often the innocent victims of pesticide poisoning from accidental ingestion or exposure. This is one of the primary reasons why chemicals should always be stored in their original containers. The EPA numbers, safety warnings, emergency numbers, and other information will be readily visible.

Another piece of information that can be found in the safety portion of the label will be treatment procedures. They will give specific instructions as to what to do should accidental exposure occur. There are usually treatment suggestions in case of ingestion or exposure. A warning might read: "If swallowed, call a physician or the poison control center immediately." If the pesticide label

FIGURE 11.6 Interpreting warnings on pesticide containers.

Signal Word	Toxicity	Approximate amount needed to kill the average person
DANGER	Highly Toxic	A taste to a teaspoon
WARNING	Moderately Toxic	A teaspoon to a tablespoon
CAUTION	Slightly Toxic	An ounce to more than a pint

contains the word "warning," then an 800 number must be provided on the label for you or a physician to be able to contact someone immediately.

There should also be a warning that will state how toxic the product is to humans and animals. This part of the label should provide specific information about possible allergic reactions from coming into contact with product. It will list possible symptoms of a reaction to the product and should have suggestions of what to do and not do with the chemical. This may be valuable if you are not sure whether you are suffering from chemical exposure.

The next section of a label that should be closely read is a description of the result of misuse of the product to the environment. This section will have many environmental warnings and give specific information of the threats the product poses to ground water, plants, and animals in the area. Some labels might even mention additional threats to endangered species.

There are smaller parts on a pesticide label that many do not understand and others simply ignore. There is a reason for every component of the pesticide label, and all the information in the label provides a "short story," so to speak, about the particular product. By not reading the label thoroughly or reading only the mixing instructions you do pose a risk to yourself and others (Figure 11.7).

The use classification is another required part of the label, which classifies the pesticide for either general or restricted use. Restricted-use pesticides are more toxic to people and the environment, and they have many additional restrictions for use and handling. A restricted-use product may cause injury to non-target plants. The retail purchase and any use of these products should only be performed by certified applicators or people who are being supervised by those who either have a license or are covered by a certified applicator's certificate.

You are starting to see just how much information is on a pesticide label and why it is important that the label be read very thoroughly

Pesticide Application 479

TURF & ORNAMENTAL PRODUCTS

Super TRIMEC®
BROADLEAF HERBICIDE

Controls dandelion, chickweed, knotweed, plantain, henbit, spurge, and many other species of broadleaf weeds.

ACTIVE INGREDIENTS:
Isooctyl (2-ethylhexyl) ester of
  2,4-dichlorophenoxyacetic acid 32.45%
Isooctyl ester of 2-(2,4-dichlorophenoxy)
  propionic acid 31.80%
Dicamba: 3,6-dichloro-o-anisic
  acid 5.38%
INERT INGREDIENTS: 30.37%
THIS PRODUCT CONTAINS: TOTAL 100.00%
2.0 lbs. 2,4-dichlorophenoxyacetic acid
  equivalent per gallon or 21.54%.
2.0 lbs. 2-(2,4-dichlorophenoxy) propionic
  acid equivalent per gallon or 21.54%.
0.5 lb. 3,6-dichloro-o-anisic acid equivalent
  per gallon or 5.38%.
Contains Petroleum Distillates.
Isomer Specific by AOAC Methods.
TRIMEC® is a registered trademark of
PBI/Gordon Corporation.

785/11-2000 AP081800
EPA REG. NO. 2217-758
EPA EST. NO. 2217-KS-1

Manufactured By

An Employee-Owned Company
1217 West 12th Street
Kansas City, Missouri 64101

KEEP OUT OF REACH OF CHILDREN

WARNING- AVISO

Si Usted no entiende la etiqueta, busque a alguien para que se la explique a Usted in detalle. (If you do not understand the label, find someone to explain it to you in detail.)

Statement of Practical Treatment
IF IN EYES: Hold eyelids open and flush with a steady, gentle stream of water for 15 minutes. Get medical attention.
IF SWALLOWED: Call a physician or Poison Control Center immediately. Contains petroleum solvent. Do not induce vomiting because of danger of aspirating liquid into lungs, causing serious damage and chemical pneumonitis. If spontaneous vomiting occurs, keep head below hips to prevent aspiration, and monitor for breathing difficulty.
IF ON SKIN: Wash with plenty of soap and water. Get medical attention.
IF INHALED: Remove victim to fresh air. If not breathing, give artificial respiration, preferably mouth-to-mouth. Get medical attention.

See below for additional Precautionary Statements.

READ THE ENTIRE LABEL FIRST.
OBSERVE ALL PRECAUTIONS AND
FOLLOW DIRECTIONS CAREFULLY.

PRECAUTIONARY STATEMENTS
Hazards to Humans and Domestic Animals
WARNING: Causes substantial but temporary eye injury. Do not get in eyes or on clothing. Wear protective eyewear. Harmful if swallowed. Harmful if absorbed through skin. Avoid contact with skin. Harmful if inhaled. Avoid breathing vapor or spray mist.

Personal Protective Equipment (PPE):
Some materials that are chemical-resistant to this product are listed below. If you want more options, follow the instructions for category E on an EPA chemical resistance category chart.
Applicators and other handlers must wear: • Long-sleeved shirt and long pants • Chemical-resistant gloves such as barrier laminate, nitrile rubber, neoprene rubber, or viton • Shoes plus socks • Protective eyewear • Chemical-resistant apron when cleaning equipment, mixing or loading.
Discard clothing and other absorbent materials that have been drenched or heavily contaminated with this product's concentrate. Do not reuse them.
Follow manufacturer's instructions for cleaning/maintaining PPE. If no such instructions for washables, use detergent and hot water. Keep and wash PPE separately from other laundry.

Engineering Control Statements for WPS Uses:
Containers over 1 gallon and less than 5 gallons in capacity: Mixers and loaders who do not use a mechanical system (probe and pump) to transfer the contents of this container must wear coveralls or a chemical-resistant apron in addition to the other required PPE.
Containers of 5 gallons or more in capacity: Do not open-pour from this container. A mechanical system (such as a probe and pump or spigot) must be used for transferring the contents of this container. If the contents of a nonrefillable pesticide container are emptied, the probe must be rinsed before removal. If the mechanical system is used in a manner that meets the requirements listed in the Worker Protection Standard (WPS) for agricultural pesticides [40 CFR 170.240(d)(4)], the handler PPE requirements may be reduced or modified as specified in the WPS.
When handlers use closed systems or enclosed cabs in a manner that meets the requirements listed in the Worker Protection Standard (WPS) for agricultural pesticides [40 CFR 170.240(d)(4-6)], the handler PPE requirements may be reduced or modified as specified in the WPS.

USER SAFETY RECOMMENDATIONS:
Users should:
• Wash hands before eating, drinking, chewing gum, using tobacco or using the toilet.
• Remove clothing immediately if pesticide gets inside. Then wash thoroughly and put on clean clothing.
• Remove PPE immediately after handling this product. Wash the outside of gloves before removing. As soon as possible, wash thoroughly and change into clean clothing.

ENVIRONMENTAL HAZARDS:
This product is toxic to aquatic invertebrates. Drift or runoff may adversely affect aquatic invertebrates and nontarget plants. For terrestrial uses, do not apply directly to water, or to areas where surface water is present, or to intertidal areas below the mean high water mark. Do not contaminate water when disposing of equipment washwater. When cleaning equipment, do not pour the washwater on the ground; spray or drain over a large area away from wells and other water sources.
Most cases of groundwater contamination involving phenoxy herbicides such as 2,4-D and 2,4-DP have been associated with mixing/loading and disposal sites. Caution should be exercised when handling 2,4-D and 2,4-DP pesticides at such sites to prevent contamination of groundwater supplies. Use of closed systems for mixing or transferring this pesticide will reduce the probability of spills. Placement of the mixing/loading equipment on an impervious pad to contain spills will help prevent groundwater contamination.

Physical or Chemical Hazard
Do not use or store near heat or open flame.

DIRECTIONS FOR USE
It is a violation of Federal law to use this product in a manner inconsistent with its labeling.
Do not apply this product in a way that will contact workers or other persons, either directly or through drift. Only protected handlers may be in the area during application.
For any requirements specific to your State or Tribe, consult the agency responsible for pesticide regulation.

AGRICULTURAL USE REQUIREMENTS
Use this product only in accordance with its labeling and with the Worker Protection Standard, 40 CFR part 170. This standard contains requirements for the protection of agricultural workers on farms, forests, nurseries, and greenhouses, and handlers of agricultural pesticides. It contains requirements for training, decontamination, notification, and emergency assistance. It also contains specific instructions and exceptions pertaining to the statements on this label about personal protective equipment (PPE) and restricted-entry interval. The requirements in this box only apply to uses of this product that are covered by the Worker Protection Standard.

DO NOT enter or allow worker entry into treated areas during the restricted-entry interval (REI) of 48 hours.

PPE required for early entry to treated areas that is permitted under the Worker Protection Standard and that involves contact with anything that has been treated, such as plants, soil, or water, is: • Coveralls • Chemical-resistant gloves such as barrier laminate, nitrile rubber, neoprene rubber, or viton • Shoes plus socks • Protective eyewear.

FIGURE 11.7 A pesticide label.

480 *Turfgrass Chemicals and Pesticides: A Practitioner's Guide*

NON-AGRICULTURAL USE REQUIREMENTS
The requirements in this box apply to uses of this product that are NOT within the scope of the Worker Protection Standard for agricultural pesticides (40 CFR Part 170). The WPS applies when this product is used to produce agricultural plants on farms, forests, nurseries, or greenhouses.

Do not allow people (other than applicator) or pets on treatment area during application. Do not enter treatment area until spray has dried.

FOR USE ON RESIDENTIAL AND ORNAMENTAL TURFGRASS SITES AND SOD FARMS (COOL SEASON GRASSES OTHER THAN BENTGRASS)

Do not apply this product through any type of irrigation system. Avoid drift of spray mist to vegetables, flowers, ornamental plants, shrubs, trees and other desirable plants. Do not pour spray solutions near desirable plants. Do not use on carpetgrass, dichondra, St. Augustinegrass, bentgrass, nor on lawns or turf where desirable clovers are present. Use only lawn type sprayers. Avoid fine sprays; coarse sprays are less likely to drift. Do not spray roots of ornamentals and trees. Do not exceed specified dosages for any area; be particularly careful within the dripline of trees and other ornamental species. Do not apply to newly seeded grasses until well established.

Do not spray when air temperatures exceed 85°F. Seed can be sown 3 to 4 weeks after application. Care should be taken not to make applications where runoff could carry the chemical to food crops or grazing lands where cattle, sheep, goats, swine or poultry would be exposed.

SPRAY PREPARATION
Add one-half the required amount of water to the spray tank, than add this product slowly with agitation, and complete filling the tank with water. To prevent separation of the emulsion, mix thoroughly and continue agitation while spraying.

INSTRUCTIONS
Maximum control of weeds will be obtained from spring or early fall applications when weeds are actively growing. Avoid spraying during long, excessively dry or hot periods unless adequate irrigation is available. Do not irrigate within 24 hours after application.

APPLICATION RATES: Apply 2 to 3 pints of product in 20 to 260 gallons of water per acre (0.75 to 1.1 fl. oz. of product in 0.5 to 6 gallons of water per 1,000 square feet). Use higher rates when using the higher volume of water per acre.

The maximum application rate to turf is 0.8 pounds 2,4-D acid equivalent per acre per application per site. The maximum number of broadcast applications per treatment site is 2 per year.

CONTROLLED DROPLET APPLICATOR (CDA): Add 1½ pints of product to the HERBI container and fill with water. Spray contents over 33,000 square feet. Avoid overlapping between spray patterns.

SMALL AREA APPLICATIONS
Not recommended for hose-end sprayers.
Spray anytime during the growing season when weeds are actively growing. On new lawns — wait until the grass has hardened off — usually after it has been mowed at least three times. Poor weed control may result if spray is applied during drought or just before rain. Do not water within 24 hours after treatment.

SPRAY PREPARATIONS FOR HAND OPERATED SPRAYERS

AMOUNT OF SUPER TRIMEC® TBS.		AMOUNT OF WATER IN SPRAYER, GALLONS	AREA TO BE SPRAYED, SQ. FT.
	FL. OZ.		
1½ Tbs.	¾ fl. oz.	1 Gallon	1000 Sq. Ft.
3 Tbs.	1½ fl. oz.	2 Gallons	2000 Sq. Ft.
4½ Tbs.	2¼ fl. oz.	3 Gallons	3000 Sq. Ft.

WEEDS CONTROLLED

Aster, white heath & white prairie	Black medic	Buttercup, creeping
Bedstraw	Broadleaf plantain	Carpetweed
Beggarweed, creeping	Buckhorn plantain	Chickweed, common
Bindweed	Bull thistle	Chicory
	Burclover	Cinquefoil
	Burdock, common	Clover

Cocklebur
Compassplant
Curly dock
Dandelion
Dayflower
Deadnettle
Dock
Dogfennel
English daisy
False dandelion
(*spotted catsear & common catsear)
Field bindweed
(*morningglory & creeping jenny)
Field oxeye-daisy (*creeping oxeye)
Filaree, whitestem & redstem
Florida pusley
Ground ivy
Groundsel
Hawkweed
Healall
Henbit
Jimsonweed
Kochia
Lambsquarters
Lawn burweed

Lespedeza, common
Mallow, common
Matchweed
Mouseear chickweed
Mustard
Nettle
Oxalis (*yellow woodsorrel & creeping woodsorrel)
Parsley-piert
Pennsylvania smartweed (*smartweed)
Pennywort (*dollarweed)
Pepperweed
Pigweed
Pineappleweed
Plantain
Poison ivy
Poison oak
Prostrate knotweed (*knotweed)
Puncturevine
Purslane
Ragweed
Red sorrel (*sheep sorrel)

*Synonyms

Shepherdspurse
Spotted spurge
Spurge
Sunflower
Thistle
Velvetleaf
(*pie marker, buttonweed, Indian mallow, butter print)
Veronica
(*corn speedwell)
Virginia buttonweed
White clover
(*Dutch clover, honeysuckle clover, white trefoil & purplewort)
Wild carrot
Wild geranium
Wild garlic
Wild lettuce
Wild mustard
Wild onion
Wild strawberry
Wild violet
Yarrow
Yellow rocket

STORAGE & DISPOSAL
Do not contaminate water, food, or feed by storage or disposal.

STORAGE: Keep from freezing. Store in original container in a locked storage area inaccessible to children and pets.

PESTICIDE DISPOSAL: Pesticide wastes are toxic. Improper disposal of excess pesticide, spray mixture, or rinsate is a violation of Federal law and may contaminate groundwater. If these wastes cannot be disposed of by use according to label instructions, contact your State Pesticide or Environmental Control Agency, or the Hazardous Waste representative at the nearest EPA Regional Office for guidance.

CONTAINER DISPOSAL: Plastic Containers: Triple rinse (or equivalent). Then offer for recycling or reconditioning, or puncture and dispose of in a sanitary landfill, or incineration, or, if allowed by state and local authorities, by burning. If burned stay out of smoke. Metal Containers: Triple rinse (or equivalent). Then offer for recycling or reconditioning, or puncture and dispose of in a sanitary landfill, or by other procedures approved by state and local authorities.

LIMITED WARRANTY AND DISCLAIMER
The manufacturer warrants only that the chemical composition of this product conforms to the ingredient statement given on the label, and that the product is reasonably suited for the labeled use when applied according to the Directions for Use.

THE MANUFACTURER NEITHER MAKES NOR INTENDS ANY OTHER EXPRESS OR IMPLIED WARRANTIES, INCLUDING ANY WARRANTY OF MERCHANTABILITY OR FITNESS FOR A PARTICULAR PURPOSE, WHICH ARE EXPRESSLY DISCLAIMED. This limited warranty does not extend to the use of the product inconsistent with label instructions, warnings or cautions, or to use of the product under abnormal conditions such as drought, excessive rainfall, tornadoes, hurricanes, etc. These factors are beyond the control of the manufacturer or the seller. Any damages arising from a breach of the manufacturer's warranty shall be limited to direct damages, and shall not include indirect or consequential damages such as loss of profits or values, except as otherwise provided by law.

The terms of this Limited Warranty and Disclaimer cannot be varied by any written or verbal statements or agreements. No employee or agent of the seller is authorized to vary or exceed the terms of this Limited Warranty and Disclaimer in any manner.

© 1983, PBI/GORDON CORPORATION

FIGURE 11.7 *(continued)* A pesticide label.

before the chemical is applied. One of the most important parts of labels is the statement of the brand name as well as the active ingredients. This is an area that many people confuse because chemicals are often referred to by their brand name instead of their chemical name. I will give a common example, the chemical known as Roundup™. This chemical has name recognition worldwide and is known to be a nonselective chemical that will kill most any nonwoody vegetation that it comes into contact with. However, the actual chemical name of this product is glyphosate and is sold under many other brands. Roundup™ is what you will hear more often than not when the chemical glyphosate is referred to. There are many other products that have been marketed so well that they are commonly known by their trade names. I am simply trying to point out that there are many brands of the same products, and you should be sure that you are buying and applying the product you think you are. You can save considerable amounts of money in some cases if one brand of the same material is cheaper than a more recognized counterpart.

I briefly touched on the EPA registration information, which is a must on every label. It is a state and EPA requirement that a product must be acknowledged and registered in each state and not just nationwide. I have had a couple of instances where I was able to purchase a product that the state did not recognize or had no record of at the time and I had to call the company to advise that there was no registration in my particular area. Both times that chemical was recognized and I was able to go ahead and apply and sell the product, but when the inspectors cannot find the product in their books of registered products, it can be a pain to get that product recognized. You are also in a sense liable for proving how and why you have the product until the registration is confirmed.

The next part of the label is another one that I think is quite important: the directions for use. This is where the type of product is described in detail, and the way it should be applied is laid out very specifically. There will be information provided as to which pests the product will control, mixing information, rates, and limitations and restrictions.

The last element I want to mention that can be found on the pesticide label is information pertaining to the storage and disposal of the product. I stated earlier that accidents can happen when handling or even storing a pesticide product. Any material that is discarded after use will contain pesticide residues and can be harmful if it is not taken care of correctly. Make sure that when you dispose of pesticide containers you are following federal, state, and local laws regulateing the disposal of pesticides and containers.

TYPES OF PESTICIDES

I want to outline the different types of pesticides that can be used and the modes of action that are possible for them. There are two ways a pesticide can work: through contact modes of action or through systemic modes of action. Almost every pesticide group has both systemic and contact action products available. Systemic modes of action mean that the product will spread through the entire specimen, giving a response in the entire plant even though the product may have been absorbed in an isolated area. Contact products can still move throughout the plant, but they provide a more limited means of absorption. Contacts may need to be used to saturate the plant to provide a desirable result in some cases.

PESTPOINTER

Systemic modes of action mean that the product will spread through the entire specimen, giving a response in the entire plant even though the product may have been absorbed in an isolated area.

PESTPOINTER

Contact products can still move throughout the plant, but they provide a more limited means of absorption.

The first group of products that are used are known as herbicides. Herbicides kill weeds in most situations by impairing the plants' metabolic processes in some way. Some herbicides have multiple ways of doing this, which means that resistance is less likely, while other herbicides have only one mode of action. Some herbicides are actually derivatives of plant hormones and cause a forced response that results in death. There are nonselective herbicides, which will kill most vegetation, and selective herbicides, which will only target a specific type of plant.

A herbicide can also be a pre-emergent or post-emergent product. Pre-emergent products control weeds before they actually germinate and work on the premise of providing a soil barrier. A post-emergent product controls weeds that have already germinated and sometimes can keep additional weeds from germinating (Figure 11.8).

The insecticide family is more complex than many other groups of pesticides and is currently under a lot of scrutiny. The first group of insecticides is the organophosphates, which are the most widely used type. Over 40% of registered insecticides fall into this category. This group is losing registration and use privileges. The overall safety of this group is now being questioned, and many medical studies have shown that these pesticides cause illness. It is uncertain whether they will be completely banned, but several products have already been banned and others are on a list to be phased out over the next several years. Long-term exposure can

FIGURE 11.8 Weed-Out® (a post-emergent-selective herbicide that kills only germinated broadleaf weeds).

be a serious risk to people, but the effectiveness of these products has been very good over the years when targeting insect pest control issues. This will mean that replacement products might be quite difficult and pest control may need to be addressed differently in the future.

The next group of insecticides is the carbamates, which are widely used in and around homes. As with organophosphates, there are health issues, but the risk at this point is much lower than the organophosphate group. Carbamates are an older group of insecticides and once again have been used for many years.

Boric acids and borates are used primarily in and around livestock areas. The use of these products is not as common as the other insecticide groups, and exposure is unlikely for most people unless they are involved in farming operations. These products are used to kill cockroaches and larvae in livestock confinements. There is a concern about the animals that are exposed to this group of insecticides as well as people confined in livestock areas.

The last group of insecticides I want to mention is the pyrethroids. These are some of the newest available and are actually synthetic products. It is believed that this group is safer than the organophosphates, and I feel that many new synthetics will be developed over the next few years. This will be necessary to offset some of the other insecticides that are being taken off the market. There will also be more organic types of products created in the future and others that will use insect pheromones as an ingredient.

The next group of pesticides that you might encounter in the turf industry is the fungicides. Fungicides can be either systemic or contact. There have been many strides in fungicide development in the last few years. Systemic availability is one of the biggest improvements, as well as the length of time that some of the products will last. A lot of the older fungicides were composed of organic mercury and hexachlorobenzene. These two components can cause health problems and only provide limited protection to turfgrass. There was also a problem of limited suspension in the spray tank. Fungicides now have the capacity to be used for preventive treatments, control treatments, or both. Products that can do both are a recent trend. Fungicide use is much more common in the golf course compared to the commercial lawn care business. Unless there is a catastrophic disease in a lawn, it is fairly uncommon to use fungicides. This is largely due to their cost.

Fumigants are another group of pesticides that might be encountered in the turfgrass industry. They are highly dangerous and are

PESTPOINTER

Any restricted-use pesticides will require a license to purchase and to use to any degree.

often used only in the golf course segment of the industry or perhaps in nursery and greenhouse production.

The last group of chemicals is the rodenticides, which is another high-risk category. The reason this group is so high-risk is because the target pests are usually mammals, and what will harm them can also harm humans and pets. Many of the rodenticides that are available are restricted-use pesticides because of their deadly nature and should only be purchased and used by licensed commercial applicators. Many of these products contain such chemicals as warfarin, zinc phosphate, aluminum phosphate, and even strychnine. If you are familiar with any of these components, you will quickly realize the deadly nature of this pesticide group.

Although most pesticides can be purchased by American homeowners for use on their own property, not all chemicals can be bought by just anyone. Any restricted-use pesticides will require a license to purchase and to use to any degree. Many chemicals can be purchase at large shopping centers and are fine for the general public for personal use as long it is not for hire and the labels carry no restrictions. The important fact to remember is that a chemical license will allow a person to treat other properties for hire. Even products bought at a farm, home store, or large shopping center cannot be applied by nonlicensed people for profit. "For hire" is the key phrase that requires the applicator's license. It is illegal to use even a "weed and feed" type of product for hire, but it is fine in most cases for you to apply it to your own yard. Once again, it is important to research all the local and state laws for your particular area before making any decisions on pesticide usage.

APPLICATION EQUIPMENT

There are numerous types of equipment that are used for the application of pesticides. Some of these pieces of equipment are designed for dry products and others are used for spraying. All spraying equipment needs to be accurate and to be able to be calibrated as well. Good-quality application equipment will always be easy to calibrate and should be very accurate with several adjustments. Equipment used for pesticide applications can be very large for farming operations or fit in one hand when spreading very small areas. There is a type of equipment specially designed for application to any terrain and area, and you should know which ones you'll need to use.

The smallest type of application equipment is handheld equipment (Figure 11.9). Handheld equipment consists of small spreaders and sprayers. Backpack sprayers are one of my favorite pieces of handheld application equipment because they are inexpensive and very accurate to use. A backpack sprayer is easy to carry and gives you very good control of the liquids you apply. You can get backpack sprayers and other handheld sprayers in either a piston pump or a diaphragm pump. The piston pump will usually give you higher

PESTPOINTER

Backpack sprayers are one of my favorite pieces of handheld application equipment because they are inexpensive and very accurate to use.

FIGURE 11.9 Small handheld sprayers are versatile enough to do small tasks but do not have enough pressure to do larger jobs.

pressure for spraying, but the diaphragm sprayer will provide less clogging and is less likely to leak over time. All small sprayers can even be used for alternative applications like liquid ice melts in the winter months. There are also handheld rotary spreaders, which can be used to spread dry product accurately in small quantities. There are several brands of both backpack sprayers and handheld spreaders that you can purchase; buy one that is adjustable, accurate, and very comfortable to use.

The next step up would be an electric pump sprayer, which usually will mount on some type of equipment with a 12-volt battery. These sprayers most often hold less than 50 gallons and sometimes do not have a boom. Booms are more commonly found on engine-mount sprayers. There is often a wand that usually attaches to these types of sprayers so they can be used as spot sprayers. Spot spraying is the most common use of electric sprayers, but booms can easily be attached to many models.

There are now a few electric rotary spreaders on the market, which are becoming seemingly more popular every day. The idea behind them is to have a "ride on" type of broadcast spreader that eliminates the walking. The only complaints I have heard about these units is a difficulty in calibration and an inconsistency of ground speed while applying.

The most common type of sprayer on the market for turfgrass maintenance is the gas-powered sprayer. This sprayer can be set up a couple of different ways. The first way is to have a tank with a boom mounted in some sort of transport vehicle. Some of these units are self-contained, run off the vehicle engine, and are designed specifically for spraying. They often have features such as a cab and filters for a safer application of product. The problem with these units is that they are very expensive and limited to golf course use in most cases. Other more commonly used sprayers may mount in the back of a utility vehicle. These are usually skid-mounted so they can be moved in and out of the vehicle very easily (Figure 11.10). They are very common on golf courses, athletic fields, and some lawn care applications. The larger counterparts of skid-mounted sprayers are found in the agriculture industry. Agricultural sprayers are designed

PESTPOINTER

The larger counterparts of skid-mounted sprayers are found in the agriculture industry. Agricultural sprayers are designed to fit in the back of a pick-up truck and are not recommended for any turf use.

FIGURE 11.10 Skid type sprayers are typically used for commercial lawn care operations).

to fit in the back of a pick-up truck and are not recommended for any turf use. Most turf sprayers will have tanks that range from 25 to 250 gallons. Anything over 250 gallons is probably more suited for crop and agricultural spraying.

The other type of skid-mounted sprayer is used more for lawn care but also works off an engine-driven system. These systems seldom have booms but instead use a long hose reel, which contains 100 to 200 feet of garden-type hose. The applicator will have a hose-end sprayer that disperses the pesticide while he or she walks over an area. The problem with this type of system is the high amount of exposure that results from dragging around the hose and picking up residues as you are applying the product. Then the hose is handled while reeling and unreeling it at each job site. You also tend to walk directly through the areas you are spraying as the pesticides are applied. It is important to wear personal protective equipment and be aware of the exposure you are receiving.

DISPOSAL OF CHEMICALS

It is not uncommon to have leftover pesticides in your tank when a spraying project is finished. Their disposal can cause problems to our drinking water and the environment. Unused pesticides that are disposed of incorrectly cause as many or more problems than applied pesticides. Many people who use pesticides do not know how to correctly dispose of containers or the leftover pesticides in the application equipment.

It is pretty easy to accidentally mix up too much product or to miscalculate the square footage of an area. After all, none of us is perfect, and sometimes plans can change after the product is mixed up. The first rule of thumb is to NEVER put a pesticide product in anything other than its original container. This includes any extra pesticide that you might have left over in a tank and that you are considering saving. This is the fastest way to expose someone who has no idea that there is a pesticide present in that particular container. The best way and safest way to get rid of extra product after it is mixed up is to apply it. I am not suggesting that an applicator should ever double up coverage in an area to simply "use up" what is mixed. What I am suggesting is to have a designated area where

PESTPOINTER

The first rule of thumb is to NEVER put a pesticide product in anything other than its original container.

extra product can be applied until the tank or spreader is completely empty. By applying the product you are complying with the label and at the same time you are keeping this extra product from being dumped in a ditch or area where it will cause off-site exposure.

Pesticides should never be poured out in ditches or sprayed out in one spot until the sprayer is empty. This is a very fast way to cause the pesticide to end up in the ground water or even in sewer systems. It could also move off-site to a pond, lake, or stream, where it could end up in someone else's drinking water. Make sure that extra product is applied to an area where it will not cause any problems. You should also avoid spraying the extra chemical out over a concrete parking lot, where it might be more likely to move off-site in runoff water.

If a dry product is used, you can just pour the excess back into the original bag. I have heard stories of people dumping it out in the street instead. This is another bad idea and once again will just cause problems as the product is carried away in runoff water.

Rinsing your equipment or spray tank can be just as important as any part of the cleanup and disposal process. The runoff from washing out a spray tank or rinsing the sprayer off can contain residues that can be very harmful. It is important, when a sprayer is rinsed off, that the water be collected whenever possible and excess runoff avoided. It is possible to put in a collection dam around your rinse

PESTPOINTER

Rinsing a tank should never be done around a well or water source. That is a direct way to get chemicals into a drinking-water system.

area, but you should check with your local laws on pesticide disposal and first and foremost always follow the label. Rinsing a tank should never be done around a well or water source. That is a direct way to get chemicals into a drinking-water system.

If you find that you are using a sprayer for multiple uses, you might need to neutralize your tank during the rinse process before changing chemicals in the tank. A triple rinse should be performed and bleach added to the tank on the third rinse cycle. When you do this, make sure that the rinses have made it through the entire system including the pump. Pesticides can remain in the lines and the pump if you do not do a thorough job of rinsing the system out. This process applies to all sprayers, but is much easier to do with smaller sprayers.

I do want to mention that certain pesticides can actually adhere to some spray-tank compositions, and complete removal of the prior chemical can be difficult to accomplish. I do suggest when using smaller sprayers especially that you try to keep a different sprayer for each type of chemical you are using. I know that this practice is cost-prohibitive on a larger scale, so make sure that you do a thorough rinsing each time you change chemical types.

The next issue to consider after you are finished spraying is what to do with an empty chemical container. Disposal of the container is another critical part of the pesticide application process. You should only dispose of chemical containers in approved locations. This will ensure that they are properly disposed of at a safe landfill area. There is disposal information on the label which can tell you specifics about any particular product you might be applying.

When an applicator has finished a treatment, a record needs to be made of that application. All of the following information should be recorded each time an application is made and immediately after you are finished. It is important to have all the information, because in many states an inspector will stop by your facility to check the accuracy of your records from time to time.

Items that should be recorded by a licensed applicator for each application of product:

- *Name of noncertified applicator or technician (if applicable)*
- *License number of applicator or technician*
- *Date of application*
- *Name of person requesting the application*
- *Address of requested person*
- *Address or location of application site*
- *Target pests to be controlled*
- *Pesticide trade name*
- *EPA registration number or documentation*
- *Reasonable estimate of amount of pesticide used*
- *Time, air temperature, wind speed, and wind direction*

A chemical inspector will visit you to make sure you are recording the appropriate information, usually once a year. This inspection is given to all commercial applicators at some point during the cal-

PESTPOINTER

A chemical inspector will visit you to make sure you are recording the appropriate information, usually once a year. This inspection is given to all commercial applicators at some point during the calendar year.

endar year. You should be aware that the inspector may want to follow you for the day and do an on-site application inspection. This is done at random by the inspectors and is studied to provide statistical information and other data for the state. The inspector will follow you throughout the day while you are doing your applications to make sure you are doing everything correctly. He or she may ask you many questions and even photograph the mixing and application process. You will not get in trouble for doing something wrong; it is more for your education and for documentation of what and how applications are being made. This is also a good time for you to ask questions so you can find out anything you may have been doing wrong. It is unfortunate that people view the state inspectors as the enemy when in fact they are there to educate you as a chemical applicator and to improve the general use of pesticides.

No matter what role chemicals play in your day-to-day operation, the concepts behind pesticide applications are fairly simple. If we all follow rules and guidelines set up by our state and local governments, we are making an effort to improve the pesticide application process. We must view the pesticide label as an instruction manual and realize that all the information that is printed on those labels is there for an important reason. Lastly, we must make good judgment calls as applicators and realize that safety is the most important issue when using and handling pesticides. The pesticides that are being used by the public are relatively safe when used properly, and the control they provide will be appreciated if everything else is handled in the right way.

SUMMARY QUESTIONS

- *What safety measures are already in place at your facility regarding the use and disposal of pesticides?*
- *After reading this chapter, are there any procedures that should be put into place at your facility?*

- *How can biotechnology affect the use of chemical pesticides?*
- *What steps do you and your facility have to complete in order to legally use and dispose of pesticides in your state?*

Chapter 12

Best Management Practices and Integrated Pest Management

This chapter represents the culmination of a tremendous amount of information that has already been presented to you in this text. We have already discussed the various chemicals and pesticides you will encounter, how to apply them, and what regulatory issues surround chemical and pesticide use. This final chapter discusses Best Management Practices and Integrated Pest Management, both of which are concepts that pull together components of each of these topics discussed earlier in the text. The best way to introduce Best Management Practices (BMPs) may be simply to define them. In the context of turfgrass and landscape management, BMPs are techniques used to minimize the movement of unwanted materials into sensitive areas of the surrounding environment. This definition, of course, includes the chemicals and pesticides we have discussed at length in this text but can also pertain to fertilizers and even sediment. The concept of Best Management Practices (BMPs) is one that originated in field crops but has since found its way into the turfgrass industry. This transition is in part due to a proactive effort on the part of turfgrass professionals to optimize their productivity and efficiency. It is, however, also in response to the growing criticism of turf for what is seen by many as aggressive and irresponsible fertilizer and chemical use.

BMPs embrace a number of conservation-oriented techniques, including Integrated Pest Management, but are only useful if they are planned and implemented. These practices are designed to protect the numerous natural resources present at any turfgrass or landscape facility. These resources include things like soil, native vegetation, water, air, and wildlife. Proper design and execution of a BMP program at your facility should focus on the following objectives:

- *BMPs should minimize the movement of displaced soil, fertilizers, or chemicals (including pesticides) into unwanted areas.*

- *BMPs should result in reduced and more efficient use of chemicals and pesticides.*

- *BMPs should result in greater discretion as to what types, what formulations, and what quantities of chemicals and pesticides are applied to different areas.*

- *BMPs should result in greater use of techniques that help conserve the natural resources at a particular facility.*

- *BMPs should result in greater awareness of your staff, coworkers, and clients as to the environmentally conscious approach you are assuming.*

As you can see, the BMP concept is not simply one that follows anti-pesticide sentiment. It is a multi-layered series of strategies that will make your approach to turfgrass and landscape management more responsible, more efficient, and more accessible to those who may judge what you are doing.

A truly in depth discussion of BMPs could command an entire text of its own. Within the scope of this text, my goal is to introduce you to some of the key concepts and strategies that fall into the broader scope of BMPs. You will find that the strategies and concepts discussed in this chapter are not only practical but also usable, such that you can take them into the workplace and begin benefiting from them immediately. The goals of this chapter are to take these BMP strategies and build upon those already discussed in earlier chapters, thereby giving you the tools to manage chemical and pesticide use successfully and responsibly at your facility.

REDUCED PESTICIDE AND CHEMICAL USE THROUGH STRATEGIC PLANTINGS

We will be discussing two different categories of BMPs in this chapter. The first category includes practices called land use BMPs. These, by definition, are practices that are built into the design of a particular facility. Introduction of land use BMPs can occur during construction of a new facility or can also be incorporated into facilities that are already there. The second category of BMPs is broadly referred to as a source prevention approach. Simply defined, this approach is more active in its pursuit of resource conservation and is an excellent supplement to land use BMPs that are essentially permanent once they are introduced. Source prevention BMPs primarily include the strategies and steps taken within an Integrated Pest Management program, which will be discussed in detail later in this chapter. Two key land use BMPs that will be discussed first are strategic plantings and turfgrass/landscape design techniques. These techniques are not mutually exclusive, meaning they are not independent of one another and therefore their benefits can often overlap.

Strategic plantings include landscape beds, vegetative filter strips, trees, shrubs, and grassed swales or berms. Where these plantings are used will vary but all of these types of plantings share a common benefit. They are designed and intended to require relatively low maintenance. This means that, once they are installed, they should not require much mowing, fertilization, irrigation, or pesticide and chemical use. Selection of plantings should adhere to this desired outcome, meaning plants selected for these low input areas should be well adapted to your area and your climate. If they are not, these so-called low maintenance areas will require more input than was originally intended. Where to use these types of plantings will depend of course on the site characteristics of your facility, but also upon

> **PESTPOINTER**
>
> Strategic plantings, by reducing inputs required where they are planted, will effectively reduce the acreage at your facility that will require care.

what your facility is used for and where sensitive environmental areas are located.

Strategic plantings, by reducing inputs required where they are planted, will effectively reduce the acreage at your facility that will require care. There will usually be costs associated with installing strategic plantings but, once they are in place, reduced water, fertilizer, chemicals or pesticides, and labor time in these areas will quickly recover any costs of installation. Strategic plantings can also be easily incorporated into the design of your facility, providing natural buffers between areas that you want to keep separated. Golf courses may use large amounts of such plantings to separate golf holes and to minimize necessary inputs between them (Figure 12.1).

Lastly, strategic plantings can be used to buffer maintained areas from sensitive ones. The classic example of this is the use of a

> **PESTPOINTER**
>
> The filter strip does exactly what its name implies by filtering nutrients, sediment, and chemicals out of water moving across the turf surface before they reach the water source.

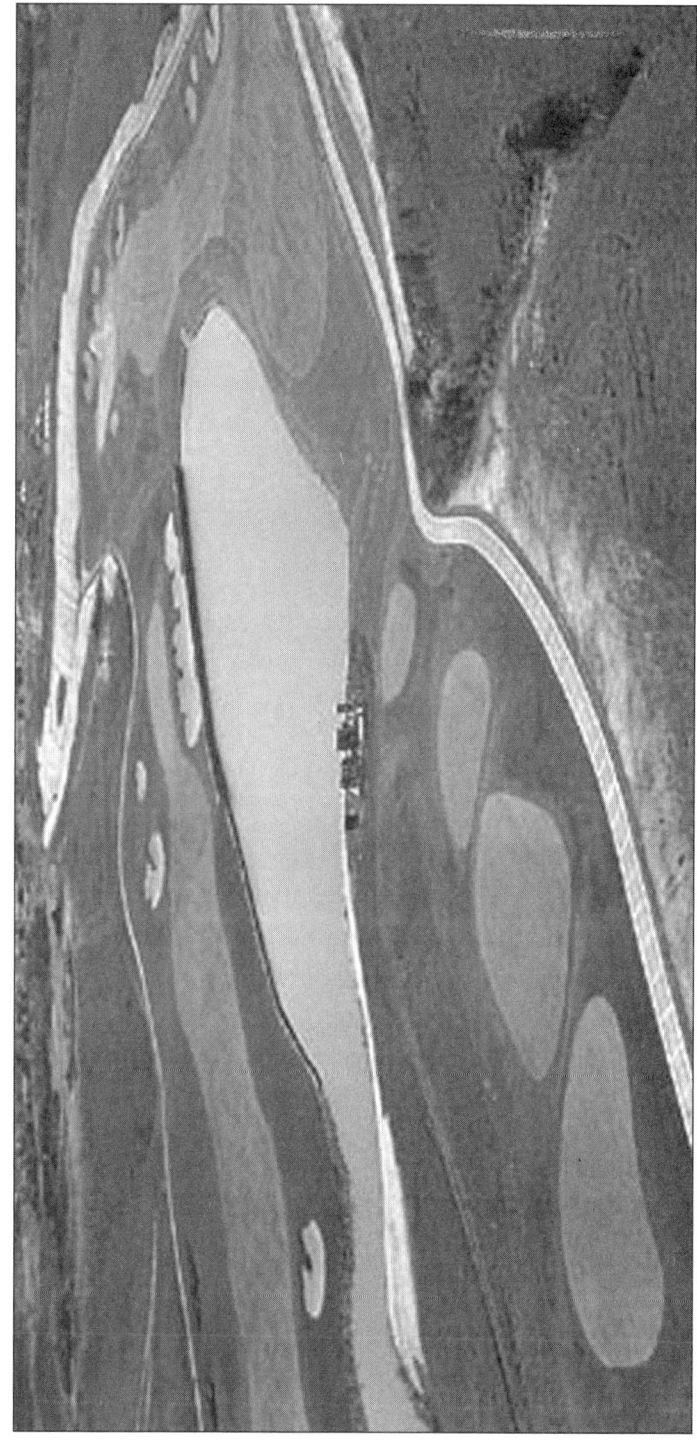

FIGURE 12.1 Note how both water and native grasses can be used in golf course design to frame a golf hole.

502 Turfgrass Chemicals and Pesticides: A Practitioner's Guide

FIGURE 12.2 A classic example of a forested riparian buffer, separating a water source from managed land.

vegetative filter strip between managed areas and sources of water (Figure 12.2). The filter strip does exactly what its name implies by filtering nutrients, sediment, and chemicals out of water moving across the turf surface before they reach the water source. Installation of a filter strip can be as simple as allowing native vegetation near the water to grow unimpeded until they have reached a more natural size. Consider all of these advantages when you are exploring the use of strategic plantings and the benefits should outweigh any potential hassles.

Turfgrass and Landscape Design Influences on Pesticide Use

We've been discussing some of the key types of plantings that can help a facility pursue a BMP approach to resource conservation. Since these types of plantings must be incorporated into the design of a turfgrass or landscape facility, the design process is critical to optimizing the impact of these plantings. For example, a common strategic planting like a vegetative filter or riparian buffer between a maintained area and a water source can produce a lot of benefits to the ecosystem. However, if the planting disrupts the function of the facility, these benefits become more questionable. Let's use a golf course as an example. Vegetative filters are excellent tools to preserving aquatic ecosystems at this type of facility. However, this type of planting should not disrupt the playability of the golf holes it is adjacent to. Proper design should focus play away from these areas so they aren't disruptive to a patron's round of golf (Figure 12.3).

Design issues are most easily implemented during construction of a new facility because the impact of the BMP design practices can be envisioned before the facility is put to use. Examples include proper design of golf holes around sensitive areas and subsequent installation or incorporation of strategic plantings in areas where they

FIGURE 12.3 Vegetative filter strips that protect water sources ideally should be placed out of the way of normal play on a golf course.

PESTPOINTER

Design issues are most easily implemented during construction of a new facility because the impact of the BMP design practices can be envisioned before the facility is put to use.

should not require extensive maintenance or traffic. If you have a wetlands area at your golf course, you will want to protect it (Figure 12.4). Many areas mandate that wetlands be preserved, either by avoiding maintenance around them or by prohibiting entry into the area altogether. With this in mind, you would not want the wetlands to be anywhere near a landing area on a golf hole. This will require both maintenance in close proximity to the wetlands and also will encourage golfers not to enter the area looking for lost balls. Proper design should situate wetlands either off to the side of a golf hole or in an area where golf shots are not likely to go. This same strategy can be applied to parks or commercial facilities by situating sensitive wetlands areas away from where patrons are likely to go.

FIGURE 12.4 Wetlands areas are highly sensitive and their preservation at a facility is essential.

PESTPOINTER

One of the fundamental ways to minimize the impact of maintenance practices on sensitive environmental areas is to reduce or eliminate the need for maintenance near these areas. Landscapes can help you achieve this goal as they will not require as many inputs as turfgrass areas.

Landscape plantings are also one of the best design tools that can be used for BMPs. Landscaped areas can be used both to beautify a facility and to buffer sensitive areas from maintained areas. One of the fundamental ways to minimize the impact of maintenance practices on sensitive environmental areas is to reduce or eliminate the need for maintenance near these areas. Landscapes can help you achieve this goal as they will not require as many inputs as turfgrass areas. This will potentially reduce the use of fertilizers and chemicals or pesticides in close proximity to sensitive areas. With this in mind, design these types of plantings to be situated where they will not only provide aesthetic value, but where they will also be most useful. A great example of this is the use of landscaping near driveways and parking lots. The potential for inputs to reach impermeable surfaces like pavement is high in these areas so the use of low input landscapes can help keep applied inputs where they belong, in addition to beautifying the area.

INTEGRATED PEST MANAGEMENT PRINCIPLES

Integrated Pest Management (IPM) is one of the most powerful concepts you can employ to optimize your chemical and pesticide use efficiency. As you have seen from the previous sections, BMPs include numerous strategies other than those that pertain directly to pest control. IPM is simply another component of BMPs that can help you achieve an efficient and environmentally responsible management program. IPM has been mistakenly perceived as an anti-pesticide program, to the point where I have humorously described the IPM acronym to mean "impulsive panic mongering" to clients I have taught in the past. As you will see, the true objectives of IPM do not suggest abolition of pesticides but rather to incorporate pesticide use into a more comprehensive plan of attack to ideally manage and control pests. Table 12.1 lists and describes six key approaches to a good IPM program.

Note that chemical use is only one of the six approaches that IPM includes but that it is indeed a key component, despite the contentions of some that IPM is against the grain of conventional pesticide use. Of course, chemical use within an IPM program is recommended to be timely, judicious in terms of where chemicals are used according to label directions, and in adequate rotation to prevent pest resistance. The other approaches to IPM are more fun-

TABLE 12.1 Different approaches to IPM and how to best use them.

Approach	Description
Genetic	Use of recommended turfgrass species and varieties that are adapted to your area and can thus best handle pest and environmental challenges
Regulatory	Use of high quality plant material that is guaranteed to be pure and free of contaminants like weed seeds
Cultural	Use of recommended turfgrass cultural practices to optimize the natural ability of our desired plants to withstand pests
Physical	Use of clean equipment to avoid unwanted transfer of pests from one place to another or physical removal of pests from a location (e.g. handweeding)
Biological	Use of biological control agents or installing structures to promote natural pest predators
Chemical	Use of traditional pesticides in a timely, judicious, and responsible manner to supplement other IPM approaches

PESTPOINTER

IPM is simply another component of BMPs that can help you achieve an efficient and environmentally responsible management program.

damental, with respect to managing your facility. They all have some pertinence to pest control but may not necessarily target control of pests. Recall the adage mentioned in the chapter on weed control that stated the best defense against weeds is a strong, healthy stand of turf. That concept epitomizes the non-chemical approaches to IPM listed in Table 12.1. Use of quality plant material, cleaning your equipment properly, managing your turf and landscape plants properly, and promoting natural predators are all techniques that can improve a pest control program without venturing into the arena of chemical applications. Many of these techniques you may already have in place. If this is the case, you are to be commended. You are well on your way to having a polished and refined IPM program in place. If not, these are fairly simple concepts and are easy to implement to move in the direction that IPM suggests. More will be discussed later in this chapter on how to execute an IPM program in a defined, stepwise fashion.

INTERPRETING PESTICIDE RISK

It is worth spending a bit more time talking about the risks associated with pesticide use, especially since many of your customers and clients may view these materials as risky. To the untrained member of the public, risk can be a rather subjective issue. Clients or patrons at your facility may arbitrarily assume they are being put at risk when you are using pesticides. This fearful presumption usually has

> **PESTPOINTER**
>
> Clients or patrons at your facility may arbitrarily assume they are being put at risk when you are using pesticides. This fearful presumption usually has nothing to do with the particular pesticide in question and may represent an overall opinion of all chemicals.

nothing to do with the particular pesticide in question and may represent an overall opinion of all chemicals. This brings us to the question of how to assess risk properly and how to educate those around you about these risks. Many of the readers of this text are trained professionals in the turfgrass and landscape industry. With this recognition comes the expectation that they not only have expertise in the application of chemicals and pesticides but also in the understanding of how these materials can affect their surroundings. This chapter section is intended to show you how risk can actually be defined for a particular chemical or pesticide and what that definition means to you and to the people you answer to.

Risk from chemicals or pesticides can be quantified as a combination of two key elements: hazard and exposure. Hazard is the likelihood of danger that will result, either to the environment or to other organisms, from an encounter with a chemical or pesticide. As you might expect, hazard levels can vary considerably among chemicals and pesticides so this criterion alone cannot accurately define the risk potential. That is why exposure is also factored in when assessing the potential for risk. Exposure relates to the total amount of contact that an organism, or perhaps a sensitive environmental area, has with a particular chemical or pesticide. Let's now look at some examples in Table 12.2 of how the combination of hazard and exposure can result in different defined levels of risk.

TABLE 12.2 Risk associated with different hazard and exposure levels for chemicals and pesticides.

Hazard Level	Type of Exposure	Level of Risk
Low	Brief, little contact	Very low
Low	Little contact, repeated on several occasions	Low
Low	Brief, thorough contact	Low
Low	Thorough contact, repeated on several occasions	Moderate
Moderate	Brief, little contact	Low
Moderate	Little contact, repeated on several occasions	Low/Moderate
Moderate	Brief, thorough contact	Moderate
Moderate	Thorough contact, repeated on several occasions	High
High	Brief, little contact	Low/Moderate
High	Little contact, repeated on several occasions	Moderate/High
High	Brief, thorough contact	High
High	Thorough contact, repeated on several occasions	Very high

PESTPOINTER

It is important to inform your staff, coworkers, and clients that pesticides may not be that dangerous if direct exposure can be limited.

As you can see, risk is definitely a function of both hazard and exposure. Chemicals or pesticides that have low hazard potential may represent moderate risk if exposure is thorough and somewhat regular. Similarly and perhaps even more importantly, materials with high hazard potential may not be that risky if exposure is slight and brief. It is therefore important to inform your staff, coworkers, and clients that pesticides may not be that dangerous if direct exposure can be limited. Proper execution of appropriate safety precautions should limit the kinds of exposure that would prove most risky to those who may encounter the materials you apply.

Risk assessment can vary among the different types of materials we apply at our facilities. To expound, we can be more specific with our determination of risk potential, based upon some characteristics of these different materials. For example, we don't need to get too caught up worrying about risk to human beings or other organisms with herbicide applications. These materials tend to have very low toxicity. However, herbicides may present risk to the environment so our analyses of hazard and exposure might be best suited from an environmental contamination standpoint. Refer to Table 12.3 for a brief outline of how to determine herbicide risk to the environment. Note that our criteria for herbicide risk include issues like longevity (usually determined by the half-life of the material in question) and mobility,

TABLE 12.3 Criteria used to determine herbicide risk to the environment.

Longevity/half-life	Mobility	Level of Risk
Low	Low	Very low
Low	Moderate	Low/moderate
Low	High	Moderate
Moderate	Low	Low
Moderate	Moderate	Moderate
Moderate	High	High
High	Low	Moderate
High	Moderate	High
High	High	Very high

the combinations of which give us a good assessment for potential to contaminate areas away from where the application is being made. In fairness to the herbicide industry, there are few, if any, herbicides on the market today that represent high environmental risk. Those that may or did fit into this category have been removed from the market in recent years to promote environmental responsibility.

By contrast to herbicides, fungicides and insecticides are generally more toxic materials so a proper risk assessment would place greater emphasis on toxicity to humans and other organisms. Certainly, the same sorts of environmental criteria that are listed in Table 12.3 are also relevant for these pesticides but toxicity parameters should also be factored in to optimize safety to both clients and applicators. Potential toxicity of a fungicide or insecticide can be classified in one of two ways: acute or chronic. Acute toxicity is an immediate harmful effect that follows contact with a pesticide while chronic toxicity is a harmful effect that can result after more prolonged exposure or a series of brief exposures. Naturally, either type of toxicity can factor into a determination of risk but high potential for acute toxicity can elevate the risk potential of a fungicide or insecticide more so than for chronic toxicity. Pesticide labels and MSDS are the best ways to determine the potential toxicity of a material you may want to apply. Use this information to not only optimize safety measures used when making an application but also to determine areas where to apply the material. Areas that are environmentally sensitive should

PESTPOINTER

By contrast to herbicides, fungicides and insecticides are generally more toxic materials so a proper risk assessment would place greater emphasis on toxicity to humans and other organisms.

be avoided to minimize risk to both the environment and to the organisms that may be affected.

DEVELOPING AN INTEGRATED PEST MANAGEMENT PLAN FOR TURFGRASS AND LANDSCAPES

Now that we have covered some of the fundamental principles of BMPs and IPM, how do we take these strategies to the field and implement them? Hopefully, from the earlier discussions in this chapter, you will see that BMPs and IPM do not necessarily represent a dramatic departure from techniques and strategies you are currently using. Rather, the objectives of these programs are for you to visualize a series of steps to follow, determine what, if any, changes or additions you may need to make, and then integrate them into your current management procedures. There are six key steps to implementing and executing an IPM program at your facility. These are:

- *Monitoring or evaluating what pests you have at your facility and the effects of pesticides*

- *Identifying what pests are the greatest threats and how much of each can be tolerated*

- *Making decisions as to which, if not all, of the six IPM approaches to use and how to best incorporate them into your management schedule*

- *Training and/or informing your staff, coworkers, and clients of what IPM strategies you are using, why they are being used, and what impact they will have*

- *Properly selecting chemicals to use, where to use them, and how frequently they need to be used*

- *Assessing the results of your IPM strategies and using these results to evaluate strengths and weaknesses of your program so you can make informed decisions in the future*

The six different approaches to IPM discussed earlier in the chapter did not rank equally in terms of their importance and practicality. By contrast, these six steps to carrying out an IPM program are each very important and need to be performed for IPM to be as successful as it can be. Let's look at each of these steps in more detail.

Monitoring pest populations is one of the cornerstones of modern IPM. We now have the ability to conduct many types of monitoring and even better ways to track the results. Four primary types of monitoring exist, as pertains to IPM: reconnaissance, record keeping, subjective, and objective. Reconnaissance is often better known as pest scouting. This form of monitoring involves you or someone else at your facility periodically evaluating the incidence of different pests, where they are, and when the problem appears. This is highly proactive so is only successful if your facility is committed to scouting. Several members of the maintenance staff should be skilled in scouting as this will allow for earlier detection of pests and more timely execution of control measures. Recording exactly where and when pests are found is a critical aspect of scouting and can help reduce total pesticide use by allowing you in some cases to target only those affected areas. Scouting can and should be performed while doing other tasks to optimize worker efficiency, but a log or journal should accompany the scout to optimize record keeping.

Record keeping as a form of monitoring goes beyond pest scouting. It is also a critical aspect of adhering to regulatory guidelines.

PESTPOINTER

Monitoring pest populations is one of the cornerstones of modern IPM. We now have the ability to conduct many types of monitoring and even better ways to track the results.

> **PESTPOINTER**
>
> It is now mandated that all chemical and pesticide applications be documented. This documentation should include date of application, total amount of pesticide or chemical used, time of day for the application, rate of application, and specific areas of the facility that are treated.

It is now mandated that all chemical and pesticide applications be documented. This documentation should include date of application, total amount of pesticide or chemical used, time of day for the application, rate of application, and specific areas of the facility that are treated. These records not only satisfy legal obligations but are prudent in case of a complaint or environmental issue that may arise following the application. Poor record keeping can result in fines or even loss of job so take this very seriously. The same record keeping can also be a valuable tool within IPM. Knowing when pests were treated, where, and with how much of a particular product can help you effectively judge the success of applications and potentially identify ways to reduce pesticide or chemical use and therefore save money. In this way, record keeping should be cross-referenced with scouting reports to help add precision to your chemical and pesticide application program.

Subjective monitoring is really an extension of pest scouting. A trained scout, who scouts and then makes an immediate decision to act towards control of a particular pest, usually performs subjective monitoring. For example, a golf course superintendent not only is an effective scout at his or her own facility but also has the discretion to allocate resources and effort to take action on the results of the scouting effort. This form of monitoring can result in very efficient control of pests and is termed subjective only because it is not based upon data that is generated. Rather, subjective monitoring relies upon the educated opinion of the person doing the monitoring. There is an implied margin of error with subjective scouting but education of personnel can reduce the potential for mistakes.

The last form of monitoring is termed objective because it relies exclusively upon the collection and interpretation of data. Objective monitoring is not used as much to determine if and when chemicals or pesticides need to be applied but it can be useful for assessing any environmental impact of these applications. Common forms of objective monitoring include measurements of sediment, nutrient, and chemical or pesticide levels in sensitive areas like ponds, lakes, or streams (Figure 12.5). Objective monitoring often has little to do with the actual pests themselves but can be a critical part of an environmentally responsible management program. Objective monitoring is an extremely valuable tool in determining the impact of both turfgrass installation and maintenance on the surrounding environment.

FIGURE 12.5 A stream running through a golf course would be a common place to conduct objective monitoring.

To be most effective, objective monitoring should be conducted multiple times each growing season and, when possible, at times just following fertilizer or chemical applications to determine the immediate impact, if any, of these practices. Understandably, monitoring streams or ponds at a facility six weeks after any types of applications were last made might produce a false sense of security. Therefore, if we can determine a lack of impact under the most vulnerable of circumstances, the effectiveness of monitoring is increased. Results of objective monitoring can produce information at any given time of sampling but can also be used to determine trends over time, which are perhaps the most useful when using these data to make decisions about a particular type of application. Objective monitoring is best achieved by taking random water samples at defined times and having them analyzed professionally to determine their contents.

Once monitoring results have been collected, it is important to determine what levels of pest control are necessary. This usually involves establishment of pest thresholds, which dictate at what level of population a particular pest needs to be controlled in order to preserve turfgrass quality and health. For example, many weeds might have reasonably high thresholds, due to their relative lack

PESTPOINTER

Results of objective monitoring can produce information at any given time of sampling but can also be used to determine trends over time, which are perhaps the most useful when using these data to make decisions about a particular type of application.

> **PESTPOINTER**
>
> Once monitoring results have been collected, it is important to determine what levels of pest control are necessary. This usually involves establishment of pest thresholds, which dictate at what level of population a particular pest needs to be controlled in order to preserve turfgrass quality and health.

of aggressiveness, while others may be less tolerated. Established thresholds for diseases may actually dictate what type of control measure is used. Highly destructive and fast-acting diseases may require preventive treatment, where fungicides are applied when conditions are suitable for disease outbreaks. Others can be handled on a curative basis, where control is only used when disease symptoms first become evident. As was discussed in Chapter 2, thresholds can be affected by the characteristics of the pest, the type of turf that is at stake, and the economic capabilities at a facility. Establishment of thresholds and using them to optimize control measures are valuable ways to potentially avoid unnecessary applications and to target those pests that pose the greatest threat to you and your facility.

When pests have been identified and their thresholds have been reached or surpassed, control measures only then become necessary. As was mentioned earlier in the chapter, IPM includes several approaches that can be employed to control pests. Pursuit of all of these approaches must be at least considered in a good IPM program. Planning stage IPM approaches like genetic or regulatory methods are more pertinent at the planning stage or prior to planting turf. However, cultural, physical, and biological approaches are all valid considerations, in addition to the use of chemicals or pesticides. For example, physically removing dew from putting greens might be an identified alternative way to diminish one's reliance on fungicides for disease control. Chemicals and pesticides may ultimately be your best weapon against a particular pest but consideration and/or incorporation of these other approaches may help reduce your pesticide needs.

Education is as fundamental a part of IPM as any because it intimately affects all of the other steps. Scouting or other forms of monitoring are only as effective as the people who are doing the work involved. Decisions made concerning pest thresholds and pest control strategies must come from an educated source or mistakes can and will be made. And, of course, the execution of applications must be handled and supervised by skilled and trained personnel who know what they are doing. Education is a superceding part of all of these IPM steps and, while often overlooked, is perhaps the easiest phase of IPM to implement. No decisions need to be made during education, although good training will enable a staff member to make good decisions. Education can be as simple as spending 30 minutes a day to train a new staff member or as involved as encouraging staff members to enroll in courses or seminars to broaden their knowledge. Any person can be trained to complete a task, based upon a series of steps that they follow. We must take education beyond sim-

PESTPOINTER

Any person can be trained to complete a task, based upon a series of steps that they follow. We must take education beyond simply training a person to perform a duty for IPM to be most effective.

ply training a person to perform a duty for IPM to be most effective. Some practical examples of educational activities on the job that pertain to IPM include:

- *Allowing staff members to lead and direct certain tasks after they have acquired experience*
- *Allowing staff members to observe and listen when decisions are discussed and made*
- *Asking staff members for their thoughts and input during a decision-making process*
- *Involving as many staff members as possible to participate when a pest is identified or diagnosed*
- *Allowing less trained staff members to scout with more experienced individuals to boost their skills*

Education is as much about what students do with their knowledge as it is about the learning process. Staff members that are never given opportunities to apply their learning on their own will tend not to use what they have learned. I can't emphasize enough that this process at your facility must include giving responsibility to staff members who have made the effort to learn. Given this opportunity, most people will rise to the challenge and may surprise themselves, if not their supervisor. Allowing all of the staff members at your facility to become familiar with IPM and how it works in your environment will only increase how effective the program can be because they will all become scouts, analysts, and perhaps even decision-makers.

A key component of any IPM program is the actual use of pesticides. Pesticides always have been and still are one of the most useful tools we have to target pests. IPM doesn't discourage chemical use but does promote their use as one of many tools available in a total pest control program. How do we refine chemical and pesticide use so that it is a tool, rather than the only approach taken to control pests? One way to achieve this objective is to build chemical and pesticide applications around the previous steps taken. Using the results of pest monitoring and, after employing alternative IPM approaches, chemical pest control can be directed towards areas that are in the greatest need for the applications. Once areas have been identified to receive pesticide applications, the next step is to properly select a pesticide. Criteria that should be used to select pesticides in an IPM program are as follows:

- The efficacy of the pesticide against the target pest
- The potential environmental impact of the pesticide
- The characteristics of the location(s) where the pesticide is to be applied
- The potential for risk to applicators or clients associated with the pesticide
- The cost effectiveness of the pesticide

It would be very difficult to assess the performance of an IPM program accurately without an evaluation of the program. For this reason, the last key step of IPM is to look back on what was done during the course of a growing season or other defined period during which IPM strategies were implemented. Some things you try may work extremely well, some things may be marginal, and some may be disappointing. Hopefully, you can see that it can be critical to determine which strategies work and which ones don't so your efforts can be focused on what is working for you. Let's look at some examples.

Last year, you decided to apply a fungicide only to areas where disease problems had been most rampant. By evaluating where disease outbreaks were located and when they occurred, you were able to determine that this particular strategy kept disease incidence within acceptable levels and also reduced your fungicide costs by 40 percent. Other situations may not produce the same good results. During the same year, you decided to approach weed control in a similar fashion to the fungicide example described above. You decided to apply only preemergence herbicides for crabgrass control in areas where weed pressure had been the worst in past years. This approach, however, backfired, and you ended up having to apply a postemergence herbicide later in the year to crabgrass that was thriving in the untreated areas. In this case, you would likely determine that use of a preemergence herbicide across your entire facility was both more time efficient and more cost effective. By performing a simple evaluation in each of these two cases, you can effectively determine which approaches best fit the needs of your facility and which ones might be worth modifying or outright eliminating. Either way, you have a stronger plan of attack for the following growing season. In this light, the evaluation step of IPM, even though it is the last of the six listed steps, is really the first key step you will take in later growing seasons once you have started an IPM program.

Much of this section has been devoted to the specifics of IPM programs. However, especially when it comes to the evaluation step, the same techniques can be used for the broader scope of BMPs. Earlier in this chapter, we talked about using strategic plantings to minimize the environmental impact of our management practices. Say, for example, you decided to allow plants to grow taller around the perimeter of a pond at your facility and thus create a natural buffer strip. This technique can be very useful but only if you evaluate how it has helped you. If algae had always been a problem in this pond in summer months and you noticed a significant decline in algae after allowing the buffer strip to grow, you could accurately determine that the buffer strip was responsible for this positive change. This change could then be interpreted in a number of ways. You could conclude that fertilizers applied near the pond were not reaching the water as much as before, thus reducing the algae. Similarly, you could also conclude, because fertilizer had reached the water in the past, that your fertilization program could be modified to be more precise near the water or tapered down to reduce total fertilizer use. Regardless of the conclusions drawn or how they impact your future decisions, you can see how the use and proper evaluation of BMP strategies can impact what you do at your facility. Nature can tell us a lot about what we are doing within it so pay attention to your surroundings and use both BMP and IPM strategies to optimize your facility's beauty and your performance.

CONCLUDING REMARKS

Congratulations! You have successfully completed reading one of the most comprehensive and practical texts that is available regarding this most important of topics. I sincerely hope that your journey through this text has been both informative enough to expand your base of knowledge and practical enough to bring immediate relevance to the tasks you perform on a regular basis. However, reading this text is only the beginning. You must now take what you have learned and apply it. As an author, I can only hope that you will take the contents of this text and use them to bring your skills and awareness to the next level. I have mentioned them on numerous occasions in this text but feel that certain things should be briefly stated one more time. Make sure that this text is only one of many tools that you will use during the course of your activities as a turfgrass practitioner. Many other tools are available to you and will be essential to your successful career development. Among these are extension resources (both written and verbal), label and MSDS information for the chemicals and pesticides you use, technical and sales representatives for these products, and your peers. Your peers may be some of the most valuable sources of information you will encounter. They have access to the same forms of information you do but they also have an additional and often understated resource to offer: experience. To read this or any text and presume it to offer you everything you will need on the job would be a tremendous disservice to your own professional development. Attend seminars, stay current with changes in your industry, consult with your peers, and don't be afraid to try different things to achieve success. All of these things will be excellent supplements to a written resource like this text and

I, as the author, hope you will use each and every one of them to make the most of the career path you have chosen.

SUMMARY QUESTIONS

- *As an introductory question, how would you rate the conservation practices currently used at your facility? What immediate steps could you reasonably take to improve this rating if there is perceived need for improvement?*
- *What types of BMPs have you employed at your facility? Did you think about them as conservation practices when you started or has your perception now changed?*
- *Are there strategic plantings BMPs that would be practical to use at your facility? If so, think of where you would use them and for what purpose.*
- *How well does the design of your facility adhere to the principles of BMPs? What could you do to improve this situation?*
- *What IPM approaches do you currently use on a regular basis? Are there others that you would find reasonable and practical to start using?*
- *Do you, your staff, or your co-workers scout for pests in areas where you traditionally apply pesticides? If so, what procedures do you follow, how do you document your findings, and how often is the scouting performed?*
- *If IPM strategies are currently being used at your facility, how do you evaluate the results of these efforts?*

Index

2,4-D,
  discovery of, 11,187
  injury symptoms, 177, 200
MSDS,
  content of, 59-68
  importance of, 30, 58, 68
  safety training using, 68

Active ingredient,
  definition of, 32
  discovery of, 33-34
Adjuvant, 362
  activator, 363
  buffers, 369-370
  compatibility agents, 368
  definition of, 362
  dyes, 371
  list of, 363
  spreaders, 369
  stickers, 369
  utility modifiers, 370
Algae, 385
  control of, 387-389
  impact of turfgrass
    management practices, 386
  pictures of, 385, 387-388
Applicator testing, 468-470
Aquatic weeds, 384
  Alligatorweed, 391
  Cattails, 389
  control of, 393
  Horsetails, 390
  Purple loosestrife, 392
  Water hyacinth, 390
  Water lilies, 391
Atrazine,
  injury symptoms, 179
  introduction of, 11
  resistance to, 205

Best Management Practices
  (BMPs), 497
  design issues, 503
  evaluation of, 521
  landscaping and, 505
  land use, 499
  objectives of, 498
  source prevention, 499
Biostimulants, 372
  hormone ingredients, 373-374

525

micronutrient ingredients, 372–373
pigment ingredients, 373
plant defense activator, 375
uses of, 375–376
Biotechnology and turf, 17, 466
Bird pests, 394–396
Broadleaf weed species, 117
  Black medic, 122, 126
  Broadleaf plantain, 135, 138
  Bull thistle, 108, 129–130
  Carpetweed, 121
  Chickweed, 126–127
  Curly dock, 136, 138
  Dandelion, 132, 137
  Field bindweed, 133, 138
  Henbit, 105–106, 127–128
  Horseweed, 123, 125
  Lespedeza, 126
  list of, 119–120
  Prickly lettuce, 124
  Prostrate knotweed, 122–124
  Prostrate spurge, 99, 107, 123, 125
  Speedwell, 137, 139
  White clover, 111, 134, 138
  Wild carrot, 130
  Wild geranium, 131
  Yellow rocket, 128–129
  Yellow woodsorrel, 135, 138

Contact herbicides, 188–190

Deer, 396–398
Diseases, 210
  Anthracnose, 212–213
  Brown patch, 213–216
  causal organisms, 211
  diagnosis of, 263–264, 277
  Dollar spot, 224–225
  Dutch Elm disease, 267–269
  effects of cultural practices, 245–246
  Fairy ring, 232–233
  Gray leaf spot, 226–227
  Gray snow mold, 220, 222
  Helminthosporium, 230–232
  Large patch, 229
  leaf spots, 274
  Pink snow mold, 219–221
  plant health and, 261, 276
  Pythium blight, 218–219
  Red thread, 216–217
  root/crown rot, 274
  rust, 217, 273–274
  signs of, 211
  spreading of, 255–256
  Spring dead spot, 226–228
  St. Augustine decline, 229–230
  Summer patch, 222–223
  symptoms of, 212
  Take all patch, 223
  Verticillium wilt, 272–273
  Yellow patch, 213, 215
Disease triangle, 210, 241–245, 251
DDT, 324–325
  banned use of, 14, 324–325
  marketing of, 12
  use of, 4, 7, 19

EPA, 15
  formation of, 15
  product registration, 476, 481
  signal words, 59

FDA pesticide regulation, 13
Fungicides, 234, 485
  application of, 247, 253, 259–261
  contact, 234–235, 278
  diseases controlled by, 248
  formulations, 238
  injections of, 256–258

injury to turfgrass, 239–240
list of, 236
programs for use, 249
resistance to, 237
systemic, 235, 237

Genetically modified organisms, 18, 187
Glyphosate,
and herbicide resistant crops, 18, 466
and herbicide resistant turf, 17, 186–187
herbicide use of, 185
Grass weed species, 139
Annual bluegrass, 104, 144, 146
Annual ryegrass, 144
Bahiagrass, 151
Bermudagrass, 112, 149–150
Crabgrass, 97, 141–143
Dallisgrass, 151–152
Field sandbur, 142–143
Foxtail, 144–145
Goosegrass, 101, 143
Johnsongrass, 155–156
list of, 140
Little barley, 144, 147
Orchardgrass, 149–150
Quackgrass, 153–154
Rough bluegrass, 153–155
Torpedograss, 153
Tufted hardgrass, 146, 148

Herbicide families, 173
Herbicide mode of action,
amino acid disruption, 180–181
cell division inhibition, 176
grass-killing, 175
hormone disruption, 176–177
photosynthesis disruption, 176–179

Herbicide names and products, 193–194
Herbicide resistance, 201
definition of, 202
in annual bluegrass, 202, 205
Herbicide selection criteria, 192, 194–195
Herbicide selectivity, 185, 187–188

Insect control, 328
cultural, 341
diagnosis and, 329
evaluation of, 334
plant selection and, 340
steps in, 329
Insecticides, 324
activity of, 333
application of, 351–352
banning of, 339–340, 483
biological, 326
Borate, 484
Carbamate, 326–327, 484
Chlorinated hydrocarbon, 327
list of families, 326
mode of action, 325
Organophosphate, 327, 483
products, 331–332
Pyrethroid, 328, 485
resistance to, 333–334
Insect pests,
Armyworms, 305–309
Asian long horned beetles, 354–356
Bagworms, 343–344
Billbugs, 316–320
Bores, 347–348
Chinch bugs, 292–294
complete metamorphosis, 280–281
Crane flies, 314–315
Cutworms, 309–311
damage from, 337

Fire ants, 320–323
Frit flies, 312–313
Grasshoppers, 338
Ground pearls, 284–286
Gypsy moths, 353–354
incomplete metamorphosis, 280, 282
Leafhoppers, 288–289, 348
list of, 283
Mealybugs, 286–288
Mites, 323, 346–347, 350
Mole crickets, 294–297
Mosquitoes, 336
Scale, 344–345
Sod webworms, 301–305
Spittlebugs, 289–291
White grubs, 298–301
Wireworms, 316
Integrated Pest Management (IPM),
approaches to, 506
birth of, 17
evaluation of, 520–521
implementation of, 209, 507
objective monitoring, 515
program development, 512
record keeping and, 493, 514
role of education, 518–519

Methyl bromide, 36, 485
Moss, 139

National Fire Protection
Association, 66–67
Nematodes, 233
Nicotine, 6, 8
Nonselective herbicides, 185, 483
Nuisance pests, 398
Chiggers, 401
Mosquitoes, 400
Reptiles, 399
Ticks, 400–401

Paraquat,
activity of, 179, 185
injury symptoms for, 180
Personal protective equipment (PPE), 474, 476
Pesticide fate, 75
leaching, 80–81
summary list of, 76
surface runoff, 80
volatility, 78
Pesticide formulations,
baits, 40
definition of, 33
discovery of, 34–35
dry flowable, 42, 45
dusts, 37, 39–40
emulsifiable concentrate, 46, 49–51, 53
flowable, 46–47, 49
gaseous, 35–36
granule, 40–42
list of, 38
microencapsulated concentrate, 46, 51–52
numbering schemes for, 52–56
powders, 42, 44–45
ready to use (RTU), 46–47
solution concentrate, 46, 52–53
wettable granules, 43–44
Pesticide labels,
content of, 71–75, 475–482
importance of, 70
Pesticide risk, 507
assessment of, 510–511
exposure and, 475, 508, 510
hazard and, 508, 510
levels of, 509
Pesticides,
banning of, 25
disposal of, 482, 490–493
history of use, 8–10
public perception of, 18

regulation of, 15, 21
selection of, 520
storage of, 69
Pesticide safety, 470
Pest scouting, 513
Pest thresholds, 82, 516–517
    how to determine, 84
Plant growth regulators (PGRs),
    371–372
    classes of, 377–378
    herbicidal, 378
    plant hormones and, 374
    selection of, 380
    uses of Type I and Type II,
    377
Practitioners' rights, 73
Preemergence herbicides,
    activity of, 178, 182, 184
    and turfgrass establishment,
    196–199
    examples of, 175, 178
    injury caused by, 197
    resistance to, 206
    target species for, 182

Rabbits, 396, 398
Rachel Carson
    influence of, 20, 324
    Silent Spring, 13–14, 20–22
Rodents, 380
    damage caused by, 383
    example pests, 381–383
Rodenticide, 485

Sedge and sedge–like species,
    Kyllinga, 158, 160
    list of, 141
    Purple nutsedge, 158–159
    Star of Bethlehem, 160,162
    Wild garlic, 114, 159, 161
    Yellow nutsedge, 109, 118,
    155–157

Spray equipment, 486–490
    booms, 421
    cleaning of, 455–456, 492
    guns, 432–434
    hoses, 419–420
    manual, 404–407
    power, 407
Sprayer calibration, 434
    calculations for, 440, 442–445,
    447–448, 454
    checking for accuracy, 452–455
    flow rate and, 436–437
    for manual sprayers, 435–436
    for power sprayers, 436
    ground speed and, 437–438
    spray width and, 438
    steps in, 446–449, 452–453
Sprayer flow control, 415
    agitation, 415–417
    flow control valves, 417
    pressure gauge, 418
    ways to regulate, 417–418
Spray nozzles, 422
    components of, 430
    extended range flat fan,
    422–423
    flooding flat fan, 423–425
    selection of, 443–444,
    450–451
    tip materials, 431–432
    Turbo flat fan, 426–427
    Turbo flood, 425–426
    Turbo turf flood, 427–428
    Venturi, 428–430
Spray pumps, 408
    centrifugal, 411–413
    characteristics, 410
    diaphragm, 413
    piston, 413
    roller, 411
Spray tanks, 413–415
Spreaders, 457

calibration of, 461–464
drop, 457
rotary, 458
use of, 459–461
Steps in weed control, 171
Strainers, 418–419
Strategic plantings, 499, 501
benefits of, 500
filter strips, 502–503
Surfactants, 364
crop oils, 366
detergents, 364–365
how they work, 364
organosilicone, 366–367
wetting agents, 365
Systemic herbicides, 190–191

Turfgrass,
cultural practices, 245
industry, 1, 17, 22, 27
species origins, 164

USDA,
formation of EPA, 15
pesticide regulation, 13–14

Weed biology, 114
broadleaf, 115
grass, 116
sedge, 116–117
Weed life cycles, 94
biennial, 105–106
characteristics of annuals, 95
daylength and, 99–101
germination requirements for annuals, 183
perennial, 107–109
perennial storage structures, 110–113
summer annual, 96–98
temperature and, 100
winter annual, 100–105
Wetlands, 504